Tidal Friction and the Earth's Rotation

Edited by
P. Brosche and J. Sündermann

With Contributions by
M. Bonatz P. Brosche O. Calame H. Enslin
K. Lambeck L. V. Morrison J. D. Mulholland
J. D. A. Piper C. T. Scrutton F. R. Stephenson
J. Sündermann W. Zahel J. Zschau

With 81 Figures

Springer-Verlag Berlin Heidelberg GmbH 1978

PROFESSOR DR. PETER BROSCHE
Observatorium Hoher List
Universitätssternwarte Bonn
5568 Daun/Eifel

PROFESSOR DR. JÜRGEN SÜNDERMANN
Universität Hamburg
Institut für Meereskunde
Heimhuder Straße 71
2000 Hamburg 13

ISBN 978-3-662-39200-3   ISBN 978-3-662-40203-0 (eBook)
DOI 10.1007/978-3-662-40203-0

© by Springer-Verlag Berlin Heidelberg 1978
Originally published by Springer-Verlag Berlin Heidelberg 1978

Offsetprinting: Beltz Offsetdruck, Hemsbach/Bergstr., Bookbinding: Brühl-
sche Universitätsdruckerei, Gießen.
2131/3130-543210

Contents

Historical Background and Introduction

P. Brosche

The development of the ideas and observational techniques related to
the subject of our meeting "Tidal friction and the Earth's rotation",
Bielefeld, September 1977 is one of the most fascinating books — not
merely chapters! — of the modern history of science. Its genealogical
tree is as intricate as that of mankind itself: There are dead ends
and superfluous re-discoveries. Due to these circumstances and to the
pure extent of the topic, it is impossible to give more than a few
highlights here.

The first relevant observational fact was discovered by the famous
English astronomer E. Halley in 1695 (Berry, 1961). He simply could
not arrive at an agreement between ancient and recent eclipses using
a constant mean angular motion of the Moon. Instead, he had to introduce an empirical acceleration term in the mean motion. Known as the
"secular acceleration", it has ever since been a most challenging subject of celestial mechanics and a main branch of the genealogical tree
already mentioned.

In 1754, completely independently and almost certainly in ignorance of
those specialists' activities, the German philosopher Kant established
the idea of tidal friction as a decelerating mechanism for the rotation
of the Earth (Felber, 1974). Although he made some errors in his rough
computations, the majority of the constitutive elements of his concept
have survived to the present day (Brosche, 1977).

For us, the connection between the lunar secular acceleration and the
behaviour of the Earth's rotation is taken for granted. However, to
seek for the first time the reason for a seemingly purely celestial
phenomenon in the Earth itself must be admired as an autonomous scientific contribution of a specific kind, namely, the establishment of
a relationship between existing but hitherto unconnected pieces of
knowledge, thereby transforming the meaning of the information. In
contrast to the dialectic leap arising from the accumulation of quantity, this case may be called a catalytic leap. It is still not completely clear who was the first to perform this catalysis. A probable
candidate is Tobias Mayer, astronomer, cartographer, and geodesist in
Göttingen. In a letter of 1753, November 25th, to Euler, dealing with
periodic effects in the lunar orbit (the "inequalities") and also with
the secular acceleration, he expresses the opinion that inequalities
in the Earth's rotation should be studied at least so carefully (Kopelevich, 1959; Forbes, 1971).

I shall pass over the glorious achievements of celestial mechanists
in the eighteenth and early nineteenth centuries, such as Laplace,
Delauny, and Adams. This is justified because it is the best documented
part (Berry, 1961) of the history of our topic and because it concerns
primarily the establishment of the *non*-tidal part of the secular acceleration. I must admit, however, that only in this way can the tidal
part be determined at the same time.

I would identify the beginning of the modern period of tidal friction
research with the work of Robert Mayer (1927), better known for his

formulation of the theorem of the conservation of energy. He gives a synopsis of the whole process, which, although or because it is written without any formulae, can be regarded as one of the highlights of scientific prose. At this point I have to leave the path of the historical development and shall now try to take a briefe glance at the *observational situation* in an increasing order of the time scales of the applied techniques:

OA 1. In *recent years*, the development of the lunar laser ranging (LLR) technique led to an unrivalled accuracy in measuring distances between stations on Earth and on the Moon and thus to the corresponding determination of their relative geometry. There are well-founded hopes that radio interferometry will supplement LLR with regard to absolute orientation in space.

OA 2. During the least few *decades* the advent of atomic clocks has made possible direct and independent measurements of the rotation of the Earth as well as of the motion of the Moon. However, due to fluctuations with a time scale longer than that of those precise observations, they do not enable us to determine values of any tidal friction parameters.

OA 3. For modern times, that is, for a few *centuries*, the observed motions in the solar system define an appropiate time scale for studying the Earth-Moon system in respect to those motions.

OA 4. For historical times — a few *millenia* — there exist records of solar eclipses. Taking the orbital motion of the Earth as granted, they give us information on a combined effect of changes in the Earth's rotation and in the lunar orbit.

OA 5. In an unforeseeable way, namely, from periodical growth in fossil animals, the time scale of observations has been extended to at least several *hundred million years*.

While the preceding observations refer to changes in the Earth-Moon system, we certainly also need observations of what we believe is the reason for those changes, that is, of tides:

OT 1. Present-day tides of the solid Earth can be measured either locally (by gravimeters) or globally with the aid of satellites.

OT 2. Likewise the tides of the oceans can be measured at specific sites or by satellite altimetry. Unfortunately until now, the first possibility was restricted to shelf areas and the second is only beginning to be used.

OT 3. It seems that there is not only palaeontological evidence of the period of oceanic tides but also of some kinds of amplitudes and phases, thus extending the time scale of observations by 10^6!

OT 4. Since oceanic tides depend crucially on the shape of the oceans, we have to consider information on continental drift as an indispensable contribution to our general topic.

Theory is applied in two opposite directions: Firstly, for reducing and clarifying the observations, secondly, to explain the observed effects. This burden has to be carried by at least the following branches of science:

T1. From celestial mechanics we expect the long-term behaviour of the Earth-Moon system, including the nonconservative process of tidal friction.

T2. The realm of rheology of the solid Earth is the response of the solid part of our planet, including its interaction with the fluid part.

T3. Theoretical oceanography has to deliver models of tides accurate enough to determine the average tidal friction. Correspondingly to T2, the interaction with the solid Earth should be included.

Having carried out all possible analyses, we have to envisage the *consequences* of the picture at which we arrive. This is true even if we cannot take our state of understanding as firmly established, because only in this way can we confront our view with otherwise overlooked facts. Until now, consequences have been discussed for the following scientific areas:

C1. The origin and early evolution of the Earth-Moon system

C2. The origin of large topographic features on the Moon and Earth

C3. The origin of life on Earth

Although considerable progress has been made in several areas of our theme, we are still left with many of the basic *questions*. Without anticipating the conclusions of our assembled experts, I would like to ask a few of these questions to which we may get an answer here:

Q1. Has the process of tidal friction varied measurably within historical time?

Q2. How important are the tides of the solid Earth as compared with oceanic tides?

Q3. Are backwards-integrations, which attempt to give the changes in the Earth-Moon system on the basis of a schematic picture of the tidal friction process, acceptable?

In contrast to Q1 - Q3, I do not expect a final answer to even the most simplified version of our main problem:

Q4. What was the minimum distance between Earth and Moon and when did this situation occur?

At least we, the initiators of this meeting, admit that we feel we are not aware of all the relevant information. This is especially since it comes from so many disciplines. May I therefore ask all the participants to address us as non-specialists, to sum up the essential results in as quantitative a manner as possible and with a balanced evaluation of the uncertainties, and finally to inform us of the main problems in their view, and of any proposals for their solution if it seems they can be tackled in the near future.

References

Berry, A.: A Short History of Astronomy. New York: Dover Publ. 1961, p. 254 (originally published 1898)
Felber, H.J.: Die Sterne 50, 82 (1974)
Brosche, P.: Die Sterne 53, 114 (1977)
Kopelevich, Y.K.: Istoriko-astronomitsheskye issledovanya, V, Moskva, 1959, p. 370
Forbes, E.G.: The Euler-Mayer correspondence (1751-1755), New York: MacMillan 1971, p. 21 and 77

4

Mayer, R.: Beiträge zu Dynamik des Himmels, originally published Heilbronn 1848.
Reprinted in: Ostwald's Klassiker der exakten Wissenschaften Nr. 223, Leipzig,
1927, chapter 8 = p. 33-43

Pre-Telescopic Astronomical Observations

F. R. Stephenson

1. Introduction

The ideas that the Earth's rotation was non-uniform and that the orbital motion of the Moon was accelerated (two of the themes central
to the present conference) were conceived more than two centuries ago.
However, a sound grasp of the whole problem has only been realised in
the last few decades.

Before the time of Halley, there had been no reason to suspect that
the motion of the Moon was anything but uniform. However, in 1695
Halley published a paper on the longitudes of ancient cities in which
he hinted that the Moon's orbital motion might be accelerated. He ar-
rived at this conclusion by comparing ancient astronomical observations with those made in his own time, but never made any quantitative
deductions. At the end of his paper Halley expressed the conviction:
"And I could then pronounce in what Proportion the Moon's Motion does
accelerate, which that it does, I think I can demonstrate, and shall
(God willing) make it appear to the Publick".

Unfortunately, Halley never seems to have realised his objective, and
it was left to Dunthorne (1749) to make the first numerical determination of the lunar acceleration using ancient, medieval, and (what were
then) recent astronomical data.

The discovery of the acceleration of the Moon was a purely observational result; there had been no reason to suspect its existence. On
theoretical grounds, Kant (1754) reasoned that tides raised by the
Moon and Sun in the seas and solid body of the Earth must retard the
rotation of the Earth by friction. This idea found some favour in the
nineteenth century, but it was not until the current century that its
significance was clearly recognised.

In the present paper, ancient and medieval astronomical observations
(all pre-telescopic) are analysed in order to study the variation in
the rotation of the Earth and the mean lunar motion during the historical past. Useful telescopic observations go back as far as about A.D.
1630 and historical data extend the time scale only to about 1000 B.C.
The earliest known usable observation was made in 1375 B.C., but there
are very few data prior to 200 B.C. It should be emphasised just how
minute this period is compared with the time scale of palaeontologial
data. Extrapolation into the geological past based on observational
data, whether "modern" or "ancient", should thus at best be regarded
as only a first approximation.

Institute of Lunar and Planetary Sciences, School of Physics, University of Newcastle
upon Tyne, NE1 7RU, England

2. Historical Development

The early pioneers in the analysis of pre-telescopic observations —
among whom Halley, Baily, Airy, Hansen, Ginzel, Newcomb, Nevill, Cowell, and Fotheringham deserve special mention (more or less in chronological order) — worked exclusively within a Universal Time (UT)
framework. Universal Time was then the only time scale conceived. In
the earlier stages, attention was largely confined to deriving better
values for the lunar acceleration, but in 1905 Cowell discovered a
solar acceleration. This later proved to be purely apparent — a direct
reflection of the retardation of the terrestrial rotation. It became
the practice to express the lunar and solar "secular accelerations"
measured on U.T. as coefficients of T^2, where T was measured in Julian
centuries from a standard epoch (usually 1800.0 or 1900.0). If these
secular accelerations are designated m and s, then the true accelerations are obviously 2m and 2s. Present-day convention assigns the symbols ν and ν' for m and s.

De Sitter in 1927 obtained values for the accelerations from ancient
observations which remained definitive until the last few years with
the publication of the exhaustive studies by Newton (1970, 1972) and
Muller and Stephenson (1975). De Sitter made a least squares analysis
of the ancient observations, rediscussing the investigations by Fotheringham and Schoch. He obtained:

$$m = + 5\overset{''}{.}22 \pm 0\overset{''}{.}30 \text{ cy}^{-2} \tag{1}$$

$$s = + 1\overset{''}{.}80 \pm 0\overset{''}{.}16 \text{ cy}^{-2} \tag{2}$$

Spencer Jones (1939) analysed in detail modern (telescopic) observations since about A.D. 1680. These were principally of occultations,
solar declinations, and transits of Mercury. Because of the short period (less than three centuries) covered by these data, it was impossible to determine the lunar acceleration directly, and he had to
assume De Sitter's result for m of + 5$\overset{''}{.}$22 cy^{-2} in order to derive a
result for the solar acceleration. For the first time, Spencer Jones
was able to fully separate out the irregular fluctuations in the mean
motions of the Moon, Sun, and Mercury; these had long been a source
of controversy. He derived a result for the "modern" solar acceleration:

$$s = + 1\overset{''}{.}23 \pm 0\overset{''}{.}04 \text{ cy}^{-2} \tag{3}$$

The significant discord between the ancient and modern values of s
led Spencer Jones to the view: "There seems to be no escape from the
conclusion that the effects of tidal friction are appreciably less at
the present time than the average effects over the past two thousand
years". The truth (or otherwise) of this statement is one of the subjects dealt with in this paper.

A thorough geophysical understanding of the results obtained from the
observational data became possible with the work of Clemence (1948).
He showed (in the notation of the present paper) that if $\dot{n}$ is the lunar acceleration on Dynamical Time, TD (formerly known as Ephemeris
Time, ET), which is independent of the rotation of the Earth, then

$$\dot{n} = 2 \ (m - ks) \tag{4}$$

where the factor k is the ratio of the mean lunar and solar motions
(13.37). The quantity $\dot{n}$ is a true acceleration (twice the coefficient
of T^2).

Using this last equation, De Sitter's values of m and s lead to:

$$\dot{n} = -\ 37\overset{''}{.}6 \pm 4\overset{''}{.}1 \ cy^{-2} \tag{5}$$

Spencer Jones' figures from modern data give

$$\dot{n} = -\ 22\overset{''}{.}44 \pm 0\overset{''}{.}88 \ cy^{-2} \tag{6}$$

Comparison between these two results seems to bear out Spencer Jones' inference that the mean lunar tidal torque has decreased substantially during historical times.

3. Inter-Relation of Parameters

An important quantity is the rate of recession of the Moon from the Earth. It is readily shown that if r is the present distance of the Moon from the Earth then

$$\frac{\dot{r}}{r} = -\ \frac{2}{3} \frac{\dot{n}}{n} \tag{7}$$

Thus a typical value for $\dot{n}$ of $-\ 30''\ cy^{-2}$ leads to a rate of tidal recession of the Moon of

$$\dot{r} = 4.4 \ cm \ per \ year \tag{8}$$

On TD the acceleration of the Sun and planets is zero. Instead, we are concerned with the (negative) acceleration of the Earth's rotation. Two common ways of expressing this latter quantity are:

1. $\dot{e}$ in units of seconds of time cy^{-2} and

2. $\frac{\dot{\omega}}{\omega}$ in units of cy^{-1}, where ω is the present rate of spin of the Earth.

The smoothed expressions for ΔT (= TD - UT), with fluctuations averaged out, may be written:

$$\Delta T = a + bT - 1/2\dot{e}T^2 \tag{9}$$

where a and b are constants and T is measured in Julian centuries (of 36525 days) from the epoch 1900.0. The additional constants a and b are necessary because 1900.0 is an arbitrary epoch for which neither ΔT nor its derivative is zero.

Direct conversion factors from s (on UT) to $\dot{e}$ and $\frac{\dot{\omega}}{\omega}$ are:

$$\dot{e} = -\ 48.7 \ s \tag{10a}$$

$$\frac{\dot{\omega}}{\omega} = -\ 15.4 \cdot 10^{-9} s \tag{10b}$$

A useful quantity is the rate of change in the length of the day ($\dot{d}$) expressed in milliseconds per century. We have:

$$\dot{d} = 1.33 \ s \tag{11}$$

A typical value of $\dot{d}$ is around 2 ms cy^{-1}.

.4. Recent Investigations

Recently the application of historical astronomical observations in the study of the Earth's rotation and the motion of the Moon has been revived, with extensive studies of the raw data. Newton (1970) made a thorough survey of ancient and medieval observations of various kinds, and from these he determined the following results:

(1) Mean epoch 200 B.C.

$$\dot{n} = -41\overset{\shortmid\shortmid}{.}6 \pm 4\overset{\shortmid\shortmid}{.}3 \ cy^{-2}; \quad 10^9 \ \frac{\dot{\omega}}{\omega} = -27.7 \pm 3.4 \ cy^{-1} \tag{12}$$

(2) Mean epoch A.D. 1000

$$\dot{n} = -42\overset{\shortmid\shortmid}{.}3 \pm 6\overset{\shortmid\shortmid}{.}1 \ cy^{-2}; \quad 10^9 \ \frac{\dot{\omega}}{\omega} = -22.5 \pm 3.6 \ cy^{-1} \tag{13}$$

The values quoted represent averages over the interval between the mean epoch and the present day.

Comparing the above results for $\dot{n}$ with the value derived by Spencer Jones (1939) from recent (telescopic) data and his own result deduced from a study of near-Earth satellites ($\dot{n} = -20\overset{\shortmid\shortmid}{.}1 \pm 2\overset{\shortmid\shortmid}{.}6$) led Newton to conclude: "Thus there is a strong presumption that $\dot{n}_m$ (= $\dot{n}$) has changed by a factor of 2 within historic times".

From a detailed investigation of ancient and medieval solar eclipse data Muller and Stephenson (1975) obtained similar results:

$$\dot{n} = -37\overset{\shortmid\shortmid}{.}5 \pm 5" \ cy^{-2} \tag{14}$$

$$10^9 \ \frac{\dot{\omega}}{\omega} = -29\overset{\shortmid\shortmid}{.}0 \pm 3\overset{\shortmid\shortmid}{.}2 \ cy^{-1} \tag{15}$$

An important conclusion of this paper is that the ΔT curve over the past three millennia can be closely approximated by a simple parabola. At no time during this period is there evidence of a departure of more than 100 s from this form. This leads us to suspect that $\dot{e}$ has remained constant — i.e., that tidal friction has not changed significantly. The current paper is a development of our 1975 analysis.

5. Remarks on the Selection of Suitable Observations

Pre-telescopic observations, which have been the subject of studies of the Earth's rotation and lunar motion in the historical past, fall mainly into the following categories:

1. Solar eclipses which were either total or near-total (untimed)

2. Large solar eclipses of unspecified magnitude (untimed)

3. Estimates of solar and lunar eclipses magnitudes (untimed)

4. Times of solar eclipse contacts

5. Times of occultations of stars by the Moon

6. Times of lunar eclipse contacts

7. Times of planetary conjunctions

8. Times of spring and autumn equinoxes

Until as late as the seventeenth century, the accurate measurement of time proved very difficult. The first astronomers to attempt to measure time with what could be called reasonable accuracy were the Babylonians. From about 700 B.C. to 50 B.C. there exists a series of timed contacts for both solar and lunar eclipses. These are in general estimated to the nearest us or even 1/2 us (1 us = 4 min, the time taken for the Earth to rotate through 1°). The *intrinsic* accuracy of the measurements is high, but the author (Stephenson, 1974) has shown that there is reason to believe that systematic instrumental errors could be very large. An analysis of a late Babylonian astronomical text (323 – 319 B.C.) describing measurements of the interval between the rising and setting of the Sun and Moon around new and full Moon (made with some kind of clepsydra) led to the conclusion that over an interval of the order of one hour systematic drifts of up to 25 % were possible. Eclipse times are invariably measured with reference to sunrise or sunset so that quite possible an observation made near midday of midnight could be as much as an hour in error.

The Greeks were less ambitious, measuring time to the nearest third of an hour only. After the Babylonians until the Renaissance in Europe only the Arabs achieved any degree of accuracy in time measurement. The most accurate timed observations are of eclipses made between about A.D. 800 and 1000 and these are of similar intrinsic precision to the Babylonian data. However, they are probably much more reliable, for most of the data are in the form of altitude measurements (to the nearest degree). With these, the very uncertain question of systematic instrumental errors still remains.

It was not until the mid-seventeenth century, by which time the use of the telescope was well established, that certain astronomers, notably Hevelius, began to time their observations to a fraction of a minute. However, even at this late period many prominent observers were still content with quoting times to no better than a minute.

An accuracy of a few minutes in the time of an eclipse seen 1000 or even 2000 years ago is not as useful as might first appear, for in analysing such an observation the two fundamental unknowns ($\dot{n}$ and $\dot{e}$) are highly correlated. We believe that the only really useful pre-telescopic timed observations are of equinoxes. Although these are of much lower precision (an hour or even several hours) only $\dot{e}$ is to be found. The extant Greek and Arabian data will be analysed in another paper.

Of the untimed material, the least promising relates to eclipse magnitudes. There is no reason to suppose that the few magnitudes which have survived from various parts of the world — Greece, Babylon, the Arab Lands and China — represent more than crude guesses to the nearest twelfth or fifteenth of the solar or lunar diameter.

We have elsewhere criticized the use of untimed observations of large solar eclipses where the magnitude is not specified (Muller and Stephenson, 1975). With these kinds of data, the best that can be achieved is a least squares fit to the observations of each eclipse. Unfortunately, such a method does not allow for the biased distribution of population centres. In medieval Europe, where many of the observations were made, this effect is particularly marked, as illustrated by Figure 1. The total eclipse of A.D. 1239 June 3 was observed extensively in Italy, but only at scattered sites elsewhere. The diagram shows the position of the zone of totality in the vicinity of Italy, computed on the basis of values of $\dot{n}$ and $\dot{e}$ similar to those finally adopted in this paper. On any "reasonable" values of these quantities, a displacement of the track northwards or southwards by only a few km would be expected; the track happens to reach its northernmost limit in this region. On the other hand, the observations taken

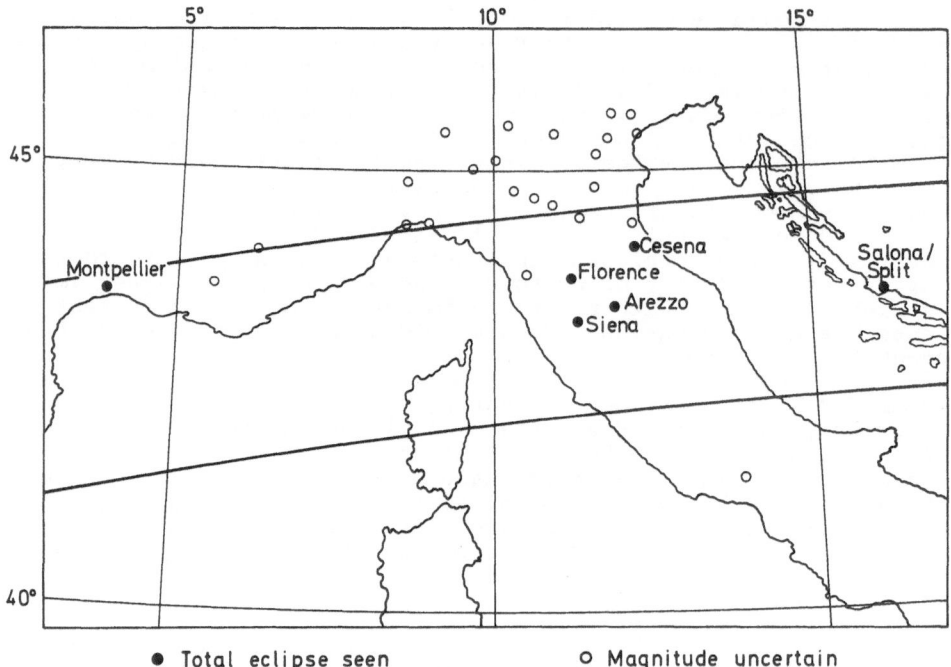

Fig. 1. The total solar eclipse of A.D. 1239 June 3

as a whole seem to require a northerly displacement of some 200 km, which is impossible. The explanation is that in the thirteenth century — as at the present time — most of the population centres of Italy were situated in the north of the country and therefore very few records would be expected to come from the midlands or south. However, it is significant that all of the places where totality was witnessed (often described in vivid terms) lay within the computed central zone.

Numerous other similar instances could be quoted. As far as we can judge there seems to be no satisfactory way of making allowance for population bias unless an observation makes it clear whether or not the central phase was witnessed. This is the standpoint adopted in the present paper.

6. Observations of Total and Near-Total Solar Eclipses

"There was an eclipse of the Sun .... some time after midday, with stars appearing and the Sun completely hidden from our sight; and the sky was so clear that no clouds appeared in the air".

So runs an account of the eclipse of A.D. 1241 October 6 in the chronicle of the monastery of Stade in Germany. Descriptions of this form, which give only an approximate indication of the time of day, are very common, and the above account is fairly typical in this respect. In such cases, all that can usually be inferred is that the observer was *somewhere* within the zone of complete shadow. Just how accurate an observation does this represent?

Because the zone of totality is very narrow on the Earth's surface — typically 200 or 300 km wide — it is clear that prospects are at least reasonably good. As a rough indication of the *potential* accuracy of such an observation, we can note that the semi-duration of totality is seldom more than about 2 or 2 1/2 min and is often substantially less. Thus without any equipment for measuring times or altitudes it is possible for an observer who witnesses a total eclipse to provide information at least as accurate as the Arabian timed contacts — but free from systematic errors — and far superior to the Babylonian timed data.

What makes observations of totality especially valuable is the very hard limits of the zone of complete shadow. Because the solar photosphere has a very sharp edge (on the $0''\!.1$ scale) and the chromosphere is comparatively faint, an observer without optical aid who by chance is situated near the northern or southern limits of totality can help define the exact limit with very high precision. This is well exemplified by experiments performed in New York City during the total eclipse of 1925 January 24. The southern limit of totality on this occasion passed across Manhattan Island and 149 pairs of untrained observers (employees of the public utilities) were stationed at one-block intervals along Riverside Drive and other streets roughly normal to the expected southern limit of totality. (For details see T.I.E.S., 1925.) The two observers in each pair were assigned different tasks: One was instructed to note the oncoming shadow; the other was asked to describe the appearance of the eclipsed Sun and to decide whether or not the eclipse appeared to be total.

The results were impressive. All observers except one fell into two classes: Those who reported totality and those who denied it. One single *shadow* observer (very near the southern limit) expressed doubt. This remarkable consistency permitted the definition of the limit to within 200 metres ($0''\!.1$) or so. It would thus appear that although the lunar limb is very irregular (on the km scale) any beads of sunlight shining down the clefts of the limb are so brilliant that even to the unaided eye it is clear that the eclipse is not quite complete. This was noted at the eclipse of 1715 May 3 (Halley, 1715).

The $0''\!.1$ referred to above is the resolution normally obtained by a 1-metre telescope! Thus if we take an ancient or medieval account, which clearly describes the total phase — albeit in simple terms — although in most cases we cannot tell just where within the total zone the observer was situated, the limits are extremely well defined.

So far we have discussed only the advantages of total eclipse reports. There are several disadvantages that were not fully appreciated by the early investigators (particularly Cowell and Fotheringham), and this has led to very valid criticism of their usage by Newcomb — at the beginning of the present century — and by Newton very recently. Most of the difficulties arise from the fact that, by and large, because of the spectacular nature of major solar eclipses, the records are found in literary and historical sources rather than astronomical works. This is no way prevents an astronomically accurate description of the phenomenon from being given, but very few early descriptions of total eclipses trouble to mention *both* the date and place of observation: In fact, several give neither. Newton (1970) has with justification drawn attention to the very real possibility of a striking eclipse being retrospectively linked with a major historical event, e.g., a battle. Fotheringham's famous diagram in relating solar and lunar accelerations in his 1920 paper ("A solution of ancient eclipses of the Sun") attracted much attention and formed the basis of numerous geophysical arguments a decade or so ago. However, present opinion (that of Muller, Newton, and the author) is that probably all of Fotheringham's

Table 1. Usable observations of total and near-total eclipses made before the introduction of the telescope

Julian date	Place	Lat. N.		Long. E/W[*]		Observations
B.C. 1375 May 3	Ugarit	35°	47'	35°	47'	Sun put to shame; went down in daytime
709 Jul.17	Chu-fu	35°	32'	117°	01'	Total
601 Sep.12	Ying (?)	30°	20'	112°	15'	Total
549 Jun.12	Chu-fu	35°	32'	117°	01'	Total
198 Aug. 7	Ch'ang-an	34°	21'	108°	53'	Annular
188 Jul.17	Ch'ang-an	34°	21'	108°	53'	Almost complete
181 Mar. 4	Ch'ang-an	34°	21'	108°	53'	Total; dark in daytime
147 Nov.10	Ch'ang-an	34°	21'	108°	53'	Almost complete
136 Apr.15	Babylon	32°	33'	44°	17'	Total; 4 planets seen
89 Sep.29	Ch'ang-an	34°	21'	108°	53'	Not complete; like a hook
80 Sep.20	Ch'ang-an	34°	21'	108°	53'	Almost complete
35 Nov. 1	Ch'ang-an	34°	21'	108°	53'	Not complete, like a hook
28 Jun.19	Ch'ang-an	34°	21'	108°	53'	Not complete, like a hook
2 Feb. 5	Ch'ang-an	34°	21'	108°	53'	Not complete, like a hook
A.D. 2 Nov.23	Ch'ang-an (?)	34°	21'	108°	53'	Total
65 Dec.16	Kuang-ling (?)	32°	26'	119°	27'	Total
120 Jan.18	Lo-yang	34°	47'	112°	41'	Almost complete; on Earth it was like evening
360 Aug.28	Chien-k'ang	32°	02'	118°	47'	Almost total
516 Apr.18	Chien-k'ang (?)	32°	02'	118°	47'	Annular
522 Jun.10	Chien-k'ang	32°	02'	118°	47'	Total
840 May 5	Bergamo (?)	45°	42'	9°	40'	Sun hidden from world, then shone again
912 Jun.17	Cordoba (?)	37°	53'	4°	46'[*]	Total; darkness covered the Earth
968 Dec.22	Constantinople	41°	01'	28°	59'	Sun deprived of light (very clear account of corona)
968 Dec.22	Farfa	42°	13'	12°	42'	Sun in darkness
975 Aug.10	Kyoto	35°	02'	135°	45'	Total; sun colour of ink
1124 Aug.11	Novgorod	58°	30'	31°	20'	Sun perished completely
1133 Aug. 2	Salzburg	47°	48'	13°	04'	Sun disappeared suddenly (allusion to corona ?)
1133 Aug. 2	Heilsbronn (?)	49°	21'	10°	50'	Sun in a moment became black as pitch
1133 Aug. 2	Reichersberg	48°	20'	13°	22'	Sun entirely hidden
1133 Aug. 2	Vysehrad	50°	04'	14°	25'	Small crescent at S. limb of sun
1133 Aug. 2	Prague (?)	50°	06'	14°	25'	Sun appeared small
1133 Aug. 2	Cambrai (?)	50°	11'	3°	20'	Sun disappeared suddenly

Table 1 continued.

Julian date	Place	Lat. N.		Long. E/W		Observation
A.D. 1176 Apr.11	Antioch	36°	12'	36°	10'	Sun totally obscured
1178 Sep.13	Vigeois	45°	23'	1°	31'	Sun like 2- or 3-day-old Moon
1185 May 1	Novgorod	58°	30'	31°	20'	Description of chromosphere
1221 May 23	Kerulen River	48°	11'	115°	54'	Total
1239 Jun. 3	Coimbra	40°	13'	2°	26'*	Sun as black as pitch
1239 Jun. 3	Cerrato (?)	41°	57'	4°	31'*	Total
1239 Jun. 3	Toledo	39°	52'	4°	02'*	Sun lost all its strength
1239 Jun. 3	Montpellier	43°	37'	3°	53'	Sun entirely covered by Moon
1239 Jun. 3	Arezzo	43°	28'	11°	54'	Whole of sun covered for several minutes
1239 Jun. 3	Cesena	44°	08'	12°	15'	Sun completely black (prominence ?)
1239 Jun. 3	Florence	43°	46'	11°	15'	Whole sun obscured
1239 Jun. 3	Siena	43°	19'	11°	20'	Sun completely obscured
1239 Jun. 3	Salona/Split	43°	32'	16°	28'	Whole sun obscured
1241 Oct. 6	Reichersberg	48°	20'	13°	22'	No part of sun visible
1241 Oct. 6	Stade	53°	37'	9°	29'	Sun completely hidden
1241 Oct. 6	Cairo (?)	30°	02'	31°	15'	Sun completely darkened
1267 May 25	Constantin. (?)	41°	01'	28°	59'	Total
1275 Jun.25	Lin-an	30°	15'	120°	10'	Total
1292 Jan.21	Ta-tu	39°	55'	116°	25'	Sun like a golden ring
1330 Jul.16	Konigsaal	49°	59'	14°	24'	Sun like a 3-night-old Moon
1406 Jun.16	Braunschweig	52°	15'	10°	30'	Sun stopped shining
1415 Jun. 7	Altaich	48°	46'	13°	02'	Sun entirely lost its light (for several minutes)
1415 Jun. 7	Prague	50°	06'	14°	25'	Whole sun eclipsed
1415 Jun. 7	Krakow	50°	07'	19°	55'	Darkness fell suddenly
1485 Mar.16	Melk	48°	14'	15°	21'	Complete eclipse
1560 Aug.21	Coimbra	40°	13'	2°	26'*	Moon covered whole sun
1567 Apr. 9	Rome	41°	53'	12°	33'	Narrow circle of light surrounding whole sun

data is valueless. The main reason is dating difficulties. Like several of his contempories, Fotheringham was not above playing what Newton (1970) has called the "Identification Game". This amounts to attempting to date large eclipses on the basis of astronomical computation and then using them to improve the accuracy of subsequent computation! I have found it necessary to completely re-investigate the historical data.

Table 1 contains a list of all the seemingly reliable pre-telescopic observations that I have been able to isolate as a result of a careful search through the literature of China, Europe, and Babylon, supplemented by a few observations made in other parts of the world. There are many medieval European data, essentially as a result of the vast number of monasteries, each of which kept independent chronicles. The Chinese records are also fairly numerous because of the extended period covered and also because professional astronomers were employed to maintain a regular watch of the sky. The Babylonian astronomers may well have been just as diligent as the Chinese, but their observations cover a much shorter period, and the material which has managed to survive the ravages of time is in a very fragmentary condition. At the present time, data from the Arab world represents something of an unknown quantity; there *ought* to be much more in view of the interest in astronomy cultivated by the Arabs. It seems that real progress can be made here in the discovery of further data.

It is well to explain here just what is meant by "near-total" eclipses. A few records quite carefully state that the Sun did not completely disappear, although we know that the eclipse was total on the Earth as a whole. To give an example, the following observation was made at Vysehrad (Prague) in A.D. 1133:

"An eclipse of the Sun appeared in a wonderful manner; this, gradually decreasing, diminished to such an extent that a crown like the crescent Moon proceeded to the south part, which afterwards turned round to the east, henceforth to the west; at length it was transformed to its original state".

Here we have a very clear statement that the path of totality passed to the north of Prague. A small number of annular eclipses are reported either as central or partial in the Chinese annals. We have rejected all European references to annular eclipses since descriptions of these seem to be particularly vague and unintelligible. However, the Chinese records, being the work of professional astronomers, seem fairly precise. The Chinese used the same term *chi* ("complete") to describe both total and annular eclipses when they were central where seen. The pictogram originally portrayed a man seated near a bowl of food and turning his head away, indicating that he was replete, so that the astronomical meaning is clear enough, whether applied to ring or total eclipses.

Table 1 gives the Julian date, place of observation (with geographic co-ordinates), and a brief description based on the original text. This latter is often a mere summary giving only the salient points: Many of the European records are extremely detailed. Only in a single example (the very first) is there *any* doubt regarding the date. In every other instance there is an error of no more than a year (which readily allows the correct date to be established) and frequently the exact day is specified. Most of the descriptions are non-technical (a notable exception is the Babylonian report in 136 B.C.), but in their own way the various statements are quite precise, clearly differentiating between central and partial eclipses.

Unfortunately, in a number of cases the place of observation is in doubt. Although very few reports state directly where an eclipse was seen, usually the place is implicit in the text. Thus a monastic chronicle may be very much concerned with local affairs. Again, in the case of the Chinese records there are sound reasons for believing that the capital was the place of observation. However, occasionally it is apparent that the place where a particular eclipse was seen can be no more than surmised. In order to keep the quality of the data fairly high we have included only those observations for which we estimate the probability of the assumed location being correct as at least 0.50.

Most of the observations are of very high reliability in this regard; there is scarcely any doubt with respect to location.

None of the data in the table can be considered as accurately timed. However, among the large number of Babylonian timed contacts of solar eclipses there exists a single observation that was made very close to sunset. This is reported on one of the late Babylonian tablets in the British Museum (BM 34093). The inscription reads: "On the 28th day (of the month Ulul) at $3us$ before sunset (there was a) solar eclipse .... It set eclipsed". The UT of sunset at Babylon on this day (which corresponds to B.C. 322 September 26) can readily be calculated. The principal uncertainty probably lies in the inevitable delay between true first contact and the actual sighting. Still, as the timing error is likely to be very small, we have here what could prove to be a very useful observation.

7. Method of Analysis

The selection of only a comparatively small number of reliable sightings of central and near-central eclipses from among a much more extensive set of "large" eclipses may seem questionable. However, there appears to be no satisfactory way of utilizing the numerous eclipse observations of unspecified magnitude; these are merely of academic interest. One of the major difficulties in applying a least squares analysis to a large body of unsorted data of this type has already been pointed out; this is the effect of population bias. Additionally, assigning weights to individual observations is of doubtful validity, especially where the less reliable sightings are concerned; very low-weighted data can materially affect the final solution. On the other hand there is a definite advantage in selecting only a smallish data set (as in the present instance). It becomes possible to enquire in detail into the historical circumstances relating to each individual observation. Obviously this would be impracticable where a vast bulk of data was concerned.

Use of linear inequalities seems to be by far the most promising method of analysing total and near-total eclipse observations. This was developed in its present context by Fotheringham (1920). For such a technique to be successfully applied there must be only two unknown parameters: $\dot{e}$ and $\dot{n}$. This requires a reliable value for the motion of the lunar node. Newcomb (1912) and Brown (1919) derived essentially the same figure for the centennial motion of the node, but Spencer Jones (1925) obtained a correction of $+ 3\overset{.}{.}74$ to Brown's result. (Such a significant correction was confirmed by Martin and Van Flandern (1970) who deduced $+ 4\overset{.}{.}30$.) More recently Muller et al. (in preparation) obtained $+ 4\overset{.}{.}39$ from an integration of the solar system emphemeris. Pending the investigation by Morrison currently in progress, this will be taken as definitive.

Figure 2 shows schematically the path of a central (total or annular) eclipse in the vicinity of a particular observing station. A small change $\delta\dot{e} > 0$ moves the path precisely westward ($\delta\dot{e} < 0$ eastward), but a small change $\delta\dot{n} < 0$ moves the path only approximately westward ($\delta\dot{n} > 0$ eastward), the precise direction depending on the circumstances prevailing. If we are confident that an eclipse was central at a given place, then we can vary $\dot{n}$ and $\dot{e}$ independently to ensure that the total or annular phase was indeed witnessed. Thus each observation provides a pair of linear equations relating $\dot{n}$ and $\dot{e}$. The "correct" solution can lie anywhere inside the solution space for $\dot{e}$ and $\dot{n}$ defined by the essentially parallel edges of the eclipse track. As the precise effect

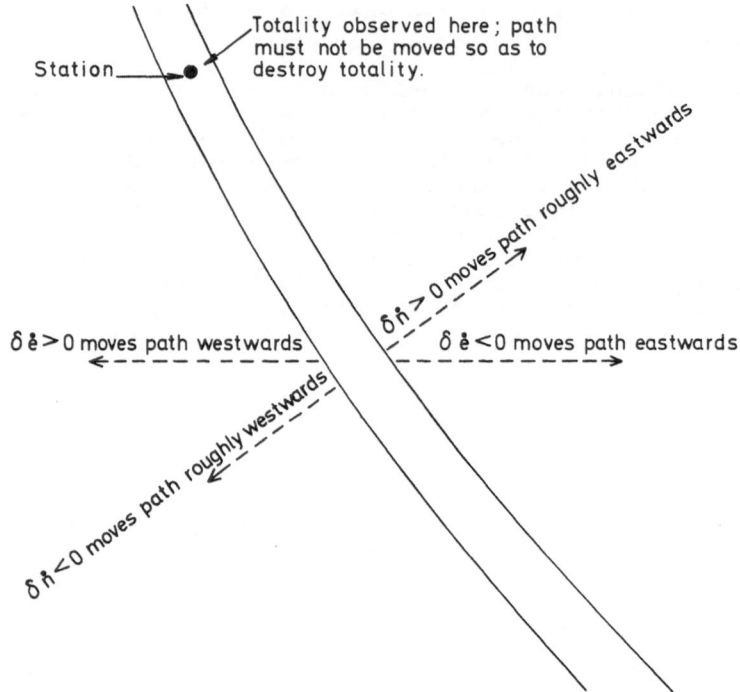

Fig. 2. Effects on eclipse track by varying the accelerations of the Earth and Moon

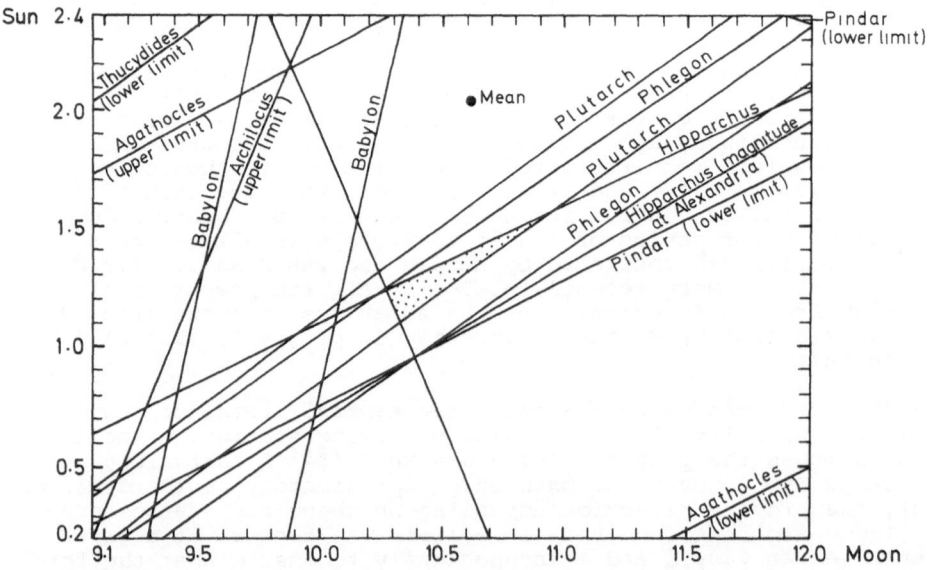

Fig. 3. Fotheringham's diagram (1920) of solar acceleration as a function of lunar acceleration for ancient observations of large solar eclipses

of changing ṅ varies for different eclipses, use of several observations allows the separation of the two parameters ṅ and ė. However, it should be emphasised that variations are smaller that might be wished; as a result the linear combination of ṅ and ė is much better known than either of the individual parameters.

The application of linear inequalities to the problem at hand can be conveniently illustrated by reference to Fotheringham's (1920) eclipse diagram. Figure 3 is copied directly from this. Fotheringham worked in U.T. and he chose as ordinate the solar acceleration (s) and as abscissa the lunar acceleration (m + 6.1) relative to the equinox — i.e., it includes the gravitational term. The solar acceleration on UT is directly proportional to ė, whereas the lunar acceleration is a linear combination of ė and ṅ.

In the diagram pairs of parallel lines represent equations corresponding to the edges of the central zone of the appropriate eclipse. In certain cases only one of these lines happens to cut the area covered by the figure — e.g., where the belt of totality was unusually wide. The individual gradients of the line pairs depend on a variety of factors such as the time of year when the eclipse occurred and the local time of day at the place of observation. Within the small triangular shaded area, all of the observations which Fotheringham deemed to be valid are satisfied. If his data had been as reliable as he supposed, his solution would have deserved the widespread attention which it attracted — particularly in later decades. In reality not a single one of the eclipses which Fotheringham used can be relied upon, either because the date, place, or magnitude (or a combination of these) is very uncertain.

To separate ṅ and ė (on TD), we have used a similar technique to that developed by Fotheringham. The complication introduced by the fact that ΔT is not just given by $-1/2\ \ddot{e}\ T^2$ (where T is measured from 1900.0), but by the polynomial $\Delta T = a+bT - 1/2\ \ddot{e}\ T^2$, has been discussed earlier. Muller in Muller and Stephenson (1975) devised an iterative procedure to deduce a and b as well as ṅ and ė. This has been simplified (Muller, 1975) to a process of single parameter solution using partials with respect to the observables. Following this method, and incorporating the revised value for the motion of the lunar node, the results of the present investigation are:

$$\dot{n} = -30\overset{.}{.}0 \pm 3\overset{.}{.}0 \ cy^{-2} \tag{16a}$$

$$\dot{e} = -76^s.6 \pm 6^s.8 \ cy^{-2} \tag{16b}$$

$$\Delta T = 20^s + 114^s \ T + 38^s.30 \ T^2 \tag{17}$$

The result for ė is equivalent to

$$10^9 \ \frac{\dot{\omega}}{\omega} = -24^s.3 \pm 2^s.0 \ cy^{-1} \tag{18}$$

8. Geophysical Discussion

The above value of ṅ is by far the smallert which has emerged from an analysis of ancient data. This results partly from the adoption of a revised value of the motion of the lunar node, but also from our selection of data (which we believe to be particularly reliable). This figure is still significantly higher than the earlier estimates of ṅ from telescopic observations, e.g.,

$$\dot{n} = -22\overset{\prime\prime}{.}4 \pm 1\overset{\prime\prime}{.}1 \; cy^{-2} \quad \text{(Spencer Jones, 1939)} \tag{19}$$

$$n = -17\overset{\prime\prime}{.}9 \pm 4\overset{\prime\prime}{.}3 \; cy^{-2} \quad \text{(Clemence, 1943)} \tag{20}$$

However, the very careful analysis of a large number of transits of Mercury observed since 1677 by Morrison and Ward (1975) yields

$$\dot{n} = -26\overset{\prime\prime}{.}0 \pm 2\overset{\prime\prime}{.}0 \; cy^{-2} \tag{21}$$

Subject to further investigations of both modern and ancient data, I feel that there is no longer any need to suppose that $\dot{n}$ has changed appreciably during the *historical* past. This is what we would expect. Changes in sea level over the past two or three millennia are unlikely to have been more than about 1 metre, which would correspond to a change in $\dot{n}$ of the order of 1 % (cf. Munk and MacDonald, 1960).

It seems particularly encouraging that recent determinations of oceanic tidal dissipation yield results for $\dot{n}$ which are compatible with the astronomically determined figures. Thus Lambeck (1975) obtained from a detailed tidal analysis

$$\dot{n} = -35\overset{\prime\prime}{.}0 \pm 4\overset{\prime\prime}{.}0 \; cy^{-2} \tag{22}$$

The earliest studies of tidal dissipation, e.g., Jeffreys (1920) and Heiskanen (1921), yielded very small values for $\dot{n}$ — around $-10'' \; cy^{-2}$. These conflicted with the astronomically determined values both from ancient *and* modern data. Because of the difficulties inherent in both the study of early astronomical observations and oceanic tidal measurements, the agreement between Lambeck's result and our own should be regarded as at least as good as we might expect. In any event, it would appear that the two distinct methods are capable of yielding results which are fully compatible — a major step forward.

Van Flandern (1975) devised a novel method for determining $\dot{G}$ — the rate of change of the universal gravitational constant — from the various values of $\dot{n}$ determined astronomically. If G is decreasing, as several cosmologies would lead us to believe — then it follows that the solar system is expanding. Obviously this will not be detectable on TD since this time scale is defined with reference to the planetary motions. However, we should expect a systematic discrepancy between values of $\dot{n}$ derived on TD and those on AT (atomic time) if $\dot{G}$ is finite. It is readily shown that:

$$\frac{\dot{n}_{AT} - \dot{n}_{TD}}{n} = 2 \frac{\dot{G}}{G} \tag{23}$$

Van Flandern combined his own result for $\dot{n}$ on AT (-65 ± 18) with that given in Muller and Stephenson, 1975 (-38 ± 4) and deduced:

$$\dot{G} = -8 \pm 5 \cdot 10^{-11} \; yr^{-1} \tag{24}$$

More recently he obtained (1977) from a more extensive study of modern observations:

$$\dot{n} = -36\overset{\prime\prime}{.}0 \pm 5\overset{\prime\prime}{.}0 \; cy^{-2} \tag{25}$$

Combining this with the result of the present study we deduce:

$$\dot{G} = -2 \pm 2 \cdot 10^{-11} \; yr^{-1} \tag{26}$$

Obviously this last figure is only marginally significant. Preliminary results from LLR (these are, as yet, unpublished) appear to be in good accord with $\dot{n} = -30"$ cy^{-2}. If this is so, $\dot{G}$ would effectively vanish. Personally I feel that at the present time the reality of a non-zero $\dot{G}$ is questionable. It would appear that the results from lunar laser ranging which are likely to emerge within the next few years could be crucial in settling this issue once and for all.

The derived value of $\dot{e}$ corresponds to a rate of change in the length of the day of

$$\dot{d} = 2.1 \text{ ms cy}^{-1} \tag{27}$$

This result is in satisfactory agreement with values obtained from palaeontological data. However, the significance of this conclusion is far from clear in view of major difficulties in the interpretation of fossil records. It is further felt that attempts to isolate and interprete the non-tidal component of $(\dot{\omega}/\omega)$ are at present of doubtful value since the uncertainties in the parameters involved are still fairly large.

Surely the most interesting aspect of the study of the Earth's rotation in the past lies in the analysis of the ΔT fluctuations. The decade fluctuations have been extensively studied in recent years by Morrison (1973). Obviously in pre-telescopic times only a very general idea of the form of the ΔT curve can be obtained, but it is possible to set some valuable bounds on the amplitude of the irregular fluctuations.

Figure 4 shows the $\delta\Delta T$ plot for the data in Table 1 using $\dot{n} = -30"$ cy^{-2}. The abscissa represents the ΔT parabola ($\Delta T = 20^s + 114^s T + 38^s.30T^2$). With the exception of A.D. 2, which is clearly discordant, all observations are in excellent accord. However, in many instances data in Table 1 are not represented merely because any value of $\delta\Delta T$ in the region covered by the diagram would yield a total (or non-total eclipse) at the place of observation in full agreement with the record. The

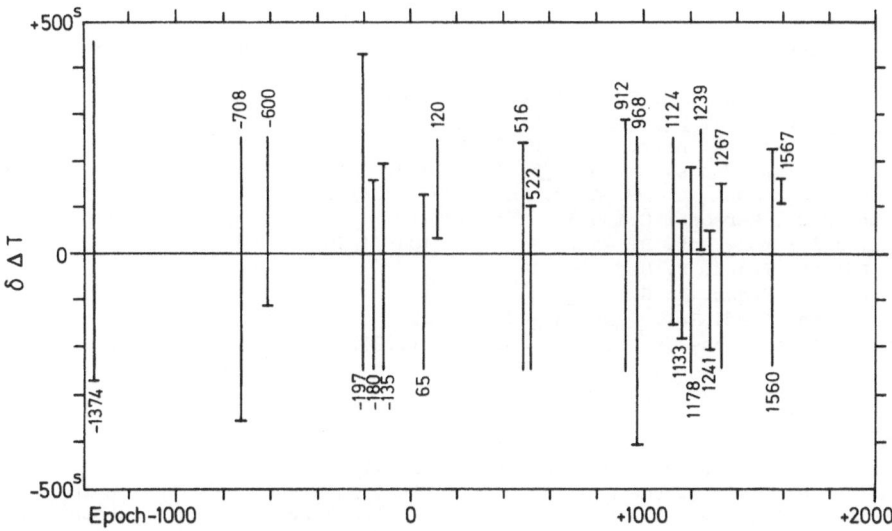

Fig. 4. Observational bounds on $\delta\Delta T$

usual reason here is that near the place concerned the path of central eclipse happened to traverse almost a line of latitude so that changes in ΔT have only a very small effect on the eclipse magnitude. Needless to say, it is disappointing to see so many good data fall by the wayside. For these, all that can be said is that the circumstances describing individual reports are amply confirmed. However, this leads us to infer that the more critical observations (these are displayed in Fig. 4), which were selected by exactly the same criteria, are indeed reliable.

For most of the period covered, a value of ΔT close to zero is acceptable. The very precise observation that the Jesuit astronomer Clavius made at Rome in A.D. 1567 (he happened to witness a total eclipse whose central zone was only a few kilometres wide) indicates a departure of ΔT from the mean curve of rather more than 100^s. There seems to be no evidence in the previous two millennia of any changes of amplitude greater than this. Thus it is possible to represent the ΔT curve back to about 1000 B.C. by a simple parabola on which are superposed irregular fluctuations of amplitude some 100^s.

9. Conclusions

The most valuable deductions from the ancient and medieval eclipse observations are the evidence that ṅ has remained essentially constant over the last three millennia and the very smooth form of the historical ΔT curve. The material presented in this paper is in the process of being expanded into a book-length manuscript in which the reliability of the individual observation and the method of analysis will be further critically reviewed. However, it is confidently anticipated that there will be no significant alterations in the results presented here.

References

Brown, E.W.: Tables of the Motion of the Moon. New Haven: Yale University Press, 1919, Vol. 1

Clemence, G.M.: Astr. Pap. Am. Eph. 11, part 1 (1943)

Clemence, G.M.: Astron. J. 53, 169 (1948)

Cowell, P.H.: Mon. Not. R. Astr. Soc. 66, 3 (1905)

Dunthorne, J.: Philos. Trans. R. Soc. 46, 162 (1749)

Flandern, Van, T.C.: Mon. Not. R. Astr. Soc. 170, 333 (1975)

Flandern, Van, T.C.: Personal Communication (1977)

Fotheringham, J.K.: Mon. Not. R. Astr. Soc. 81, 104 (1920)

Halley, E.: Philos. Trans. R. Soc. 19, 160 (1695)

Halley, E.: Philos. Trans. R. Soc. 29, 255 (1715)

Heiskanen, W.: Ann. Acad. Sci. Fenn. A 18, 1 (1921)

Jeffreys, H.: Philos. Trans. R. Soc., A 221, 239 (1920)

Kant, I.: Wöchentliche Fragund Anzeigungs-Nachrichten 23 & 24 June 1754

Lambeck, K.: J. Geophys. Res. 80, 2917 (1975)

Martin, C.F., Van Flandern, T.C.: Science 168, 246 (1970)

Morrison, L.V.: Nature (London) 241, 549 (1973)

Morrison, L.V., Ward, C.G.: Mon. Not. R. Astr. Soc. 173, 183 (1975)

Muller, P.M.: An Analysis of the Ancient Astronomical Observations with the Implications for Geophysics and Cosmology. PhD Thesis, University of Newcastle upon Tyne (1975)

Muller, P.M., Newhall, X.X., Van Flandern, T.C., Williams, J.G.: In Preparation

Muller, P.M., Stephenson, F.R.: In: Growths Rhythms and the History of the Earth's
  Rotation. Rosenberg, C.D., Runcorn, S.K. (eds.), London: Wiley 1975
Munk, W.H., MacDonald, G.J.F.: The Rotation of the Earth. Cambridge: University Press,
  1960
Newcomb, S.: Astr. Pap. Amer. Eph. $\underline{9}$, part 1 (1912)
Newton, R.R.: Ancient Astronomical Observations and the Accelerations of the Earth
  and Moon. Baltimore-London: Johns Hopkins University Press, 1970
Newton, R.R.: Medieval Chronicles and the Rotation of the Earth. Baltimore-London:
  Johns Hopkins University Press, 1972
Sitter, De, W.: Bull. Astr. Inst. Neth. $\underline{4}$, 21 (1927)
Spencer Jones, H.: Ann. Cape Obs. $\underline{8}$, part 8 (1925)
Spencer Jones, H.: Mon. Not. R. Astr. Soc. $\underline{99}$, 541 (1939)
Stephenson, F.R.: Philos. Trans. R. Soc., A $\underline{276}$, 118 (1974)
T.I.E.S.: Trans. Illum. Eng. Soc. $\underline{20}$, 565 (1925)

Tidal Deceleration of the Earth's Rotation Deduced from Astronomical Observations in the Period A. D. 1600 to the Present

L. V. Morrison

1. Introduction and Principle of Method

Under the action of tidal friction the Earth's rate of rotation decreases and the Moon's orbit expands. As the Moon's orbit expands its orbital velocity decreases, but the orbital angular momentum (the product of the orbital radius and angular velocity) increases, such that the rotational angular momentum lost by the Earth is gained by the Moon's orbit.

The decrease in the Earth's rate of rotation due to tidal friction cannot be measured directly from astronomical observations made over the past three centuries because it is obscured by the much greater fluctuations in rate occurring over periods of several decades, which are thought to be caused by the transfer of angular momentum between the core and mantle. Data over a time span of at least ten centuries are required in order to measure directly the comparatively small underlying trend in the rate due to tidal friction.

The decrease in the Earth's rate of rotation is *deduced* from telescopic observations over the past three centuries by measuring the decrease in the Moon's angular velocity and hence its increase in orbital angular momentum. By the law of the conservation of angular momentum this gain in orbital angular momentum by the Moon is equated to the Earth's loss in rotational angular momentum from which the decrease in the Earth's rate of rotation can be calculated.

2. Moon's Orbital Angular Deceleration

In order to measure the Moon's orbital angular deceleration we require a series of observations of its orbital position on a uniform time scale. Since the invention of the telescope around A.D. 1600, accurate observations have been made of the Moon's orbital position using the known positions of the stars as reference points. The observations which I shall discuss here are timings of the instants of occultation of stars in the Moon's path. From these observations a series of observed positions of the Moon is obtained relative to the star positions for certain instants of time. These instants of time are noted on clocks keeping Universal Time (UT); that is, the time scale derived from the observed period of rotation of the Earth on its axis. Since the Earth varies in its rate of rotation, the derived time scale, UT, is not uniform, and therefore the deceleration of the Moon measured with respect to this time scale is spurious. The UT scale must first be converted to a uniform time scale: There are two methods by which this can be done.

Royal Greenwich Observatory, England

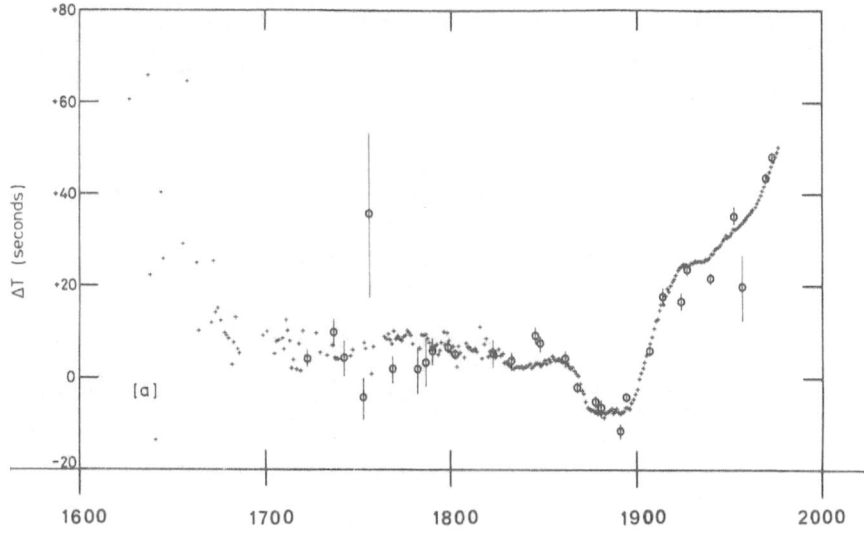

Fig. 1. Values of ΔT for the period A.D. 1627 to 1975 derived from observations of the transits of Mercury (*open circles with half-bar lengths of one standard error*) and lunar occultations (annual mean values shown as *crosses*)

2.1 Deceleration with Respect to Dynamical Time

The predicted positions of the Sun, Moon, and planets, calculated from the integration of their dynamical equations of motion, are made for instants of time on the uniform time scale which is implicit in those equations. I shall refer to this time scale as Dynamical Time (TD). Neglecting possible defects in the equations of motion and observational errors, the difference between the calculated and observed times for a particular orbital position of the Sun, Moon, or a planet is a direct measure of the difference between TD and UT. The difference in time at points along these time scales is usually denoted by ΔT and is measured in the convenient unit of seconds. Thus, UT is converted to a uniform time scale by the addition of the values of ΔT.

Because of its rapid angular motion, the Moon should provide an effective way of measuring ΔT, by forming the differences between the predicted and observed times of occultations of stars. However, the predicted times are dependent on the equations of motion, and these are not complete without an estimate of the orbital deceleration due to the tidal interaction with the Earth. This is the unknown which we are seeking to determine, so we must use some other body, independent of the Earth-Moon system, to determine ΔT.

We have chosen the motion of the planet Mercury to measure ΔT because of its rapid heliocentric motion and the circumstance that about 14 times a century it transits the disc of the Sun, thus permitting the precise instants of time at which the small disc of Mercury makes apparent contact with the limb of the Sun to be observed. There are about 2400 timings of these instants of contact for 30 of the transits in the period 1723 to 1973. Morrison and Ward (1975) have compared these observed times on the UT scale with the predicted times on the TD scale and formed mean values of ΔT for the epochs of the transits. The results are plotted in Figure 1 as open circles with half-bar lengths of one standard error. A timing made by Cassini in Paris of the transit of 1697 and timings made by Halley (see Fig. 2) on the island of St. Helena of the transit of 1677 were also used by Morrison and Ward (1975) in their analysis, but these are not shown in Figure 1: The reader is referred to Figure 2 of their paper.

With these values of ΔT from the transits of Mercury we can convert occultation observations timed on the UT scale to the uniform scale, TD, in the period 1677 to 1973 and hence measure the tidal deceleration of the Moon. In practice, the method proceeds as follows. A provisional value for the tidal deceleration is adopted in the lunar theory and this is used to calculate the expected times of occultations. The observed times (in UT) are substracted from the expected times to form a provisional set of values of ΔT which are then compared with the results from the transits of Mercury shown in Figure 1. The provisional value of the tidal deceleration is adjusted by least-squares to obtain the best agreement overall between the occultation and transit results. This adjustment has been made in Figure 1, where crosses represent the annual mean values of ΔT from occultations. The adjusted value for the tidal deceleration is found to be

$$\dot{n} = -(26 \pm 2) \ "/cy^2 \tag{1}$$

2.2 Deceleration with Respect to Atomic Time

The International Atomic Time scale (TAI) has been established since 1955.5 and this can be regarded as a uniform time scale against which to measure the tidal deceleration of the Moon. The differences, TAI − UT, published by the *Bureau International de l'Heure*, are used to convert observations made on UT to TAI. The provisional value for the tidal deceleration in the lunar theory is then adjusted until the predicted times of occultations agree overall with those observed on the TAI scale. Van Flandern's (private communication) most recent result using this method is

$$\dot{n} = -(36 \pm 4) \ "/cy^2 \tag{2}$$

The uncertainty of the result arises mainly from the fact that only observations made after 1955.5 can be used, and with only a 20-year span of data it is difficult to separate in the residuals the comparatively small affect due to a correction to the tidal deceleration from a correction to the mean motion. Van Flandern (1975) has argued that the difference between his result obtained with respect to TAI and ours with respect to TD can be taken as evidence for an extra non-tidal contribution of $-(10 \pm 4) \ "/cy^2$ arising from the decrease of G with respect to atomic time. In the rest of this discussion I shall adopt the value of $-(26 \pm 2) \ "/cy^2$ for the tidal deceleration of the Moon.

3. Tidal Deceleration of Earth's Rotation

From the conservation of angular momentum in the Earth-Moon system we find the consequential deceleration in the Earth's rotation caused by the lunar tidal torque to be given by (see e.g., Morrison, 1972)

Fig. 2a and b. Facsimiles of the first two pages of the report of Halley's observations of the transit of Mercury in 1677 contained in the appendix to *Catalogus Stellarum Australium sive Supplementum Catalogi Tychonici, Edmond Halley, 1679*

(1)

MERCURII TRANSITUS
Sub Solis disco, *Octob.* 28. Anno 1677, cum tentamine pro Solis Parallaxi.

Arum istud, & à mortalibus non nisi ter, (quod mihi scire contigit,) hactenus observatum Phænomenon transitus Mercurii sub Solis disco, mihi, in Insula *Sanctæ Helenæ* commoranti, felicius observare, quam cuivis alio Astronomo, contigit : *Gassendus* enim in transitu Anni 1631, & in hoc nostro Clarissimus *Gallet,* exitum solum spectaverunt; ingressu, huic sub densa nubium compagine, illi sub terra Orientali, latente ; Atque imperfectius adhuc Anno 1661, inclytus ille *Hevelius Gedani,* & nostrates *Londini,* qui solo situ intra faciem Solarem sumpto contenti erant : Mihi primo & ingressus & egressus momenta accuratissime confecta sunt, idque peculiari & insolito Cœli favore; erat enim nocte præcedente *Octobris* 28vum Cœli facies tristissima, cum vento valido, interdumque descendentibus Nubibus densa Nebulæ Insulæ summitates obvelavit ; luce reversa, vento licet paulo remissiore, idem manfit Cœli vultus; Juxta Solis ortum, ad instrumenta me contuli, languente jam omne spe observationis habendæ, ruboque 24 pedum in plagam Solis verso, patienter expectavi, an per Nubium aliquem hiatum conspici possit desideratissimus Phœbus : Juxta horam octavam Nubes rarescere ceperunt, ita ut 8 *b.*36 *m.* Sole clare confpecto, Mercurium nondum intrasse pronunciavi ; inde brevibus intervallis sæpius eluxit, ac sequentem habui observationem.

A

Is

(2)

In Insula Sanctæ Helenæ Anno 1677 Octobris 28. St.Vet. A.M.

9 *b.* 20 *m.* 35 *s.*			Sol purus videbatur.
9	26	17	Limbus Solis a Mercurio temeratus, facta quasi denticula, 10 grad. a Nadir Solis ad dextram circiter.
9	27	30	Erat totus ☿ intra Solem efficiens angulum contactus.

Hinc visus est magis magisque Centro Solis appropinquare, usque in mediam decimam, cum rursus Nubium densarum coalitus, spectaculum adeo jucundum oculis meis eripuit, nec iterum conspiciendum præbuit, antequam infaret hora secunda P. M. cum jam dissipatis Nubibus, videbatur ☿ brevi excessurus , itaque summa diligentia attendi ad momentum exitus , & deprehendi quod 2 *b.* 38 *m.* 39 *s.* Distantia limbi proximi Mercurii a limbo Solis non excederet Mercurialem Diametrum.

2	40	8	Limbus Mercurii attigit Solis limbum.
2	41	0	Centralis egressus 30 gr. circiter a Nadir ad dextram.
2	41	54	Solis limbus integer factus.

Ita ut a centrali ingressu ad exitum , elapsæ sunt 5 *b.* 14 *m.* 20 *s.* quod temporis spatium verissimum reputo, & ab omni exceptione liberum.

De Latitudine Mercurii ad hæc momenta nihil ausim asserere, puncta a me assignata sola æstimatione capiebantur ; Micrometri enim casu infausto inutiles reddebantur, nec in tantis Altitudinibus multum pollet id genus Instrumenti; Utamur itaque ea Latitudine, quam D. *Galleti* observatio postulat, nempe quod proxima distantia Centr.orum fit quatuor minutorum circiter ; nam in minutis secundis non est accurata vel observationum vel supputationum series, ut cuivis attendenti perspicuum est ; deia supponatur Solis semidiameter 16 *m.* 17 *s.* quod confonum est observationibus Clariss. *Cassini* & *Flamstedii* ; hinc Chorda,

$$(\dot{w}/w) \text{ lunar} = +0.93 \ \dot{n} \cdot 10^{-9}/\text{cy} \tag{3}$$

where w is the angular velocity of the Earth and $(\dot{w}/w)$ expresses the fractional change in angular velocity per century. There is a solar torque which also acts to decelerate the Earth and this ranges between 20 % and 30 % of the lunar torque, depending on the model of tidal friction used. I shall adopt a value of 25 %, thus giving a combined deceleration of

$$(\dot{w}/w)_{\text{lunar + solar}} = +1.16 \ \dot{n} \cdot 10^{-9}/\text{cy} \tag{4}$$

Substituting $\dot{n} = -(26 \pm 2)''/\text{cy}^2$ we find

$$(\dot{w}/w)_{\text{lunar + solar}} = -(30.2 \pm 2.4) \cdot 10^{-9}/\text{cy} \tag{5}$$

Adding the small *acceleration* of $+1.1 \cdot 10^{-9}/\text{cy}$ due to the solar torque on the atmosphere, the total deceleration due to tidal torques becomes

$$(\dot{w}/w)_{\text{tidal}} = -(29 \pm 2) \cdot 10^{-9}/\text{cy} \tag{6}$$

Alternatively, multiplying by $w = 47''.5 \cdot 10^9/\text{cy}$ we obtain

$$\dot{w}_{\text{tidal}} = -(1380 \pm 100)''/\text{cy}^2 = -(68 \pm 5) \cdot 10^{-23} \ \text{rad/s}^2 \tag{7}$$

The rate of dissipation of tidal energy $(-\dot{E}_{\text{tidal}})$ corresponding to this deceleration is

$$-\dot{E}_{\text{tidal}} = +(4.0 \pm 0.3) \cdot 10^{12} \ \text{Watts} \tag{8}$$

It is interesting to note that the average *observed* deceleration of the Earth's rotation over the past 300 years, which is obtained directly by fitting a parabola to the data in Figure 1, is

$$\langle \dot{w}/w \rangle \text{ observed} = -9 \cdot 10^{-9}/\text{cy} \tag{9}$$

This supports my statement earlier that the tidal deceleration of the Earth's rotation cannot be measured directly from astronomical observations made over the past 300 years.

The result deduced in this paper for the tidal deceleration of the Earth's rotation is in excellent agreement with the directly 'observed' value of $-(29 \pm 3) \cdot 10^{-9}/\text{cy}$ obtained by Muller and Stephenson (1975) in their analysis of ancient solar eclipse observations. The agreement of these two results implies that the theory of tidal friction (with a solar couple of about 25 % that of the lunar couple) provides an adequate and sufficient mechanism to explain the observed deceleration of the Earth's rotation over the past two thousand years. In particular, there is no residual *acceleration* and therefore no support for a significant change in the moment of inertia. However, there is one major problem in this argument. The ancient data also permit one to derive the orbital deceleration independently of the Earth's rotational deceleration, and Muller and Stephenson (1975) find this to be $-(37 \pm 5)''/\text{cy}^2$, which is not compatible with the Earth's 'observed' deceleration if only the mechanism of tidal friction is invoked. By the action of tidal friction an orbital lunar deceleration of $-(37 \pm 5)''/\text{cy}^2$ would imply a rotational deceleration of the Earth of $\dot{w}/w = -(42 \pm 5) \cdot 10^{-9}/\text{cy}$; but the 'observed' deceleration is only $-(29 \pm 3) \cdot 10^{-9}/\text{cy}$, so, either there is an *acceleration* of about $+ 13 \cdot 10^{-9}/\text{cy}$ due to some mechanism other than tidal friction, or the results are in error by greater amounts than their quoted errors suggest. The second explanation has proved to be the more likely one.

In the analysis of the ancient solar eclipse data, the solutions for the decelerations of the Earth and Moon are not strictly independent. They are both unknowns in the same linear equation of condition and they are highly correlated with one another. Besides these two unknowns, Muller (1975) has introduced a third for the motion of the lunar node and this has altered the solutions to: $\dot{n} = -(30 \pm 3)\,"/cy^2$ and $\dot{w}/w = -(24 \pm 2) \cdot 10^{-9}/cy$. This brings the result for the lunar deceleration into reasonable agreement with the result in this paper, but the Earth's rotational deceleration is still smaller than would be expected from the action of tidal friction alone. So, there is still some evidence to support a small accelerative component in the Earth's rotation. However, caution must be exercised in interpreting this as evidence for some internal change in the Earth, such as the moment of inertia, because Hipkin (1975) points out that some of the redistribution of angular momentum in the Earth-Moon system due to the tidal couple may be taken up by small (and as yet undetectable) changes in the eccentricity and inclination of the Moon's orbit.

4. Conclusions

The observed tidal deceleration of the Moon derived from astronomical observations made since A.D. 1600 is given by

$$\dot{n} = -(26 \pm 2)\,"/cy^2. \tag{10}$$

The tidal deceleration of the Earth's rotation deduced from this is

$$\dot{w}/w = -(29 \pm 2) \cdot 10^{-9}/cy. \tag{11}$$

References

Hipkin, R.G.: Tides and the Rotation of the Earth. Growth Rhythms and the History of the Earth's Rotation. Rosenberg, Runcorn (eds.). London: Wiley and Sons, 1975
Morrison, L.V.: The Moon 5, 253 (1972)
Morrison, L.V., Ward, C.G.: Mon. Not. R. Astr. Soc. 173, 183 (1975)
Muller, P.M.: Jet Propulsion Lab. SP43-36 (1975)
Muller, P.M., Stephenson, F.R.: The Accelerations of the Earth and Moon from Early Astronomical Observations. Growth Rhythms and the History of the Earth's Rotation. Rosenberg, Runcorn (eds.). London: Wiley and Sons, 1975
Van Flandern, T.C.: Mon. Not. R. Astr. Soc. 170, 333 (1975)

Determination of the Rotation of the Earth (at Present)

H. Enslin

1. Introduction

The determination of the Earth's rotation includes current observations of both the rotational phase of the Earth with respect to a celestial frame of reference, and the position of the axis of rotation with respect to the Earth's crust or, in other words, the regular determination of Universal Time and polar motion.

To accomplish this with the highest precision and accuracy attainable, close international cooperation is necessary. National observatories provide the results of their observations to data analysis centers working under international supervision. These centers, each of which has its particular tasks, are the *Bureau International de l'Heure* at Paris, France, and the *International Polar Motion Service* at Mizusawa, Japan.

Up to the present, the observations aimed at the regular determination of Universal Time continue to consist soleley of the measurement of directions to stars by means of optical instruments. For the past few years, polar motion has been made available on a regular basis, from Doppler satellite observations also, so that the data obtained could be included in a global solution when the Earth's rotation parameters have been computed.

Quite apart from the latter, the precision in determining the rotation of the Earth has been increased considerably during the past decade. This is due not only to instrumental improvements and increased observational efforts of the observatories, but also to the efforts of the data analysis centers, which have elaborated more refined methods for the evaluation of the observational data. Moreover, the invention of atomic standards has made possible the establishment, by the *Bureau International de l'Heure*, of International Atomic Time, an internationally agreed reference time scale which is easily available and easy to read-off, and of utmost uniformity. Comparison of atomic time and Universal Time provides any irregularities of the rate of rotation of the Earth as far as these can be separated from "noise" due to the errors of the astronomical observations.

Such comparisons since 1955.5, when atomic time became available, have shown that nonperiodic rotational accelerations of the Earth persist for several years. Values obtained are between -1.5 and $+0.3 \cdot 10^{-11}/$ day. Additional nonperiodic deviations that persist for several weeks or several months have been found to be up to $20 \cdot 10^{-11}/$day. (Secular acceleration due to tidal friction is about $-0.05 \cdot 10^{-11}/$day; maximum periodic accelerations produced by fortnightly lunar body-tide are about $200 \cdot 10^{-11}/$day.)

Deutsches Hydrographisches Institut, Postfach 220, 2000 Hamburg 4, Federal Republic of Germany

Recent studies about the polar motion occurring between about 1969 and 1976 suggest that the period and amplitude of the Chandler component as well as the amplitude of the annual component were more stable than has generally been thought.

1.1 Preview

For about two decades, the understanding of the term rotation of the Earth has included both the rate of the Earth's rotation *and* polar motion: In fact, the two subjects are closely related.

After reviewing the general features of the Earth's rotation and fundamentals, this paper explains how Universal Time (UT), which provides the rate of rotation, and polar motion are determined — at present — on a regular basis and in international cooperation. It includes a general review of the methods and instruments used at the observing stations all over the globe, and of the work of the two centers that evaluate the observational data and make the results available to different kinds of users. Some of these results are discussed.

There are several new methods for measuring the Earth's rotation, such as radiointerferometry, and range tracking of the Moon and of artificial satellites. They promise results of higher precision, and, in particular, better short-term resolution than those currently being used. However, these new methods are still in the experimental state, and no results from them are used, as yet, when computing the Earth's rotational parameters by the data analysis centers.

2. General Features of the Rotation of the Earth

2.1 Variation in the Earth's Rate of Rotation

It is customary to distinguish three kinds of variations in the rate of rotation or, in other words, in the length of day:

1. Secular deceleration due to tidal friction

2. Irregular changes of various duration caused by weather, change of see level, processes in the Earth's interior and, possibly, solar radiation

3. Periodic variations

The periodic variations — as far as a geophysical explanation has been given for evidence from astronomic observations — are due mainly either to winds, or to lunar or solar Earth tides. The well-known seasonal variation, SV, has an annual component caused mainly by zonal winds, and a six-monthly component, which is half due to solar zonal tide and half due to winds. The changes in the length of day amount to ±0.4 and ±0.3 ms/day, respectively, and the maximum semi-amplitude of the integrated effect is 30 ms. SV has been rather stable over most of the years. Another — although intermittent — term of about two year's period and of the order of 0.1 ms/day semi-amplitude has been found to be in very good agreement with zonal wind characteristics (Lambeck and Cazenave, 1973, 1977). The change of the Earth's momentum of inertia due to the effect of lunar zonal tides of 13.6 and 27.6 day periods, results in variations in the length of day of ±0.33 and ±0.17 ms/day, respectively (Munk and MacDonald, 1960). In both cases, the semi-amplitude of the integrated effect is smaller than 1 ms — which is below the noise level of the astronomic data — and even so it has been evaluated by several authors. Guinot (1974), applying spectral analysis to data over a period of seven years, obtained results with only a few percent uncertainty.

2.2 Polar Motion

The Earth's axis of rotation intersects the surface of the Earth, near the north pole, in the pole of rotation. This does *not* coincide with the pole of figure, which is the point where the principal axis of inertia pierces the surface of the Earth near the north pole. Therefore, according to dynamic theory, the pole of rotation moves about the pole of figure. The term polar motion refers to the motion of the pole of rotation with respect to a point fixed on the Earth, which is called the reference pole. The internationally adopted reference pole is the Conventional International Origin, CIO.

The position of the rotational pole with respect to the reference pole is represented in a conventional coordinate system, in which +x is along the zero meridian, and +y along the meridian 90° west. The values of x and y are usually expressed as an angular displacement of both axes in question: $0\overset{''}{.}010$ are equivalent to a linear displacement of 0.31 m of the respective poles.

The polar motion has two well-known periodic components — periods of 12 months and of about 14 months — which make the polar curve spiral in and out (see Fig. 8). The former period is due largely to a seasonal shift of air masses, which brings the rotating Earth out of balance. The 14 month's motion, often designated as free wobble or Chandler wobble, is generated by some excitation mechanism — for which satisfactory explanation has not yet been found — and is damped by internal friction.

3. Equatorial System

The Earth's rotation is measured with respect to a quasi-inertial reference system fixed in space — the equatorial system. It is determined by the celestial equator, and the vernal equinox, which is that intersection of the equator and the ecliptic where the Sun passes in spring. The places of stars are given as angular distances with respect to the equator (declination) and the equinox (right ascension when measured along the equator).

The position of the celestial reference system, with respect to the stars, changes continually. The long-term part of this motion is called precession, and the periodic parts are collectively called nutation. Star catalogs, compiled from fundamental observations with meridian transit circles, provide mean places and proper motions at a specified date. By means of these data, allowing for precession, nutation, and other effects, apparent places of stars at the time of actual observations can be computed.

4. Astronomic Longitude and Latitude

Measurements of directions to stars with respect to the vertical at an observation site provide information about the angular position of the

Earth in space. The direction of the vertical is expressed by astronomic longitude and latitude. These coordinates correspond only roughly to those furnishing the site's position in a global geometric system, because of the existence of gravity anomalies.

If L is that plane that contains the vertical of a station and is parallel to the instantaneous axis of ration, and O the respective plane at the origin of longitudes, then the angle — measured westward — from O to L, is the station's instantaneous longitude, λ. The station's instantaneous latitude, ϕ, is defined by the acute angle between its vertical and any plane perpendicular to the instantaneous axis of rotation. Northern and southern latitudes are differentiated by the signs +, or -, respectively. The conventional coordinates, λ_0, ϕ_0, refer to a particular axis fixed in the Earth; for example, to that for which the pole is the CIO.

The motion of the pole makes the values of $\lambda - \lambda_0$ and $\phi - \phi_0$ change continuously, and differently at different locations. This provides the possibility to evaluate polar motion from the variations of latitudes and/or of longitude-differences observed at some, or more, places suitably distributed over the globe.

The longitude-difference between two stations corresponds to the difference in local time as determined by astronomic observations. The local times can be compared, for example, by means of radio time signals.

5. Time Scales

5.1 Astronomic Time

There are two kinds of astronomic time (which is generally understood to be based on the rotation of the Earth), sidereal time and Universal Time, both rigorously related by formulas.

The observation of stars provides local sidereal time. By addition of the observatory's conventional astronomic longitude, λ_0, Greenwich sidereal time is obtained. Then the corresponding value of Universal Time can be computed.

The kind of Universal Time obtained in this manner is denoted UTO. After UTO has been corrected for the effect of polar motion, UT1 = UTO + $(\lambda - \lambda_0)$ is obtained. UT1 represents the true angular motion of the Earth about its axis with reference to a point that coincides nearly, but not exactly, with the mean Sun. Comparison of UT1 with a uniform time scale reveals the irregularities of the Earth's rotation speed.

For some purposes, it is convenient to remove the seasonal variation in rotation speed. The values of UT2 - UT1 = -SV, as published by the *Bureau International de l'Heure* (BIH), give UT2 = UT1 - SV.

5.2 Atomic Time

Every undisturbed cesium atomic clock provides a very uniform time scale, the scale measure of which, in the ideal case, is the second of the International System of Units. The internationally adopted reference atomic time scale is International Atomic Time, TAI. It is es-

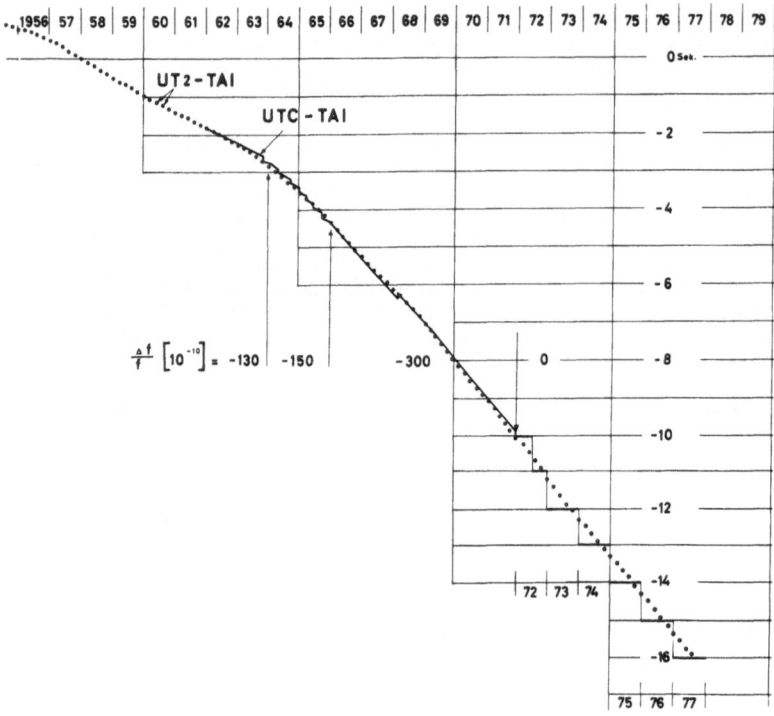

Fig. 1. Rotational time UT2, and time signal time UTC, with respect to atomic time TAI. Before 1972, the frequency of the atomic standards, which provided UTC, was offset by the amounts given in the diagram, and, if necessary, small steps were introduced in the UTC scale to keep it close to UT2. Since 1972, the frequency offset has been removed, and UTC is adjusted by the insertion of leap seconds to ensure approximate agreement with UT1 within the limits of ±0.9 s

stablished by the BIH using the data of about 80 cesium clocks located in several countries. TAI is conveniently available from time signals transmitted in the system of Coordinated Universal Time, UTC. This has, by definition, exactly the same rate as TAI and differs from it by an integral number of seconds (see Fig. 1).

The uniformity of TAI available since 1955.5 is so high that the precision of the determinations of UT1 - TAI rests entirely on the precision with which UT1 can be determined.

6. Methods and Instruments of Observation

6.1 Optical Instruments

Several types of instruments, with different methods of star observation, are in regular use providing astronomical time (with respect to a clock) and/or latitude. Table 1 lists the instrument types and quantities that contributed to the work of the BIH in 1976 (see Sect. 7).

Table 1. Contributions to the work of the BIH in 1976

Instrument type	Vert.	Observation of	Result of observation	Number 1976
Danjon astrolabe	M	the instant at which zenith distance is 30°	astronomical time (UTO) and instantaneous latitude	12
Circumzenithal				2
Photographic Zenith tube (PZT)	M	the instant of meridian transit and the zenith distance in the meridian	astronomical time (UTO) and instantaneous latitude	12
Visual transit instrument	L	the instant of meridian transit	astronomical time (UTO)	9
Photoelectric transit instr.				10
Visual zenith telescope	L	the difference of zenith distance between the two components of a pair, in the meridian	instantaneous latitude	19
Floating zenith telescope	MF			1
		Total number of instruments:		65

Fig. 2. Danjon astrolabe

Fig. 3. PZT

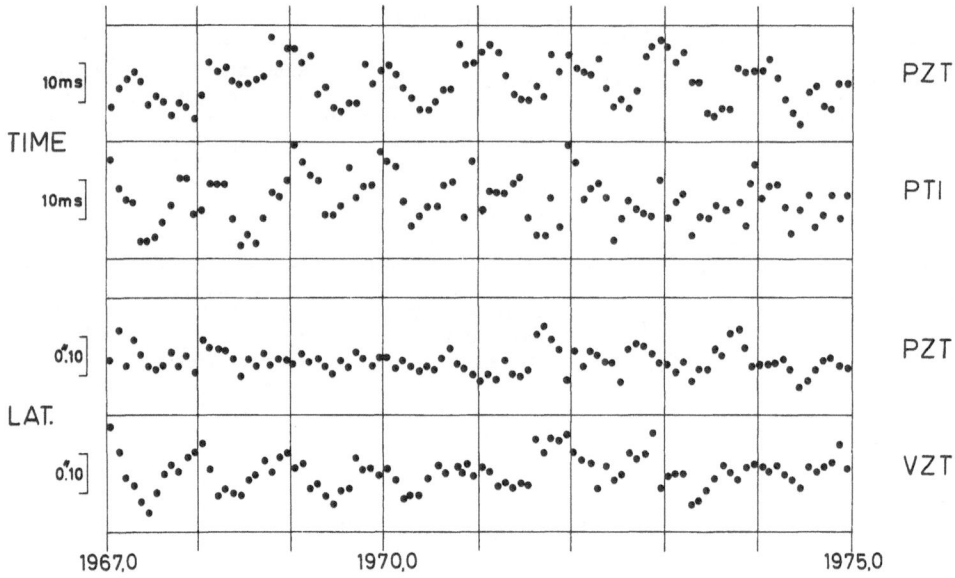

TIME

PZT

PTI

LAT.

PZT

VZT

10ms

10ms

0.10

0.10

1967.0   1970.0   1975.0

Fig. 4. Examples of instrument residuals (monthly averages) with respect to values derived by the International Polar Motion Service from astronomical observations made at many stations (see Sect. 7). *PTI* = photoelectric transit instrument; *VZT* = visual zenith telescope

An indication of the observation method and the manner of establishing the relation to the vertical is also given.

For the latter, mostly either a spirit level (L), or an "artificial horizon" formed by the surface of mercury kept in a basin (M) is used (see Fig. 2). The movable parts of the floating zenith telescope float in mercury in their entirety (MF). The observations are made visually, unless otherwise derivable from the designation of the instrument type. PZTs can be operated automatically, without the presence of an "observer" (Fig. 3).

Generally, the precision and long-term stability from PZT, astrolabe, and modern, large visual zenith telescope results are similar, and exceed those from other types. Observations using visual transit instruments can be severely affected by the observer's personal errors. In detail, the performance of instruments of the same type can often be quite different depending on local atmospheric conditions, skill of observers, or instrumental errors. Furthermore, a particular instrument's performance may change with time — for reasons which may be difficult to evaluate.

Much effort of the observatories is directed to the improvement of the observational results. This includes the improvement of the places of stars contained in the observing programs, because errors of the places used affect the time and latitude results. In particular, systematic errors of the star places reproduce themselves in the results as apparent annual variations. Internal corrections are derived by appropriately combining the results of observations. Figure 4 demonstrates the performance of some instruments.

6.2 Doppler Satellite Measurements

Since 1970, coordinates of the pole have been made available from Doppler satellite observations processed by the Dahlgren Polar Monitoring Service of the U.S. Naval Weapons Laboratory, and continued later by the Defense Mapping Agency (DMA). These polar coordinates are a by-product of the operation of the Navy Navigation Satellite System (Anderle, 1970). At present, this system includes five satellites on polar orbits, 1100 km high, which transmit frequencies, time markers, and orbit data. For orbit determination, the satellites are tracked by about 14 stations measuring the Doppler shift of the frequency emitted by the moving satellites. The position of a measuring site can be obtained analogously when the orbit of the satellite tracked is known (Mueller, 1964). The measurements are made automatically and have all-weather function.

The satellite motions are described in the equatorial system by means of six orbit constants and a number of corrections due to the many perturbations in the orbits of artificial satellites. Step-by-step modeling has led — and may continue to lead — to a gradual improvement of orbit computations through the years.

The coordinates of the DMA tracking stations have been established in a conventional three-axial coordinate system fixed in the center of, and corotating with, the Earth. Hence, these coordinates are geometrically defined and are independent of the local vertical.

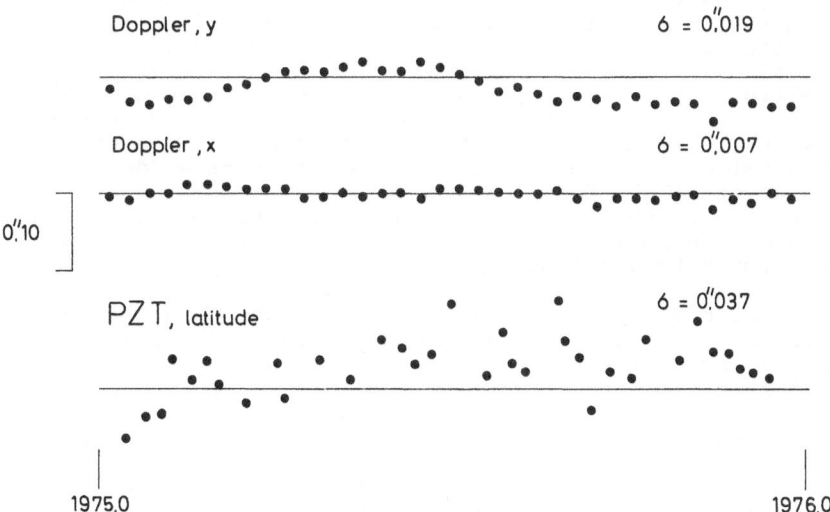

Fig. 5. Residuals of Doppler polar coordinates (10-day averages), and of latitudes obtained with one PZT (averages of equal weights), during 1975, with respect to values derived by the BIH in the global determination of the Earth's rotational parameters

The DMA reference pole has been attached empirically to that used in the astrometric work. Any deviation of the actual pole of rotation from the reference pole, the latter of which is defined by the adopted coordinates of the tracking stations, will result in apparent 24-hourly motions of the stations. The solution for the amplitude and phase of these motions yields information on the position of the pole of rotation with respect to the reference pole (Anderle, 1970).

Figure 5 demonstrates the precision of the polar coordinates as provided by the DMA. The apparent annual term in y might be due to an annual error in the combined results of the astrometric instruments, which define the BHI Reference System (see Sect. 7). It is not likely that the Doppler data are affected by annual errors. However, it has not yet been possible to confirm this using by the data of other new methods.

7. Data Analysis Centers

Bureau International de l'Heure (BIH)
International Polar Motion Service (IPMS)

From the scientific standpoint, it is indispensable to have the Earth rotation parameters on a current basis, in a homogeneous system, and with the highest precision attainable within a desired span of time. This is only possible through uniform and careful analysis of as many observational data as are available from the observatories in many countries; there is no other way to obtain final results that are affected as little as possible by changes in the numbers or types of instruments in operation, observational errors, and local anomalies. With regard to the astrometric observations — instrumental errors, atmospheric refraction, personal errors by observers, and errors of star places can introduce short-term or long-term errors in the results; changes in gravity or crustal displacements may cause nonpolar variations of the stations' astronomic longitudes and latitudes. The results of satellite observations can be affected by measuring errors as well as by model deficiencies, and crustal displacements may here also falsify the true picture of the rotation of the Earth — whatever one, in this context, may define as *the Earth*, where, strictly speaking, smaller or larger parts of masses are in continuous mutual motion. Concerning this question, the best definition of the phrase "determination of the rotation of the Earth" seems to be that contained in Robbins (1976), according to which it is "the determination of successive quasi-instantaneous positions of a unique set of terrestrial axis with respect to a non-rotating frame".

It is the task of the BIH in Paris (Director: B. Guinot) and the IPMS in Mizusawa (Director: S. Yumi) to act as data analysis centers for the rotation of the Earth. The BIH can be described as an operational service that evaluates data and publishes results with little delay; in contrast, the IPMS carries out long-term studies and therefore is able to exploit the raw material more completely. Both centers — or services — are funded mainly by the respective governments; some grants are provided from international sources. Each center is under a directing board, which includes representatives of the International Astronomical Union (IAU), the International Union of Geodesy and Geophysics (IUGG), and other international organizations.

The BIH was founded in 1919 to act as an international center on time. Nowadays, this includes atomic time as well as astronomic time.

The basic formulas used by the BIH for the computation of the polar coordinates x and y, an auxiliary unknown z (which allows for nonpolar variations common to all stations), and of the time difference UT1 – UTC, are (Guinot et al., 1970):

$$x\cos\lambda_{o,i} + y\sin\lambda_{o,i} + z \qquad\qquad\qquad = \phi_i - \phi_{o,i}, \quad \text{weight: } p(\phi)_i \tag{1}$$

$$(-x\sin\lambda_{o,i} + y\cos\lambda_{o,i})\frac{\tan\phi_{o,i}}{15} + (UT1 - UTC) = UTO_i - UTC, \quad \text{weight: } p(t)_i \tag{2}$$

A least-square solution for the unknowns is made every five days. The observed quantities in the equations, latitude ϕ_i and time difference (UTO_i – UTC), are those obtained with instrument i and transmitted to the BIH by teletype or letter, with specified corrections and appropriate interpolation between the datas of actual observations applied by the BIH before processing. Polar coordinates from Doppler measurements (included in the computation since 1972) are treated, with appropriate weights, as latitude determinations carried out at 0° and 90° longitude. $\phi_{o,i}$ ($\lambda_{o,i}$) is the conventional latitude (longitude) of instrument i.

The whole set of $\phi_{o,i}$ and $\lambda_{o,i}$ of the instruments in operation in 1967, in connection with the weights applied, form the *BIH Reference System*: The origin of longitudes and the BIH pole. The latter has been empirically attached to the CIO. Particular measures are taken to preserve the System, regardless of any changes in the weights or quantity of instruments.

In *Circular D*, the BIH publishes — with an average delay of six weeks — 5-day raw values of x, y, and (UT1 – UTC) as obtained from formulas (1) and (2), and also 5-day (one-sided) smoothed values. For the short-term prediction of the Earth's rotation, it also operates a rapid service on a weekly basis, the results of which are used especially for the navigation of deep space probes.

The precision of the 5-day raw values has improved by about 50 % from 1967 to 1976; the standard deviation of x and y is now about 0″01, and that of UT1 – UTC only slightly more than 1 ms (Feissel, 1976; Sugawa, 1976). This is due partly to the increased number of astronomic observations carried out, the increased quantity of photoelectric and photographic instruments in operation, and the introduction of Doppler satellite data, and partly to refinements in the BIH's treatment of the raw observational results (Feissel, 1976). Basic data used by the BIH, details on — and results of — computations, are published in the *BIH Annual Reports*.

The IPMS originates from the International Latitude Service, ILS, founded in 1899 to determine and study the polar motion. The ILS chain consists of five stations, with visual zenith telescope, all very near latitude 39° 08' north. The same star program is used by all. Therefore, the ILS polar coordinates are not affected by errors in star places and in proper motions. Adopted values for the latitudes of the ILS stations define the CIO. It is realized by means of the regular observations at the ILS stations and processing of the data.

Because the ILS chain consists of only five stations, there has been some uncertainty in the results. From 1962 onwards, the tasks of the ILS were redefined, and it was renamed IPMS. It continues to analyze the ILS chain's observations and, in addition, determines polar motion on the basis of long-term studies on the measurements made at many stations. Polar coordinates that were obtained from latitude observations made with 57 optical instruments from 1962 to 1972, were published in the *Annual Report of the IPMS for 1972*. The *Annual Report of the IPMS for 1974* gives polar coordinates, and also information on UT1, derived from time and/or latitude determinations carried out with optical instruments from 1967 to 1974. The IPMS has also used formula (1), or formulas (1) and (2), respectively. It is in the position to provide a surer analysis than the BIH of the data blocks, because these can be treated as closed units long after the observations were made.

Both BIH and IPMS publish residuals of the raw observational data or of unsmoothed mean values, respectively, which may be used for further studies; for example, for the study of accidental or periodic errors of the observations, or the drift of stations.

Neglecting the Doppler satellite data, which are used by the BIH only, the work of both BIH and IPMS is based on the observational results of rather the same set of stations or instruments, respectively. The ILS chain is included in both cases.

Unfortunately, the distribution of the astrometric stations is far from satisfactory: Almost two-thirds of the instruments are located in Europe, and there are only a few in the southern hemisphere. The distribution of the Doppler satellite tracking stations of the DMA network is incomparably better, reaching from the Antarctic to Greenland (Capitaine and Feissel, 1974).

8. Results on the Rotation of the Earth

Figure 6 shows annual averages of the daily rates of the speed of rotation for the past 20 years obtained from yearly differences of UT1 − TAI, which makes the effect of regular seasonal variations vanish. The values of Δ(UT1 − TAI)/365 days drawn for every 0.2 year, can be represented fairly well by segments of straight lines, the slope of which is a measure for the long-term acceleration of the rate of rotation. Maximum values obtained are −1.5 and +0.3 $\cdot$ 10^{-11}/day. For a comparison, the secular acceleration due to tidal friction, which is estimated to be −0.05 $\cdot$ 10^{-11}/day (equivalent to an increase in the length of day of 1.6 ms per cy), has also been plotted.

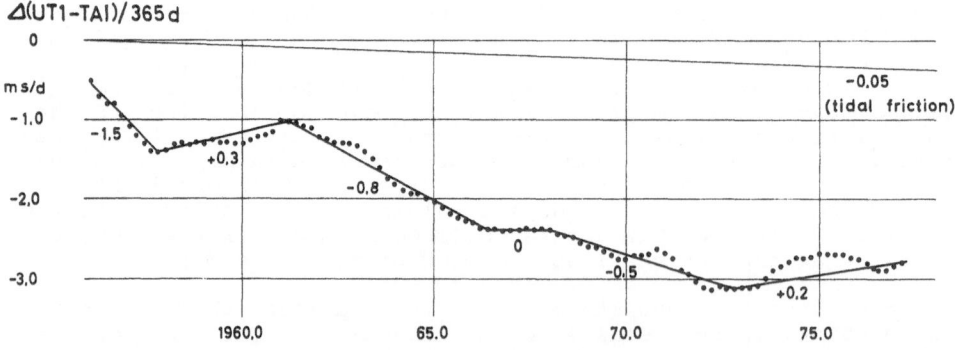

Fig. 6. Changes in the length of day from yearly differences of UT1 − TAI. The figures giving the accelerations of the Earth's rotation, according to the straight lines drawn, are in 1 $\cdot$ 10^{-11}/day

It has been suggested by Brouwer (1952) that small random changes in the speed of rotation occur, which build up to give the effect of a constant acceleration or deceleration. Another explanation would be that these fairly fast changes in accelerations are due to continuous geophysical effects, which change rather abruptly.

The long-term accelerations shown in Figure 6 are superimposed by short-term irregular and periodic accelerations that are considerably higher. In particular, the periodic accelerations produced by the two-weekly lunar Earth tide reach about $200 \cdot 10^{-11}$/day (Markowitz, 1972). Irregular accelerations continue for weeks or months. Markowitz (1972) found maximum accelerations of about $5 \cdot 10^{-11}$/day from 90-day averages, and of $20 \cdot 10^{-11}$/day from 30-day averages.

A rather abrupt change in the length of day occurred at the end of 1973. Figure 7 shows the average daily rates between 1973.5 and 1975.5 computed from 30-days differences of UT1 – TAI, and also of UT1 – UT2. The latter represent the curve that is normally to be expected because of the seasonal variations, SV. From about October 1973 to January 1974 the length of day decreased by about 0.8 ms, and then the Earth slowed down again. The maximum deviation of acceleration from normal, which can be seen in the diagram, amounts to $20 \cdot 10^{-11}$/day. It seems that these variations can be explained as anomalous seasonal variations. In fact, Rudloff (1975) has pointed out that the global distribution of winds was anomalous at that time, especially in December 1973.

The curves shown in Figure 8 have been reproduced from a paper by Markowitz (1976). He has compared ILS, IPMS, BIH, and Doppler polar motions with theoretical curves (TCs) with two, fixed magnitude vector components. The points plotted are unsmoothed monthly positions (except the computer-generated diagram of TC-2); the curves drawn through the points are to aid the eye.

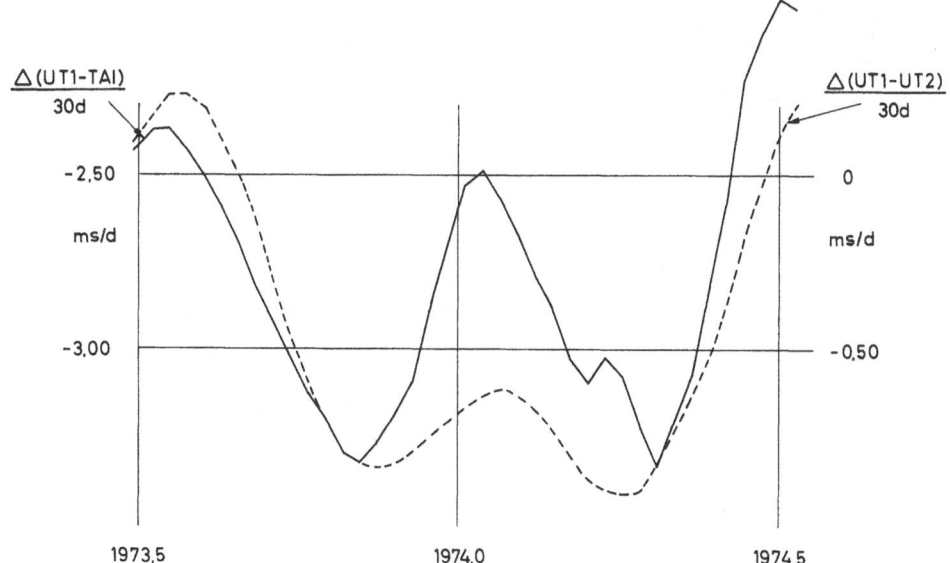

Fig. 7. Changes in the length of day during 1973.5 to 1974.5 from 30-day differences of UT1 – TAI. The dotted lines indicate the changes to be expected, due to the regular seasonal variations of the Earth's speed of rotation

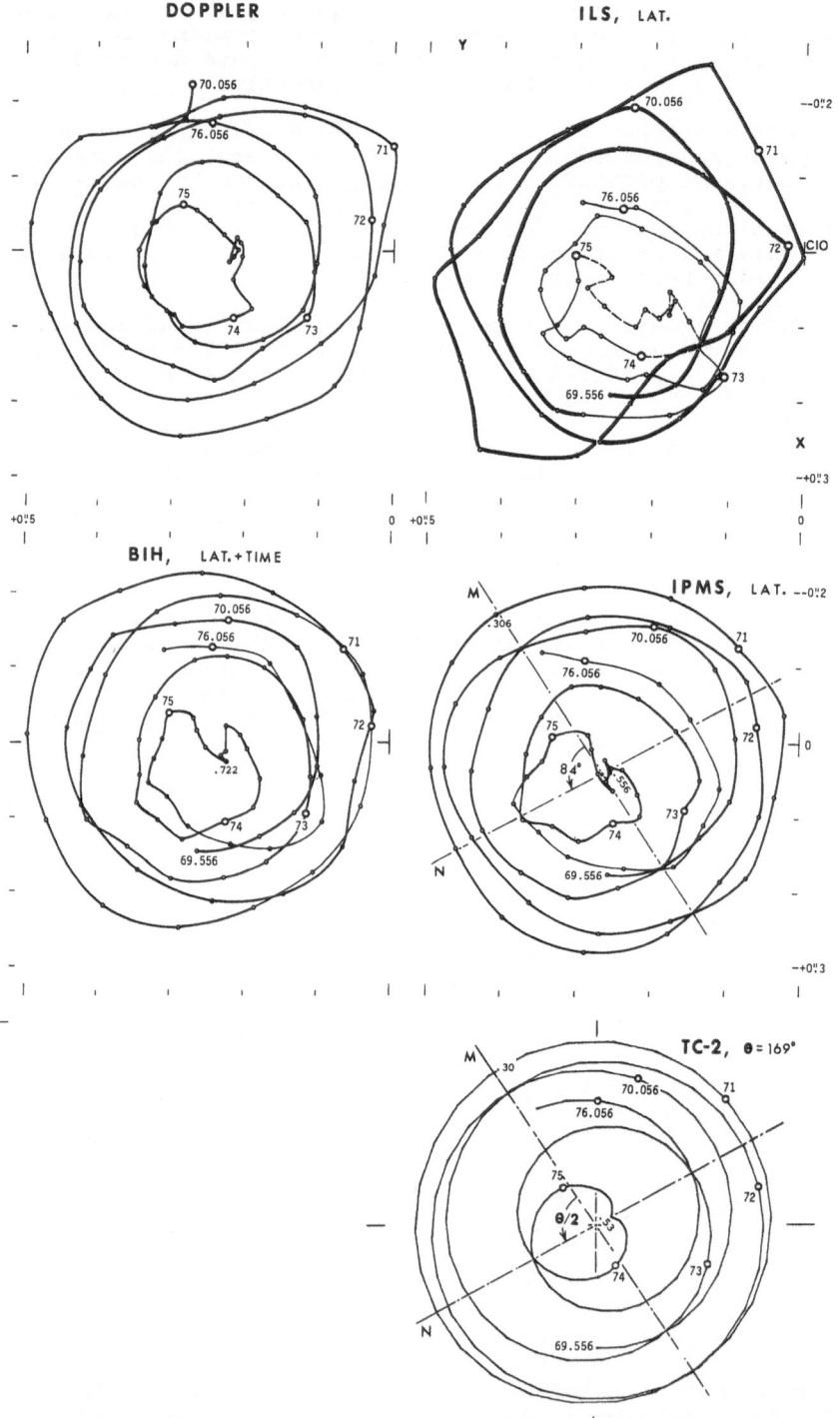

The TCs served as an external standard for the investigation of the polar motions first mentioned. Curves with two, fixed-magnitude vector components, consist of repetitive spirals that are a beat period apart, and have two axes of symmetry: The major, M, includes the maximum radius (components parallel), and the minor, N, the minimum (components antiparallel). Distinctive features of a TC are the central cardiod, and the crossings that occur on M and N. The angle between successive M and N is $\theta/2$. After θ is known, the beat period, B, and the Chandler period, H, can be computed. Let the unit of time be that assumed for the annual period, one tropical year, then $B = 6 + \theta/360°$ and $H = B/(B-1)$.

From many drawings of TCs, Markowitz found that TC-2, with $\theta = 169°$, fits best: "IPMS, Doppler, and BIH all resemble TC-2 but only IPMS shows the distinctive features of a TC well enough, the crossings, central cardiod, and maximum and minimum radii, so that a TC is determinate." His conclusions, among others, have been that (1) ILS observations cannot decide questions concerning particularities of the Chandler and the annual term; (2) IPMS is the most accurate polar motion; (3) the Chandler and the annual components have been stable for the period of investigation (the excitation energies balanced the friction losses), with the following values: Chandler period: 423.02 ± 0.15 d, semi-amplitude: 0".128; semi-amplitude of the annual component: 0".106.

Markowitz (1976) also published a diagram of the drift of the mean pole as given by the average values of x and y during a six-year cycle (Fig. 9). The ILS results indicate a drift of the mean pole since the beginning of this century of about 8 m. Doubts have been raised as to whether or not this is real, or wether one or the other of the five ILS stations have drifted. The IPMS and BIH multistation results seem to confirm that secular motion of the mean pole exists.

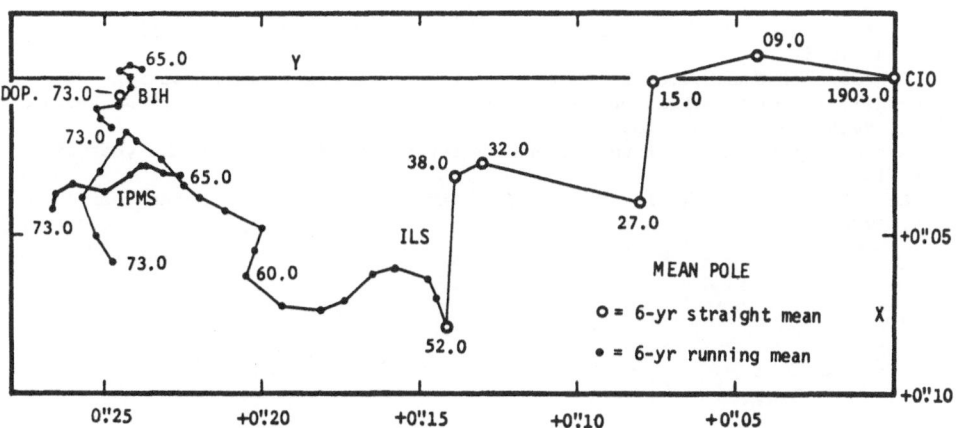

Fig. 9. Drift of the mean pole from 1903 to 1973 from ILS data, and from 1965 to 1973 from BIH and IPMS data, after Markowitz (1976). This diagram has been reproduced with kind permission of the author

◄ Fig. 8. Doppler, ILS, BIH, and IPMS polar motions between about 1969.5 and 1976.0, and theoretical curve, TC-2, after Markowitz (1976). These diagrams have been reproduced with kind permission of the author

References

Anderle, R.J.: Polar motion determinations by U.S. Navy Doppler satellite observations. NWL Technical Report T.R.2432, 1970

Brouwer, D.: A study of the changes in the rate of rotation of the Earth. Astron. J. 57, 125-146 (1952)

Capitaine, N., Feissel, M.: L'introduction des coordonnées du pôle déduites de l'observation Doppler de satellite dans les calculs du Bureau International de l'Heure. In: Proceedings International Meeting on "Earth's Rotation by Satellite Observations". Cagliari, 1973. Proverbio, E. (ed.). Bologna: Graficoop, 1974, pp. 65-73

Feissel, M.: The determination of UT1 by the Bureau International de l'Heure. In: Proceedings 8th Annual Precise Time and Time Interval Applications and Planning Meeting, Washington, 1976, pp. 47-65

Guinot, B.: A determination of the Love Number k from the periodic waves of UT1. Astron. Astrophys. 36, 1-4 (1974)

Guinot, B. et al.: Bureau International de l'Heure, Rapport Annuel pour 1969, Paris, 1970, p. 7
(The Annual Reports of the BIH are published, alternately in French and English, and began with that for 1967/French)

Lambeck, K., Cazenave, A.: The Earth's rotation and atmospheric circulation. I. Seasonal variations. Geophys. J. R. Astr. Soc. 79-93 (1973)

Lambeck, K., Cazenave, A.: The Earth's variable rate of rotation: a discussion of some meteorological and oceanic causes and consequences. Philos. Trans. R. Soc. London A. 284, 495-506 (1977)

Markowitz, W.: Rotational accelerations. In: Rotation of the Earth. Melchior, P., Yumi, S. (eds.). Dordrecht: D. Reidel, 1972, pp. 162-164

Markowitz, W.: Comparison of ILS, IPMS, BIH, and Doppler polar motions with theoretical. Report to Commissions 19 and 31, XVIth General Assembly of the International Astronomical Union, Grenoble, 1976

Mueller, I.I.: Introduction to Satellite Geodesy. New York: Frederick Ungar, 1964

Munk, W.H., MacDonald, G.J.F.: The rotation of the Earth. Cambridge: University Press, 1960, p. 98

Robbins, A.R.: A future Earth rotation service. Report to Commissions 19 and 31, XVIth General Assembly of the International Astronomical Union, Grenoble, 1976, p. 6

Rudloff, W.: Atmosphärische Einflüsse auf die Rotation der Erde. Naturwiss. Rundsch. 28, 161-164 (1975)

Sugawa, C.: Report of Commission 19 (Rotation of the Earth). In: Transactions of the International Astronomical Union. Contopoulos, G. (ed.). Dordrecht: D. Reidel, 1976, Vol. XVIA, Part 1, pp. 105-116

Yumi, S.: Annual Report of the International Polar Motion Service for the year 1972. Mizusawa, 1974

Yumi, S.: Annual Report of the International Polar Motion Service for the year 1974. Mizusawa, 1976

Effect of Tidal Friction on the Lunar Orbit and the Rotation of Earth and Its Determination by Laser Ranging

O. Calame and J. D. Mulholland

1. Introduction

E. Halley discovered that the observed longitude of the Moon shows an
acceleration by dynamic theory. Over more than two centuries, numerous
improvements have been brought to the lunar theory, but a large anomalous acceleration remains even now. The historical progress toward a
physical explanation and numerical determination has been admirably
detailed by Muller (1975). Eventually, the purely theoretical possibilities based on classical celestial mechanics were exhausted, and
it became a firmly accepted principle that the residual acceleration
was due to tidal friction. Thus, in one hypothesis, the astronomic
problem was "solved", and a new means was found for studying the physics of the elastic Earth.

Unfortunately, ad hoc solutions to observational problems do not always
lead to complete satisfaction. The recent observational determinations
of the "tidal" acceleration vary by more than a factor of two, and
there are no *theoretical* bases for establishing "reasonable" limits.

2. The Physical Problem

The physical mechanism by which it is supposed that this acceleration
is produced is briefly as follows: The deformation of the plastic Earth
under tidal stresses results in frictional dissipation. This is mani-
fested directly as a loss of spin angular momentum, that is, a slowing
of the Earth's rotation rate in space, or a lengthening of the day.
Conservation of angular momentum requires that the spin loss be com-
pensated elsewhere in the dynamic system of which the spinning Earth
is a member. The most obvious candidate is the lunar orbit; an increase
in angular momentum, transferred from the spinning Earth, requires an
increase in the mean distance to the Moon and by Kepler's third law
a corresponding decrease in the mean motion, or a deceleration in the
orbital mean longitude. The manner in which the angular momentum is
supposed to be transferred is that the internal friction in the Earth
causes a time-lag in the tidal deformation, and the difference in the
rate of rotation of the Earth and the mean motion of the Moon causes
the bulge to be displaced slightly from the center-to-center direction.
The resulting torque produces an orbital acceleration.

It is important to note that tidal friction must produce secular changes
in three parameters: The length of the day, the angular position of
the Moon, and its geocentric distance. In principle, and under standard
assumptions that we will discuss below, the tidal effect can be studied

U.S.R.A., Lunar and Planetary Institute Copernic, 06130 Grasse, France

with observations of either angular position or distance. In fact,
until recently, distance to a celestial body has not been observable,
and all available discussions of tidal effects on the lunar orbit have
been based on various kinds of angular measures.

There is a fundamental observational problem that affects all angular
observations except radio-interferometry: They cannot be related directly to the lunar center of mass, which is the point whose motion
is described by an orbital theory or an ephemeris. In general, one observes an event involving the edge of the visible disk (the "limb"),
which is irregular and situated an uncertain distance of the order of
900" from the center of mass. From such data, one deduces an acceleration of the center of mass which results in a quadratic displacement
equivalent to about 1 % - 2 % of the lunar radius after 100 years. Normally, for modern data, the problem is overcome by supposing that the
errors in limb corrections and other relevant parameters are uncorre-
lated from one observation to another, and thus their effects can be
diminished by using large numbers of observations. This is probably
true for occultations, but doubtful for transit observations, which
involve only a small portion of the limb. Solar eclipse observations,
being occultations over the entire limb, are probably least sensitive
to this problem, but are not free from other difficulties of interpretation.

On the other hand, high-precision techniques that observe points fixed
on the lunar surface are inherently related to the center of mass, because it is not only possible, but usually necessary, to use these
same measures to determine the location of the observed points. At present, such points that are *observable* consist of radio beacons (radio
Doppler and interferometry) and optical reflectors (laser ranging).
A regular observing program of laser ranging has been underway since
1969; these observations have the advantage of being highly sensitive
to *both* distance and angular position.

3. The Laser Technique and Its Application

The lunar laser ranging (LLR) technique provides measures of a nature
very different from classical astronomical observations. It is an *active*
process, in which the observer illuminates the target and observes the
illumination that he himself has provided. It is also unlike the other
new techniques, in that the observed object is entirely passive, and
thus not subject to technological failure. With the laser, a pulse of
light is transmitted by a terrestrial station toward a reflector on
the lunar surface. Because of the cube-corner design of the reflector,
the light that strikes it is retransmitted toward Earth in the same
direction from which it came. Thus, the signal is detected by a photo-
multiplier at the same station. The observation recorded is the time
delay between the transmission of the laser pulse and the detection
of the reflected signal. For convenience, one often refers to this as
a distance measurement, but it is essential to understand that it is
really a measurement of the aberration time, which is not symmetric
because of the motions of the Earth and Moon during the interval of
about 2.6 s (Calame, 1975a).

Since 1969, five laser reflectors have been deposited on the lunar
surface by the missions Apollo 11, Luna 17 (reflector on Lunakhod I),
Apollo 14, Apollo 15, and Luna 21 (reflector on Lunakhod II). The reflector on Lunakhod I has been observed only a few times by the Soviet
and French teams, with a relatively poor accuracy. The four others have

been ranged regularly at McDonald Observatory (Texas, U.S.A.) since they have been deposited. Now, more than 2000 compressed observations are available from this station, made over the past seven years. The typical accuracy of these ranges, expressed as a one-way distance, is 15 cm and the best one is about 5 cm.

Other stations are in the course of completion in Hawaii, Australia, France, and Japan. The Crimean station has yielded a few sets of observations, but does not have a regular program. Presently, an international campaign (called EROLD) has been organized, with the collaboration of the various observing groups and the *Bureau International de l'Heure*, to study the Earth Rotation from Lunar Distances (Calame et al., 1976).

The McDonald data have been analyzed by several groups, primarily in the U.S.A. and France, for the study of various subjects. Significant and interesting results have already been obtained for the dynamics of the Earth-Moon system (e.g., Calame, 1977a; Williams, 1977), cosmology (Williams et al., 1976; Shapiro et al., 1976), geodesy (e.g., Calame, 1975b), and geophysics (e.g., Stolz et al., 1976).

The analysis of such data can be conducted with various mathematical models of the orbital and rotational motions of the Earth and Moon. There exist several possible choices among the currently available models that have sufficient precision. For the lunar orbital motion, the ephemerides must be generated by numerical integration (simply because no sufficiently complete analytic theory yet exists) and adjusted to fit the laser data; several such ephemerides now exist, and the production of newer and better ones is a continuing activity. At least three usable models exist for the rotational motion of the Moon, one of which is numerical (Williams, 1974), one semi-analytic (Eckhardt, 1974), and one purely analytic (Migus, 1977). With certain modifications (Calame, 1977b) to the latter two, the accuracy obtained is nearly equivalent in all cases if one considers a global representation of the system. From these basic models, it is necessary to calculate the time-delay residuals for each observation, accounting for the various effects acting on the photon path, such as the parameters describing the motions of the station and the reflector, and the less evident effects acting on the photons themselves, such as atmospheric refraction and geodesic curvature (Calame, 1976). These residuals, in combination with the appropriate partial derivatives (e.g., Mulholland, 1977), are then used to improve the representation of the Earth-Moon system, by Gaussian least squares or an equivalent regression technique.

4. Determination of the Lunar Acceleration

The usual procedure by which one determines the value of a parameter y from an observable x is to obtain an expresssion x(y) and compute the partial derivative $\partial x/\partial y$, which is used with observed values of x to find y. If x(y) is a dynamic quantity, then it is normal to demand that dynamic theory be respected. This has not usually been the case in studies of the "tidal" acceleration. It is useful to recall that the observed angular variable is the right ascension α or the true longitude λ. In an analytic theory, it is customary to define the mean anomaly l and the mean longitude L in a fashion compatible with their Keplerian counterparts, so that there are no periodic terms:

$$L = L_o + at + bt^2 + ct^3 + \ldots \tag{1}$$

or by Taylor's series:

$$L = L_o + \dot{L}_o t + \frac{1}{2} \ddot{L}_o t^2 + \ldots \qquad (2)$$

Obviously, if there is an observed acceleration, then it must appear in the mean longitude. In most studies of the anomalous acceleration, however, this fact is treated as though the acceleration is not only constant, but that it affects *only* the mean longitude. Thus, one looks for a signal in the true longitude residuals that looks like bt^2.

Perhaps this is an adequate procedure with relatively imprecise observations but it violates dynamic theory. There is no way in which a force can be applied to the orbiting Moon that can produce this effect. Even if one supposes that the force applied is constant and normal to the radius vector, there will be effects like bt^2 in both the mean anomaly and the mean longitude, but all of the coordinates will also include secular-periodic terms like $t^2 \sin kt$. Thus, if one searches only for δL_o, a large part of the signal that exists in the observations will be lost, and this may bias or destroy the significance for tidal theory.

The situation is even worse for studies that use a numerical ephemeris, for there is no way in which the purely arbitrary separation between secular terms and periodic terms can be accomplished. In this sense, the numerical integration process forces one to abandon arbitrary mathematical formalism and to reconsider the implications of dynamic theory. The problem is more serious than the question whether $\dot{n} = \ddot{L}$ or $\dot{n} = 1/2 \ddot{L}$, which is discussed in an Appendix A. It remains convenient to use Keplerian elements to determine corrections to the epoch conditions, but these must of necessity be osculating elements. These elements cannot be easily related to "mean" elements, but are operationally defined as specific functions of the true coordinates. Furthermore, the partial derivatives are strongly dependent on the specific choice of an "independent" set of elements chosen. Thus, for example:

$$\frac{\partial X}{\partial L_o} \text{ (analytic)} \neq \frac{\partial X}{\partial L_e} (L,l,F,\ldots) \neq \frac{\partial X}{\partial L_e} (L,\Omega,\omega,\ldots) \qquad (3)$$

where the subscript e denotes the osculating value at the epoch t=0. What is perhaps more important, in the perturbed problem, the linear independence of the Keplerian elements no longer exists for the osculating elements. In particular, corrections to the mean longitude, the mean anomaly, or the eccentricity induce important interaction effects in the osculating mean motion, which are directly reflected in the average "mean motion". Thus, if one simply computes:

$$\frac{\partial X}{\partial \ddot{L}_e} = \frac{1}{2} t^2 \frac{\partial X}{\partial L_e} \qquad (4)^*$$

then the results will be totally invalid. For example, at the standard epoch of our integrations (JED 2440400.5), a correction $\Delta L_e = 1"$ will introduce automatically a $\Delta \dot{L}_e = 157"/\text{cy}$. Thus, the use of Eq. (4) will be equivalent to searching for the coefficient of:

$$t^2 + c\, t^3 \qquad (5)$$

The coefficient of t^3 will dominate the solution, which will thus be much smaller than the real value of $\ddot{L}_e$ implied by the observations. This statement seems inescapable in principle, but we have also verified it numerically.

---

* Wrong equations are marked with an asterisk.

How then does one calculate a usable partial derivative for $\ddot{L}_e$?. In fact, is $\dot{L}_e$ the quantity that one desires? To the latter question, we can respond affirmatively. If we define K as the magnitude of the transverse component of the force due to the Earth's bulge and a unit vector $\vec{K}^*$ whose direction is given by:

$$(\vec{r} \times \vec{v}) \times \vec{r} \tag{6}$$

then the force:

$$K \ \vec{K}^*$$

is constant in magnitude, contained in the orbit plane, and directed 90° from the radius vector. This is an approximation to the force supposed to arise from the tidal friction. Introduction of this force into the dynamic equations of a numerical integration produces a very sharp parabolic change in the osculating mean longitude, and a slightly noisier parabola in the osculating mean anomaly, the true longitude, and the right ascension. Thus, one can say that:

$$\delta K \propto \delta \ddot{L}_e = \delta \ddot{\lambda}_e \tag{7}$$

and that the values of these quantities at the epoch are very close to the time-averaged accelerations induced by δK in these angles. Given this information — verified numerically — it is then possible to construct explicit partial derivatives with respect to K. We have done so by finite differences of numerical integrations, but we have used these derivatives only to verify the numerical correctness of a simpler and less expensive procedure, discussed next.

It was demonstrated by Oesterwinter and Cohen (1972) that

$$3 \ K = - \ a_e \ \delta \ddot{L}_e \tag{8}$$

which we have verified numerically. As we show in the Appendix, the process of calculating the osculating elements from the state vector results in the relation:

$$\ddot{L}_e = 2 \ \dot{n}_e \tag{9}$$

so that, as with Oesterwinter and Cohen:

$$3 \ K = - \ 2 \ a_e \ \delta \dot{n}_e \tag{10}$$

Recalling that Eq. (4) was not usable because of the interaction of δL_e and δn_e, one is tempted to ask if the principle that failed with the mean longitude might be valid for the mean motion. That is:

$$\frac{\partial X}{\partial \dot{n}_e} = t \ \frac{\partial X}{\partial n_e} \tag{11}$$

As a result of numerical tests, the answer is yes; Eq. (11) is valid. How does one calculate $\partial X/\partial n_e$? The most immediate temptation is to adopt Kepler's third law, as it is given in most textbooks:

$$n^2 \ a^3 = \text{constant} \tag{12}^*$$

from which one concludes that:

$$2 \ n \ \frac{\partial X}{\partial n} = - \ 3 \ a \ \frac{\partial X}{\partial a} \tag{13}^*$$

Unfortunately, not everyone agrees that Eq. (12) is correct. On the other hand, the operational definition of the mean distance a is:

$$n^2 \ a^3 = G \ (E + M) \tag{12a}$$

where G is the universal gravitational "constant", and E + M is the total mass of the Earth-Moon system. The value of G_e is known by definition in the astronomic system of units but it is possibly a function of time. The value of E + M is only approximately known (i.e., its value is still subject to improvement), but it is still supposed to be constant. Thus,

$$2 \ n_e \ \frac{\partial X}{\partial n_e} + 3 \ a_e \ \frac{\partial X}{\partial a_e} = (E + M) \ \frac{\partial X}{\partial (E+M)} \tag{13a}$$

and

$$2 \ n_e \ \frac{\partial X}{\partial \dot{n}_e} + 3 \ a_e \ \frac{\partial X}{\partial \dot{a}_e} = G_e \ \frac{\partial X}{\partial \dot{G}_e} \tag{13b}$$

In the discussion of the high precision observations for the improvement of the numerical lunar ephemeris (and not simply the determination of abstract, nondynamic parameters that will never be put a real observational test in a new lunar theory), Eq. (13a) shows that uncertainty in the mass requires that the osculating elements be *seven* in number, including the osculating mean motion, which represents a substitute parameter for the mass. Remember that the *epoch* elements are constants, even if they are the epoch values of variable quantities. One could, of course, adopt the mass directly as a solution parameter, but it is less convenient for our purposes. Thus, we have computed explicit partial derivatives $\partial X/\partial n_e$, again by finite differences of numerical integrations. It is with these derivatives that we have verified the validity of Eq. (11) for the derivatives with respect to $\dot{n}_e$, and they were subsequently used in our solutions.

The observations used in our solutions are 2034 laser range normal points obtained at McDonald Observatory between September 1969 and October 1976. (For the definition of these data, see Abbot et al., 1973.) The physical model included the unpublished, but widely available, lunar ephemeris known as LURE2 (included in JPL ephemeris tape DE86), with the values of the astronomic constants cited by Williams (1977) as appropriate to this ephemeris. The physical libration model is a modified version of Migus' analytic solution, in which his planetary perturbations are replaced by those of Williams et al. (1973), Williams and Slade (1975), again with the LURE2 values of the constants of the lunar gravity field. The motion of the Earth rotation pole was calculated with the Woolard nutation theory modified to account for elastic effects, and the motion of the Earth's crust relative to this pole was obtained from the smoothed data of the *Bureau International de l'Heure*. The effect of geodesic curvature of the photon path was computed with a post-Newtonian expression in accord with General Relativity (Holdridge, 1967). The refractive delay was computed with an average value of the barometric pressure at McDonald Observatory (Silverberg, 1973), corresponding to 12.5 ns in the round-trip delay at the zenith.

In terms of laser ranging, LURE2 is a rather old ephemeris, produced with observations covering 1969 - 1972 only. Using the model cited above, including the LURE2 coordinates for the telescope and the reflectors, the true RMS residual (i.e., before corrections) for our observation set is 25 ns, or nearly 4 m in equivalent one-way distance. Furthermore, the residuals are not entirely random, but show a general systematic increase with time, which is quite normal in extending a numerical ephemeris beyond the interval to which it was corrected. Thus, it is not possible to use these residuals to determine ṅ alone, since they are too contaminated with other influences that are considerably larger. Thus, we were obliged to introduce ṅ as one of many parameters in a global solution for an improved physical model of the Earth-Moon system and the laser ranging network; but it was always introduced after the solution for the basic parameters was completed. In this way, the influence of this parameter on the rest of the solution could be identified.

The other solution parameters represented the coordinates of the telescope and the reflectors, the seven elements of the lunar orbit, the solar mean longitude, seven parameters representing the second and third degree lunar gravity field (β, γ, C_{30}, C_{32}, C_{33}, S_{32}, S_{33}), the fictitious "Nordtvedt" signal discussed by Williams et al. (1976), the free libration parameters (Calame, 1977a), and several global parameters representing apparent systematic errors in the coordinate systems (Calame, 1977a), which can be treated as though they are secular and periodic variations in the Earth's rotation. In one solution series, we added a correction to the origin of longitudes. Using the information given by William (1974) that the LURE2 ephemeris was constructed with an arbitrarily chosen value for the mean "tidal" acceleration of $\ddot{L}_t = -40"/cy^2$, we find:

$$\ddot{L}_t = -24.6 \pm 1.6"/cy^2$$

without this origin correction. With this parameter included, the result is:

$$\ddot{L}_t = -20.7 \pm 1.6"/cy^2$$

with an origin correction of $-0.075" \pm 0.007$; in the first case, the global post-fit residual was 3.14 ns, and in the second 3.05 ns, or about 46 cm in equivalent one-way distance. The uncertainties (approximately 10 % of the corrections) are formal standard deviations and are obviously smaller than the real uncertainties. As demonstrated by Stolz et al. (1976), these standard deviations could be reduced significantly by solving for short-period variations in the Earth rotation parameters, but this could not lead to improved values of the tidal acceleration.

It is difficult to evaluate which of these results is the more realistic. We do, however, call attention to the fact that these results are reasonably consistent with the general body of determinations made by other persons and with other techniques. It must be noted that our study used only seven years' observations, compared with determinations made with less precise observations covering time spans ranging between 18 and 3000 years.

5. Comparisons and Difficulties of Interpretation

A brief examination of the literature on the subject of the "tidal" acceleration suffices to suggest that the search for a definitive value is not converging as a function of time and depends heavily on the data and the determination methods used. Table 1 contains a sample of various results. In addition, J.G. Williams has also studied the subject using the lunar laser ranges. The value published in 1976 was

Table 1. Various determinations for tidal lunar acceleration

With ancient data	("/cy^2)	With modern data	("/cy^2)
De Sitter (1927)	-37.7 ± 4.3	Spencer Jones (1939)	-22.4 ± 1.1
Newton (1968)	-20.1 ± 2.6	Van Flandern (1970)	$-52.0 \pm 16.$
(1970)	-42.3 ± 6.1	(1975)	$-65.0 \pm 10.$
(1970)	-41.6 ± 4.3	(1976)	-35.0 ± 5.0
Muller and Stephenson (1975)	-37.5 ± 5.0	Oesterwinter and Cohen (1972)	-38.0 ± 8.0
Muller (1976)	-30.0 ± 3.0	Morrison (1977)	-26.0 ± 2.0
		Lambeck (1975)	-35.0 ± 4.0

$- 18 \pm 20$"/cy^2, but an informal communication indicates that a new value has been found and will be published after verification.

In the face of these various results, we are forced to ask if everybody is studying exactly the same thing or not. Are the divergent results due to data noise or to incommensurate models?

The most evident of the potential confusion factors is the possibility of a variation in G. It is obvious from Eq. (13b) that a nonzero $\dot{G}$ would almost necessarily induce a corresponding component in both $\dot{n}$ and $\dot{a}$. The exact relationship would depend, of course, on the exact nature of G as a function of time. With measures of angular position only, it is impossible to separate the tidal and cosmologic contributions without additional hypotheses. For this reason, we feel that such results as those by Van Flandern (1975) for $\dot{G}$ must not be taken too seriously, first because he imposes an arbitrary cosmologic model, and second because his input data are based on inhomogeneous determinations. This view seems to be supported by the large variations in the results with time. If, as with laser ranging, one can eventually measure $\dot{n}$ and $\dot{a}$ independently, then the cosmologic question can be solved explicitly by application of Eq. (13b) without other hypotheses. It is instructive to note, however, that a tidal acceleration of 25"/cy^2 will produce a change in the mean distance of only 13 cm over the span of lunar observations now available. This is well under the present noise level. Attempting thoroughness, however, we have attempted to solve for $\dot{a}$; the resulting standard deviations permit a large range of values for $\dot{G}$, both positive and negative.

The problem of inhomogeneity in Van Flandern's search for $\dot{G}$ is only a striking example of what could be a fundamental reason for the large disparities between different results. He combines determinations made by himself using one set of data, ephemerides, and partial derivatives with another determination using totally different types of data, a different ephemeris, and different partial derivatives. It is on particularly hazardous ground when he uses an analytic partial derivative in combination with a numerical ephemeris. Finally, he corrects the ephemeris for supposedly known model differences. In the end, there is no real guarantee that the two results refer to the same phenomenon, or that they are dynamically consistent. Especially when using numerical ephemerides, it is necessary to validate the partial derivatives by using them to produce new ephemerides and by measuring the degree to which the results conform to the predictions.

The final problem that has arisen in the interpretation of determinations of the tidal accelerations is the probable existence of long-period or secular defects in the lunar theory and the ephemerides, defects that are not the same in all ephemerides. It must always be remembered that what we call the "tidal" acceleration is really the sum of our ignorance of the lunar motion, part of which is surely due to tidal friction, some just as surely is not. It is only necessary to realize that the purely theoretical values of the velocities of the fundamental arguments (mean elements) are incorrect (cf. Muller, 1975) to justify doubts about the purely theoretical values of the accelerations. We present the following example:

It has long been known that the lunar gravitational coefficient C_{22} strongly perturbs the orbital motion, and this effect has thus been included in orbit theories. Until recently, this calculation was done assuming that Cassini's laws for the lunar rotation were adequate. Indeed, this seemed reasonable, because the classical libration theories produced only relatively short-period terms in the rotation. These theories are not complete. Recent improvements in the libration theory introduces significant long-period terms, one of which is due to Venus with a 270-year period. If this term is added to the acceleration model of a numerical integration of the lunar orbit, the effect on the mean longitude and the right ascension is striking. We have performed such integrations over 10^5 days, at 0.4 day step-size in 120-bit precision. The analysis of the results is not yet complete, but visual estimation from graphic differences yields an apparent (and unexplained) acceleration due to this term of about $-10.6''/\text{cy}^2$ and a periodic effect of about $3 - 7''$ amplitude with about a 270-year period (see Appendix B). The former would bias *any* result for $\dot{n}$, while the latter would influence studies based on observation intervals much shorter than 270 years. Brown's theory does not include this effect. Oesterwinter and Cohen's ephemeris does not include this effect, or even C_{22}. Ephemeris LE16 does not include this effect. Ephemeris LURE2 *may* include it, but the documentation is not explicit. Ephemeris LE25, used by Van Flandern, is implied not to have this effect. Thus, any result based on LURE2 is probably incommensurate with any results based on any other ephemeris. Tests with experimental ephemerides of our own construction, including the Venus libration term, are compatible with the LURE2 results cited above.

The important question is not just the commensurability of ephemerides, but the degree of correctness of the physical model. Even if everyone agrees on a particular value of the tidal acceleration, it will lead to erroneous geophysical conclusions if it is based on an incomplete model. We must abandon the habit of treating the unexplained acceleration as being entirely of tidal origin and search for other causes that might contribute to it. This search may lead to other interesting geophysical results: If such long-periodic effects really exist in the orbit, there must be a corresponding periodic interchange of angular momentum between the orbit and some "reservoir". The most obvious such reservoir is the rotation of the Earth, which should thus show variations with the same periodicities. This should be testable, since the second largest long-period libration term has a period of 18.6 years. This periodicity exists in UT1 also, and it is possible that the first evidence from lunar laser data of its amplitude has been obtained by Calame (1977a), although it is far from certain that this is the correct interpretation of that result. Certainly, more work needs to be done in this area.

Acknowledgments. We are grateful to the Centre d'Etudes et de Recherches Géodynamiques et Astronomiques for having provided the far-from negligible computer times for performing the calculations discussed above. A part of the program development was supported by the Office of Naval Research (USA) under contract N00014-76-C-0641 to the University of Texas at Austin.

References

Abbot, R.I., Shelus, P.J., Mulholland, J.D., Silverberg, E.C.: Astron. J. $\underline{78}$, 784 (1973)

Calame, O.: Etude des mouvements libratoires lunaires et localisation des stations terrestres à partir des mesures laser de distance, 1975a (dissertation, Université de Paris VI)

Calame, O.: C. R. Acad. Sci. Paris $\underline{280}$, 551 (1975b)

Calame, O.: Manuscripta Geodaetica $\underline{1}$, 173 (1976)

Calame, O.: In: Scientific Applications of Lunar Laser Ranging. Mulholland, J.D. (ed.). Dordrecht: Reidel, 1977a, p. 53

Calame, O.: Manuscript in preparation (1977b)

Calame, O., Chollet, F., Guinot, B.: COSPAR Bull. $\underline{77}$, 43 (1976)

Eckhardt, D.H.: Private communication (1974)

Holdridge, D.B.: Jet Propulsion Lab. SPS 37-48 III, 1967, p. 2

Migus, A.: Théorie analytique de la libration physique de la lune. 1977 (dissertation, Université de Paris VI)

Morrison, L.V.: These proceedings (1978)

Mulholland, J.D.: In: Scientific Applications of Lunar Laser Ranging, Mulholland, J.D. (ed.). Dordrecht: Reidel, 1977, p. 9

Muller, P.M.: An analysis of the Ancient Astronomical Observations with the Implications for Geophysics and Cosmology. 1975 (dissertation, University of Newcastle, published privately by the author)

Muller, P.M.: Determination of the Cosmological Data of Change of G and the Tidal Accelerations of Earth and Moon from Ancient and Modern Astronomical Data. Lunar Programs Office, NASA, 1976

Oesterwinter, C., Cohen, C.J.: Celestial Mechanics $\underline{5}$, 317 (1972)

Shapiro, I.I., Counselman, C.C. III, King, R.W.: Phys. Rev. Lett. $\underline{36}$, 555 (1976)

Silverberg, E.C.: Private communication (1973)

Stolz, A., Bender, P.L., Faller, J.E., Silverberg, E.C., Mulholland, J.D., Shelus, P.J., Williams, J.G., Carter, W.E., Currie, D.G., Kaula, W.M.: Science $\underline{193}$, 997 (1976)

Van Flandern, T.C.: Mon. Not. R. Astr. Soc. $\underline{170}$, 333 (1975)

Van Flandern, T.C.: Private communication (1974)

Williams, J.G.: Private communication (1974)

Williams, J.G.: In: Scientific Applications of Lunar Laser Ranging. Mulholland, J.D. (ed.). Dordrecht: Reidel, 1977, p. 37

Williams, J.G., Dicke, R.H., Bender, P.L., Alley, C.O., Carter, W.E., Currie, D.G., Eckhardt, D.H., Faller, J.E., Kaula, W.M., Mulholland, J.D., Plotkin, H.H., Poultney, S.K., Shelus, P.J., Silverberg, E.C., Sinclair, W.S., Slade, M.A., Wilkinson, D.E.: Phys. Rev. Lett. $\underline{36}$, 551 (1976)

Williams, J.G., Slade, M.A.: Private communication (1975)

Williams, J.G., Slade, M.A., Eckhardt, D.H., Kaula, W.M.: The Moon $\underline{8}$, 469 (1973)

Appendix A: Comments on the Perturbed Mean Motion

The principle of the variation of arbitrary constants is applied to orbital theory so that the idea of Keplerian elliptic orbits can be maintained in perturbed systems of more than two bodies. This can be done with a minimum of ambiguity when one can consider that there are only six elements, but in more complicated problems the extension is not always clear. In the case of the lunar acceleration, this is shown in a confusion over the "correct" definition of $\dot{n}$. In the two-body problem, the mean longitude is defined as:

$$L_k = \chi_0 + n\,t \tag{A}$$

where χ_0 and n are constants. There are two possible ways in which this relation can be extended to the perturbed case:

either $\quad L = \chi_0 + n_1 t$ (B1)

or $\quad\quad L = \chi_0 + \int n_2 \, dt$ (B2)

In both cases, the mean motion is a function of time. If one develops these two expresssions by Taylor's series, one finds that:

$$L = \chi_0 + (n_1)_0 \, t + (\dot{n}_1)_0 \, t^2 + \frac{1}{2} (\ddot{n}_1)_0 \, t^3 + \dots \tag{C1}$$

$$L = \chi_0 + (n_2)_0 \, t + \frac{1}{2} (\dot{n}_2)_0 \, t^2 + \frac{1}{6} (\ddot{n}_2)_0 \, t^3 + \dots \tag{C2}$$

Obviously, it is not sufficient to say that "... since n is usually defined as the lunar rate in longitude...", because at all instants, the value of $(n_1)_0$ is equal to the value of $(n_2)_0$, but:

$$(\dot{n}_1)_0 = \frac{1}{2} \ddot{L}_0 \tag{D1}$$

and

$$(\dot{n}_2)_0 = \ddot{L}_0 \tag{D2}$$

It does not appear to us that one is free to choose arbitrarily whether he will adopt relation (D1) or (D2). This choice must be considered as *imposed* by one's operational procedure. In the use of osculating elements, using Keplerian relations computed from the coordinates, it is certain that it is (D1) that is imposed, because one computes the osculating mean motion from Kepler's third law, and then the mean anomaly from Kepler's equation. Thus, it is clear that n is the coefficient of t, not the time derivative. It appears that this will also be true of any procedure in which the value (either numerical or analytic) of the mean motion is determined as a time-average over an orbital period or longer.

Appendix B: Reality of the Venus Effect

Van Flandern (1977) reports that he has found a similar secular trend in another numerical ephemeris, and that a subsequent investigation by J.G. Williams has produced an explanation. In their case, the secular deviation from the Brown analytic theory was determined to be not a constant acceleration (term in t^2), but an exponential (term in e^t), although the difference between the two functions is said to be very slight over an interval of three centuries. This deviation is thought by them to be totally spurious, due to a numerical instability triggered by the error in the initial position of the lunar C_{22} bulge. In this respect, it seems to be conceptually similar to the triggering of the spurious free librations that plague numerical integrations of the lunar rotation. If this explanation is true, then it represents a serious flaw in all numerical integrations of the lunar motion that adopt a triaxial (or higher degree) figure for the lunar gravity field, i.e., all except that of Oesterwinter and Cohen. What is worse, if this explanation is true, then the only numerical ephemerides valid even in principle for the discussion of secular effects such as tidal friction will need to be simultaneous integrations of the orbital and rotational motions, fitted to observations for short-term accuracy and subsequently fitted to analytic theories over long intervals (centuries) to avoid

the spurious numerical effects in both the orbit and the rotation.
From both a conceptual and an economic point of view, this would likely be the death-knell of numerical integration as a viable method of
analysis for such purposes. It remains to be seen if these claims are
justified by the details.

In any event, these remarks affect only the secular part of our discussion of the Venus effect. The long-periodic part remains without
modification, even if the Van Flandern-Williams findings are correct.
Perhaps more seriously, the philosophical point remains valid and important, that there may yet be important effects omitted even from the
best ephemerides of the lunar orbit, numerical or analytic in origin.
We may call the unmodeled acceleration "tidal" as a matter of convention, but one must always remember that we do not know what fraction
is tidal and what is ignorance of other factors.

Tides of the
Solid Earth from Gravimetric Measurements

M. Bonatz

1. Introduction

The 8th International Symposium on Earth Tides was held from the 19th through the 24th of September, 1977 at Bonn, Institut für Theoretische Geodäsie.

Because the present state of tidal research is documented in the proceedings of this symposium (Bonatz), to avoid duplication, this paper deals mainly with general information.

2. Basic Concepts

The gravitational forces of the Sun and Moon, acting upon the body of the Earth, produce time-variant deformations, i.e., the Earth tides. In mean latitudes, the tidal bulge is of the order of decimeters.

Terrestrial measurements for determination of the tidal deformations mean measurements on the deformed body itself. Due to the wide-ranging mode of the deformations, pure geometric techniques cannot be applied; one must therefore use an indirect method, which, in general, is to compare the tidal variations of the gravity vector, computed for a rigid Earth, with those observed in tidal stations; the differences obtained describe indirectly the tidal movements at the observation points.

To calculate the tidal deformations of the Earth from gravimetric data, the LOVE theory, which is based mainly on a radial symmetric distribution of the elastic properties of the Earth, has been applied, for practical reasons. Under this theory, one gets for the tidal variation Δr of the geocentric radius vector r of the point of observation,

$$\Delta r = h \frac{V_{M,S}}{g_o} \tag{1}$$

where $V_{M,S}$ is the lunisolar tidal potential, g_o, time-invariable part of gravity, and h, the proportionality constant (LOVE number).

The potential V_E, due to the elastic deformation of the Earth, is described by a second LOVE number k:

$$V_E = k \cdot V_{M,S} \tag{2}$$

3. The Tidal Potential

Using the polynomials of Legendre of orders, 2 and 3 for the Moon and of order 2 for the Sun, the tidal potential $V_{M,S}$ can be expressed by

$$V_{M,S} = \frac{G \cdot M}{R_M} \sum_{i=2}^{3} (\frac{r}{R_M})^i \cdot P_i \cdot (\cos\theta_M)$$

$$+ \frac{G \cdot S}{R_S} (\frac{r}{R_S})^2 \cdot P_2 \cdot (\cos\theta_S)$$

(3)

where G is the gravitation constant, M,S, the mass of Moon and Sun, respectively, $R_{M,S}$, instantaneous distance between the center of the mass of the Earth and Moon or the Sun, and $\theta_{M,S}$, instantaneous geocentric zenith distance of the Moon or of the Sun.

If the geographic coordinates of the observation point, the orbital elements of the Moon and Sun and the angular velocity of the Earth's rotation, are introduced into the basic formula (3) then the tidal potential is expressed as the sum of harmonics of different frequencies, amplitudes, and phases, a number of 505 using the Cartwright-Tayler-Edden development, which can be distinguished into long-periodic, diurnal, semidiurnal, and terdiurnal modes. The differentiation of the potential in a certain direction leads to the respective tidal accelerations, which in each case can be regarded as individual reference systems.

4. Determination of LOVE numbers from Gravimetric Data

If Δg_p is the modulus of the tidal acceleration vector, deduced from the tidal potential, i.e., associated with a rigid Earth, and Δg_o the corresponding observed value, one defines a gravimetric factor δ, which is

$$\delta = \frac{\Delta g_o}{\Delta g_p}$$

(4)

In the absence of indirect oceanic effects, δ is approximately 1.16; the maximum variation of gravity in mean latitudes is about 0.3 mGal (1 Gal = 1 cm/s^2), which is $3 \cdot 10^{-7}g$ (g $\sim$ 981 Gal). Following the LOVE Earth model, the gravimetric factor can be expressed by the equation

$$\delta = 1 + h - \frac{3}{2} k$$

(5)

The corresponding procedure for the variations $\Delta\epsilon$ of the direction of the tidal vector relative to a crust fixed-direction basis leads to a diminuation factor γ

$$\gamma = \frac{\Delta\epsilon_o}{\Delta\epsilon_p}$$

(6)

which, in the undisturbed case, is about 0.7. In mean latitudes $\Delta\epsilon_{max}$ is about 0".04.

The diminuation factor is a function of h and k too:

$$\gamma = 1 + k - h.$$

(7)

It follows that one can experimentally determine the LOVE numbers h and k, measuring the variations of the tidal acceleration vector, i.e., both modules and direction. Introducing h into Eq. (1), one gets the vertical, tidal displacement of the observation point (provided that the LOVE model describes the actual properties of the Earth sufficiently correctly!)

Since the Earth does not exactly behave as an elastic body, a phase deviation α arises between the astronomic forces and the terrestrial measurements (a positive sign of α in the tables corresponds to a phase l a g).

5. Instrumentation

The modulus of the tidal acceleration vector is measured with gravimeters, which in principal are spring balances. For the determination of the tidal tilt one uses horizontal or vertical pendulums.

The resolution of modern instruments is about $\pm$ 1 μGal and 10^{-4} arc s respectively, with non-astatized meters corresponding to displacements of the transducer point of the order of 10^{-7} mm.

6. Actual Problems

The determination of the tides of the solid Earth becomes complicated by the following main facts:

1. As lateral variations of density and elastic properties of the Earth occur, the radial symmetric model of LOVE is only an approximate description of the physical reality; thus the model has to be generalized; additional tidal information such as strain data has to be introduced.

2. The observed tidal tilt is essentially influenced by a very local effect, which is called cavity effect, i.e., the geometry of a cavity is continuously changing due to tidal forces by strain-tilt coupling effects. (To avoid thermal perturbations the tilt is measured in cavities underneath the Earth's surface.) In addition, local topography and geology plays an important role.

3. The indirect effect of the oceanic tides on the tides of the solid Earth has not been taken into account in LOVE's theory. This "disturbing" effect may, in coastal stations, exceed the primary effect; both tidal amplitudes and phases are affected. Due to the mainly semidiurnal mode of the oceanic tides, the diurnal Earth tides are less influenced than the semidiurnal ones, which is why one in general determines LOVE numbers only from the diurnal part of observations.

4. The instruments used for tidal observations not only measure tidal signals but also secondary effects due to variations of atmospheric pressure, ground-water level, temperature, etc. In addition, one has to take into account systematic instrumental effects, such as instrumental drift, calibration-, and phase errors. Thus the problem of processing the observed data is to separate the different signals and to determine band parameters for the individual tides.

Table 1. Diurnal waves — vertical component

Mean results from 22 series in which P_1 has been separated from K_1

	$\delta = 1 + h - \frac{3}{2} k$	α
K_1	1.1436 ± 0.0147	$+ 0.08° \pm 0.47°$
P_1	1.1502 ± 0.0210	$+ 0.25° \pm 0.63°$
O_1	1.1608 ± 0.086	$- 0.16° \pm 0.42°$

All these series are from Western Europe plus Hyderabad (India), Armidale and Alice Springs (Australia) and four series in North America.

Results from 26 other European stations where the group $K_1 P_1 S_1$ could not be separated

	$\delta = 1 + H - \frac{3}{2} K$	α
$K_1 P_1 S_1$	1.150 ± 0.016	$+ 0.15° \pm 1.12°$
O_1	1.160 ± 0.015	$- 0.21° \pm 0.47°$

Table 2. Smaller diurnal waves — vertical component (arithmetic means)

Wave	Group	Amplitude in μ Gals	Number of series	$\delta = 1 + h - \frac{3}{2} k$	α
$2Q1$	12– 21	0.9	8	1.166 ± 0.044	$-0.59° \pm 3.70°$
$\sigma1$	22– 32	1.1	9	1.166 ± 0.041	$-0.53° \pm 2.51°$
$Q1$	33– 52	7.3	21	1.160 ± 0.018	$-0.48° \pm 1.15°$
$\rho1$	53– 62	1.3	9	1.159 ± 0.032	$-0.10° \pm 1.31°$
$NO1$	89–103	2.3	10	1.150 ± 0.017	$-0.42° \pm 1.52°$
$\pi1$	111–113	0.9	8	1.158 ± 0.054	$-0.22° \pm 2.62°$
1	137 143	0.7	9	1.175 ± 0.092	$+3.13° \pm 3.35°$
$J1$	152–165	3.2	17	1.171 ± 0.020	$+0.24° \pm 1.45°$
$OO1$	173–183	1.8	12	1.162 ± 0.024	$+0.01° \pm 1.46°$

The nine basic series are those from Frankfurt (3 series), Hannover (1 serie), Sèvres (1 serie), Bruxelles (2 series), Ottawa (2 series).

Among the other stations, those outside of Europe are Hyderabad (India), Armidale and Alic Springs (Australia), and McDonald (USA).

Table 3. Diurnal waves — horizontal East-West component

Results from 14 European stations

Wave	Group	Amplitude 0"001	Number of series	$\gamma = 1 + k - h$
K1	124-134	5.7	14	0.7429 ± 0.0045
P1	114-120	1.7	14	0.7054 ± 0.0157
O1	63- 78	3.9	14	0.6788 ± 0.0056
Minor components				
Q1	33- 52	0.7	12	0.659 ± 0.042
NO1	89-103	0.3	3	0.711(± 0.021)
Ø1	137-143	0.07	2	0.680(± 0.007)
J1	152-165	0.3	12	0.669 ± 0.073
OO1	173-183	0.2	11	0.669 ± 0.075
ψ1	135-136	0.02	3	0.432(± 0.048)

For the very small waves the series used are two from Dourbes and one from Walferdange.

7. Recent Results

The report on the activities of the International Centre for Earth Tides, presented by P. Melchior during the 8th International Symposium on Earth Tides (Bonatz, in prep.), included some recent numerical results that are repeated in Tables 1 - 5.

With regard to the theme of the present meeting we may note that the phase lag angles of the gravimetric M_2-tide in Western Europe are around $\alpha = 2° \pm 1°$. Similar or even smaller values are exhibited by measurements in the interior of the continents. Especially the station of Talgar (24 km east of Alma Ata) in central Asia should be mentioned, which is more than 2200 km distant from the nearest ocean. There $\alpha = +0°8 \pm 0°3$ was obtained (Pariysky et al., 1976). By judging these values in the context of tidal friction problems one has well to distinguish between the phase lag α of the observed tidal gravity variations and the phase lag of the Earth's tidal bulge: In mean latitudes an observed gravimetric M_2 phase lag of 1° corresponds to a phase lag of the tidal bulge of about 7°.

8. Future Activities

Activities in tidal research to be done in the near future are summarized in the resolutions of the 8th International Symposium on Earth Tides, the main points of which are paraphrased here:

Table 4. Derivation of the diurnal LOVE numbers from the experimental values of the γ and δ factors

Wave	Doodson argument	$\gamma = 1 + k - h$	$\delta = 1 + h - \frac{3}{2}k$	h	k	k/h
K1	165.555	0.7429±0.0045	1.1436±0.0069	0.484±0.017	0.227±0.016	0.469±0.032
P1	163.555	0.7054±0.0157	1.1502±0.0210	0.584±0.070	0.289±0.068	0.495±0.090
O1	145.555	0.6788±0.0056	1.1608±0.0086	0.642±0.017	0.321±0.011	0.500±0.022
Q1	135.655	0.6590±0.0420	1.1600±0.0180	0.700	0.360	

For γ Verbaandert-Melchior quartz horizontal pendulums 14 stations, 18,056 days of registration, 433,864 hourly readings.

For δ registrating gravimeters of different kinds 22 stations, 14,346 days of registration, 344,304 hourly readings.

Table 5. Semi-diurnal waves — vertical component

Wave	Mean amplitude (μGal)	$\delta = 1 + h - \frac{3}{2}k$	α
(a) Results from the seven fundamental series in Western Europe			
Lunar waves			
M2	36.0	1.1916 ± 0.008	2.00° ± 0.95°
N2	7.6	1.1760 ± 0.007	2.43° ± 0.99°
2N2	1.0	1.1830 ± 0.037	2.43° ± 2.37°
μ2	1.1	1.1670 ± 0.019	3.27° ± 1.51°
ν2	1.5	1.1670 ± 0.036	2.90° ± 1.72°
L2	0.7	1.1950 ± 0.068	1.61° ± 2.64°
Solar waves			
S2	18.4	1.2010 ± 0.014	0.37° ± 0.83°
K2	4.4	1.1970 ± 0.013	0.80° ± 0.79°
T2	1.1	1.2130 ± 0.029	-0.93° ± 1.39°
(b) Mean results from 21 other European stations			
		δ	α
M2		1.200 ± 0.012	1.84° ± 0.97°
N2		1.178 ± 0.020	2.15° ± 1.05°
S2K2		1.211 ± 0.015	1.05° ± 1.47°
A		1.015 ± 0.028	
B		1.082 ± 0.088	
C		1.027 ± 0.027	

The principal objectives of earth tidal studies are to contribute to various branches of geodesy, geophysics, oceanography, and astronomy, more specifically for instance, satellite orbital perturbations, ocean tides and tidal currents, precession and nutations, earth structure and deformation, volcanic and earthquake prediction, secular variations of gravity field, high precision levelling etc.

During the last decade or so, the celebrated achievement of solving the global ocean tides on the basis of numerical integration of the hydrodynamical equations contributed significantly to the knowledge of the ocean tides in the open oceans. Unfortunately, the global tidal models thus far obtained do not agree among themselves to the point that precise calculations of the ocean tidal loading effect on gravity, tilt, and strain based on these tidal models can be carried out. Such calculations are urgently needed to make further progress on earth tides. Extended experimental ocean-bottom tidal observations are most urgently needed.

The global framework of tidal stations has still to be improved.

References

Bonatz, M.: Proceedings of the 8th International Symposium on Earth Tides, in preparation.
Pariysky, N.N. et al.: Tidal Gravity Variations in Talgar during Twelve Years of Observations. Proc. 7th Int. Symp. Earth Tides Sopron 1973, Stuttgart, 1976

Tidal Friction in the Solid Earth:
Loading Tides Versus Body Tides

J. Zschau

1. Introduction

Astronomical as well as palaeontological evidence suggests a secular
retardation of the Earth's rotation, which is attributed to tidal friction, i.e., mainly to the nonequilibrium and imperfectly fluid response
of the Earth's oceans, as well as to the imperfectly elastic response
of the solid Earth to tidal forces. Estimates of the rotational energy
dissipated in the oceans show that the oceanic term probably accounts
for most of the dissipated energy (Pekeris and Accad, 1969; Pariiskii
et al., 1972; Kuznetsov, 1972; Brosche and Sündermann, 1972; Hendershott, 1972), although the exact share between both, the oceanic dissipation and the dissipation within the solid Earth, is not known.
This is attributed to insufficiencies in the knowledge of the marine
tides in the open oceans, and to the fact that nothing is known about
the rheological mechanism of tidal dissipation within the solid Earth.
Measurements of tidal gravity variations at the Earth's surface, as
well as precise observations of the tidal effect on satellite orbits
have not yet revealed reliable results on imperfectly elastic body
tides of the Earth. Model calculations give also only rough estimations of the tidal energy dissipated within the Earth, mainly because
no information is available on the specific tidal dissipation function,
i.e., the quality factor Q within the Earth.

The quality factor is defined by

$$Q^{-1} = \frac{1}{2\pi \ E^*} \ \oint \frac{dE}{dt} \ dt \tag{1}$$

Here the line integral over $\frac{dE}{dt}$ is the energy dissipated during a complete cycle of sinusoidal straining, and E^* is the peak energy stored
in the system during the cycle (see Fig. 1, next page). There is some
information on the Q-factor from seismological data, i.e., from the
measurement of body waves, surface waves, and free oscillations of the
Earth. It is doubtful, however, whether the seismological Q values are
valid in the case of the Earth tide as well, because they have been
obtained for higher frequencies and for lower amplitudes of the deformation compared with those in the tidal case. Careful examination of
the frequency- and amplitude dependencies of the seismic Q, however,
has shown that the seismic Q is nearly frequency-independent, and is
also nearly amplitude-independent over a wide range of frequencies and
amplitudes (see for instance Gordon and Davis, 1968). As the frequency
of the main tidal constituent M_2 differs by only about one order of
magnitude from the lowest free oscillation frequency of the Earth, and
as the investigations of the seismic Q's amplitude dependency inlcude
the range of tidal strain amplitudes, it seems more likely now that
the Earth tidal Q is near to the seismological Q than that it is not.

Institute of Geophysics, Kiel-University, D-23 Kiel, Germany

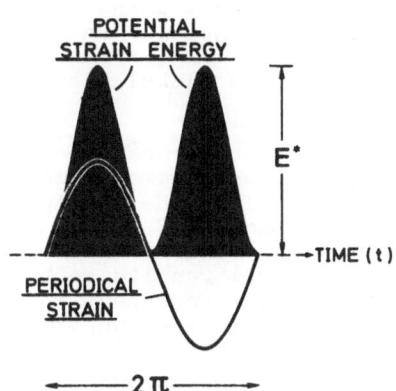

Fig. 1. Definition of the specific dissipation function Q. $Q = 2\pi$ means that the peak energy stored in the system is completely dissipated during one cycle of straining

$$\frac{1}{Q} = \frac{1}{2\pi E^*} \cdot \oint \frac{dE}{dt}\, dt$$

$\frac{dE}{dt}$ = DISSIPATION RATE

Munk and MacDonald (1960) gave an analytic expression for the body tide dissipation rate within the solid Earth by introducing the body tide Q as a free parameter. To obtain this expression, they had to assume that the Earth is homogeneous and incompressible. Kaula (1964) found an expression for the body tide dissipation in the solid Earth, which does not need the above restrictions. However, his expression is rather complicated, and he had to assume $\lambda = \mu$ for the Lamée constants to reduce the complexity of his formulas. Furthermore, his expression requires the integration of the locally dissipated energy fractions throughout the whole volume of the Earth.

In Section 2.1 we will present a theory which relates the body tide dissipation rate to the imaginary part of the complex LOVE number defined by $k^* = \psi^s/\psi^p$, where ψ^p is the tidal potential on a rigid Earth, and ψ^s is the complex secondary potential due to the Earth's tidal deformation. By merely integrating the external work function over the Earth's surface, we will obtain a final expression for the dissipated energy which is rather simple, and is valid for the inhomogeneous, incompressible Earth. This expression may easily be used to calculate the body tide solid dissipation within other planets, too, if the imaginary part of the planet's complex LOVE number k^* is known.

Additionally, the corresponding expression for the dissipation of loading tide energy within the solid Earth is derived by introducing the imaginary parts of the complex mass load coefficients $k^{**} = \psi^s/\psi^p$ and $h^{**} = gu/\psi^p$, where ψ^p is the primary potential of the applied load, ψ^s is the complex secondary potential due to the loading deformation, u is the complex radial displacement, and g is the gravitational acceleration. Loading tide dissipation within the solid Earth as a possible sink for part of the Earth's rotational energy has never been taken into account so far.

Computational results on both body tide dissipation and loading tide
dissipation within the solid Earth are given in Sections 2.2. and 2.3.

A comparison between estimated global body tide Qs and global loading
tide Qs is given in Section 3.1.

It has been argued in the past that one should be able to measure the
phase delay in the *body tide* gravity variation, and therefrom deduce
the tidal energy dissipated within the mantle as well as the mantle's
viscosity- or Q structure (Harrison et al., 1963; Lagus and Anderson,
1968; Pariiskii et al., 1972; Melchior et al., 1976). This problem will
be discussed in Sections 3.2 to 3.4, and new numerical results to be
expected on the body tide gravity phase delay will be presented. These
results will require a more exact definition of the tidal bulge angle
than given so far, and a critical examination of currently used formulas which relate the body tide gravity phase delay to the tidal bulge
delay angle, and the tidal bulge delay angle to the Earth's global Q
(Sect. 3.3).

Finally, computational results on the expected phase shifts between
the *loading tides* on an imperfectly elastic Earth and those on an elastic
Earth will be presented in Section 3.5. It will be discussed whether
and where these phase shifts could be measured, and whether they might
be used to obtain information on anelasticity within the Earth's asthenosphere.

In all sections we will need the Earth's complex LOVE numbers and mass
load coefficients as defined in Zschau (1978a). How they may be obtained, is, therefore, briefly reviewed here now:

Starting point is a general form of the linear stress-strain relation,
which for the shear components may be expressed as

$$\sum_s a_s \frac{\partial^s}{\partial t^s} \tau_{ik} = \sum_s b_s \frac{\partial^s}{\partial t^s} \varepsilon_{ik}$$

with τ_{ik} being the shear stress, ε_{ik} being the shear strain, a_s, b_s
being constants describing elasticity and anelasticity of the system,
and t being the time. After Fourier transformation, this equation
takes the same form as the corresponding relation for the purely elastic Hooke body:

$$\tau_{ik}^* = 2\mu^+ \varepsilon_{ik}^*$$

Here τ_{ik}^* and ε_{ik}^* are the Fourier transforms of the shear stress and
the shear strain, respectively. τ_{ik}^* is complex, whereas ε_{ik}^* is real.
μ^+ is a complex function of the constants a_s, b_s, and therefore depends
on the constitutive law for the body. The above formal correspondency
between the purely elastic case and the linearly anelastic cases is
known as "correspondency principle" (Biot, 1954, 1955; Lee, 1955).
Taking into account that the strain energy may be expressed as the
product of the stress and the strain tensors, and introducing the correspondency principle into Eq. (1), one easily finds that the quality
factor $Q\mu$ due to dissipation in shear may be written as

$$Q\mu^{-1} = \frac{\mu^I}{\mu^R} .$$

μ^I and μ^R are the imaginary and the real part, respectively, of the
complex quantity μ^+. For the case of a Maxwell body, one may find that

$$\mu^I = \frac{\eta \; \omega \; \mu^2}{\mu^2 + \eta^2 \; \omega^2} \; , \qquad\qquad \mu^R = \frac{\eta^2 \; \omega^2 \; \mu}{\mu^2 + \eta^2 \; \omega^2} \; ,$$

where η is the local viscosity, μ is the local rigidity, and ω is the circular frequency of the straining. These local values for μ^+ determine the local $Q\mu$ values within the Earth. If the local $Q\mu$ factors are the known quantities, μ^+ may be determined by setting $\mu^R \sim \mu$ and $\mu^I = \mu \; Q\mu^{-1}$. Finally, the correspondency principle allows us to use the same equations for the computation of the LOVE numbers and mass load coefficients in the case of an imperfectly elastic Earth as are used in the case of a perfectly elastic Earth, if merely the rigidity μ is substituted by the complex quantity μ^+. The equations for the elastic case may be taken for instance from Longman (1962, 1963) or Farrell (1972).

2. Dissipated Tidal Energy in the Solid Earth

2.1 Theory of Dissipation in a Heterogeneous Incompressible Earth

The first law of thermodynamics states that the increment of the total energy of a body is equal to the sum of the work done by external forces and the quantity of heat supplied.

We may calculate the rate at which work is done by external forces. It is expressed by the formula (see Love, 1927)

$$\frac{dE}{dt} = \iiint_V \rho \; \frac{\partial \vec{u}}{\partial t} \; \text{grad} \; \Psi \; dV + \iint_S \frac{\partial \vec{u}}{\partial t} \; \vec{t}_s \; dS \; . \tag{2}$$

Here the first integral denotes the rate at which work is done by body forces, and the integration is taken through the volume of the body in the *unstrained* state. The second integral denotes the rate at which work is done by surface tractions, and the integration is taken over the surface of the body in the *unstrained* state. ρ is the density, $\vec{u}$ the displacement vector, Ψ the potential of the disturbing body force, $\vec{t}_s$ the vector of the surface tractions, t the time and dE/dt the rate of potential energy. Ψ usually has two components, the primary potential Ψ^P of the applied force, and the secondary potential Ψ^s due to the deformation of the body:

$$\Psi = \Psi^P + \Psi^s \; .$$

We assume that the body is strained periodically. Then

$$\rho = \rho_o + \delta\rho \; , \tag{3}$$

where ρ_o is the constant part of ρ, and $\delta\rho$ is the variable part. $\delta\rho$, as well as all the other quantities in Eq. (2), are periodic functions of time. The energy dissipated during one cycle of loading may then be defined as

$$\Delta E = \oint \frac{dE}{dt} \; dt \; . \tag{4}$$

It will be shown in this and in the following two sections that in the body tide case, as well as in the loading tide case, the dissipated energy may be related to the imaginary part of the complex LOVE-numbers, and the complex mass load coefficients, respectively, as defined in

Zschau (1978a,b). We will give exact analytic expressions for the energy dissipated within the Earth, which do not require an integration throughout the whole volume of the Earth, but only the integration over the Earth's surface. The expressions are obtained without approximating the real Earth by an incompressible and homogeneous one as was necessary in former calculations, for instance by Munk and MacDonald (1960).

With Eq. (3), the volume integral in Eq. (2) may be written as

$$\iiint_V \rho \, \frac{\partial \vec{u}}{\partial t} \, \text{grad} \, \Psi \, dV = \iiint_V \rho_o \, \frac{\partial \vec{u}}{\partial t} \, \text{grad} \, \Psi \, dV + \iiint_V \delta\rho \, \frac{\partial \vec{u}}{\partial t} \, \text{grad} \, \Psi \, dV . \qquad (5)$$

From Green's 1st integral transformation we have

$$\iiint_V \rho_o \, \frac{\partial \vec{u}}{\partial t} \, \text{grad} \, \Psi \, dV = \iint_S \rho_o \, \Psi \, \frac{\partial u}{\partial t} \, dS - \iiint_V \Psi \, \text{div} \, (\rho_o \frac{\partial \vec{u}}{\partial t}) \, dV \qquad (6)$$

for a spherical Earth, where u is the radial component of the displacement vector $\vec{u}$. With

$$\text{div} \, (\rho_o \frac{\partial \vec{u}}{\partial t}) = \frac{\partial}{\partial t} \, [\text{div} \, (\rho_o \vec{u})] ,$$

and Poisson's equation (see Pekeris and Jarosch, 1958)

$$\text{div} \, \text{grad} \, \Psi = 4\pi G \, \text{div} \, (\rho_o \vec{u}) \qquad (G = \text{gravitational constant}),$$

it follows that

$$\iiint_V \Psi \, \text{div} \, (\rho_o \frac{\partial \vec{u}}{\partial t}) \, dV = \frac{1}{4\pi G} \iiint_V \Psi \, \text{div} \, \text{grad} \, (\frac{\partial \Psi}{\partial t}) \, dV . \qquad (7)$$

Applying Green's 1st integral transformation to the right-hand side of equation (7) gives

$$\iiint_V \Psi \, \text{div} \, (\rho_o \frac{\partial \vec{u}}{\partial t}) \, dV = \frac{1}{4\pi G} \, [\iint_S \Psi \, \frac{\partial}{\partial t}(\frac{\partial \Psi}{\partial r}) \, dS - \iiint_V \text{grad} \, \Psi \, \frac{\partial}{\partial t}(\text{grad} \, \Psi) \, dV] \qquad (8)$$

for a spherical Earth, where $\frac{\partial \Psi}{\partial r}$ is the derivative of the potential with respect to the radius coordinate r. With Eqs. (2), (5), (6), and (8) the rate of potential elastic energy is

$$\frac{dE}{dt} = \iint_S \left\{ \frac{\partial \vec{u}}{\partial t} \, \vec{\tau}_s + \Psi \, [\rho_o \, \frac{\partial u}{\partial t} - \frac{1}{4\pi G} \, \frac{\partial}{\partial t} \, (\frac{\partial \Psi}{\partial r})] \right\} \, dS$$

$$+ \iiint_V \left\{ \delta\rho \, \frac{\partial \vec{u}}{\partial t} \, \text{grad} \, \Psi + \frac{1}{4\pi G} \, \text{grad} \, \Psi \, \frac{\partial}{\partial t} \, (\text{grad} \, \Psi) \right\} \, dV \qquad (9)$$

As $\delta\rho$, $\vec{u}$, and grad Ψ are periodic functions varying with some frequency ω, and as

$$\oint \cos \, (\omega t + \phi_1) \cos \, (\omega t + \phi_2) \cos \, (\omega t + \phi_3) \, dt = 0$$

for any arbitrary phases ϕ_1, ϕ_2, ϕ_3, it follows that

$$\oint \delta\rho \, \frac{\partial \vec{u}}{\partial t} \, \text{grad} \, \Psi \, dt = 0.$$ (10)

The expression $\frac{\partial}{\partial t} (\text{grad} \, \Psi)$ is by $\pi/2$ out of phase with respect to grad Ψ. Hence

$$\oint \text{grad} \, \Psi \, \frac{\partial}{\partial t} (\text{grad} \, \Psi) \, dt = 0$$ (11)

From Eqs. (10) and (11) it is evident that the volume integral in Eq. (9) does not contribute to dissipation, and with Eqs. (4) and (9) the energy dissipated during one cycle of loading, therefore, is represented by a surface integral only:

$$\Delta E = \oint_t \iint_S \left\{ \frac{\partial \vec{u}}{\partial t} \, \vec{\tau}_s + \Psi \left[\rho_o \frac{\partial u}{\partial t} - \frac{1}{4\pi G} \frac{\partial}{\partial t} \left(\frac{\partial \Psi}{\partial r} \right) \right] \right\} \, dS \, dt \, .$$ (12)

All the variable quantities in Eq. (12) may be developed into spherical harmonics. Because of the orthogonality of the spherical surface harmonics on the Earth's surface the totally dissipated energy simply is the sum of the single harmonic energy terms. Hence

$$\vec{u} = \sum_n \vec{u}_n, \quad \vec{\tau}_s = \sum_n (\vec{\tau}_s)_n, \quad \Psi = \sum_n \Psi_n, \quad \Delta E = \sum_n (\Delta E)_n$$

with n being the degree of the spherical harmonic.

In the following, the variable quantities are supposed to be the nth term of the corresponding expansion into spherical harmonics, although for simplification we drop the index n in the expressions.

The nth term in the expression (12) may be reduced by applying the boundary condition

$$\frac{\partial \Psi}{\partial r} = \frac{2n+1}{R} \Psi^P - \frac{n+1}{R} \Psi + 4\pi G \, \rho_o u \quad (R = \text{Earth's radius}),$$ (13)

which holds for the disturbing potential at the Earth's surface (see for instance Longman, 1962). Eq. (12) may then be written as

$$\Delta E = \oint_t \left\{ \iint_S \left[\frac{\partial \vec{u}}{\partial t} \, \vec{\tau}_s - \frac{1}{4\pi G} \Psi \left(\frac{2n+1}{R} \frac{\partial \Psi^P}{\partial t} - \frac{n+1}{R} \frac{\partial \Psi}{\partial t} \right) \right] \, dS \right\} \, dt$$ (14)

As the expression $\frac{\partial \Psi}{\partial t}$ is by $\pi/2$ out of phase with respect to Ψ, we have

$$\oint_t \Psi \, \frac{\partial \Psi}{\partial t} = 0 \, .$$

For the same reason we have

$$\oint_t \Psi^P \, \frac{\partial \Psi^P}{\partial t} = 0 \, ,$$

and with $\Psi = \Psi^P + \Psi^s$, Eq. (14), therefore, reduces to

$$\Delta E = \oint_t \left\{ \iint_S \frac{\partial \vec{u}}{\partial t} \, \vec{\tau}_s - \frac{2n+1}{4\pi GR} \Psi^s \frac{\partial \Psi^P}{\partial t} \right) \, dS \right\} \, dt \, .$$ (15)

Eq. (15) is valid for a heterogeneous incompressible Earth.

2.2 Dissipation of Body Tide Energy

In the case of body tides, the tangential stresses are zero at the deformed surface, and may be approximately set to zero at $r = R$, i.e., at the surface in the unstrained state. However, the normal stress τ is only zero at the deformed surface, but may not be set to zero at $r = R$. There it takes the negative value of the pressure caused by the weight of the material heaped up over $r = R$ by the deformation. This term of the surface tractions is important for planetary bodies like the Earth, which are hydrostatically prestressed (see Love, 1927; Chap. 176). Hence, in the body tide case we have

$$\vec{\tau}_s \frac{\partial \vec{u}}{\partial t} = \tau \frac{\partial u}{\partial t}, \quad \tau = -\rho_o \, gu \quad (g = 980 \text{ cm/s}^2). \tag{16}$$

From Eq. (16) it is obvious that τ is by $\pi/2$ out of phase with respect to $\partial u/\partial t$, and consequently

$$\oint \tau \frac{\partial u}{\partial t} \, dt = 0. \tag{17}$$

The normal surface traction due to the hydrostatic prestress of the Earth may, therefore, be neglected in the dissipation problem, although it is important for the calculation of the potential elastic energy.

With Eqs. (15) to (17) the dissipated body tide energy reduces to one term

$$\Delta E = -\frac{2n+1}{4\pi GR} \oint_t \iint_S \psi^s \frac{\partial \psi^p}{\partial t} \, dS \, dt \tag{18}$$

The secondary potential ψ^s due to the deformation may be written as

$$\psi^s = k^* \psi^p,$$

where k^* is the complex LOVE number

$$k^* = k + jK, \quad j = \sqrt{-1}$$

(see Zschau, 1978a).

For the Earth $k \approx |k^*|$, and with

$$\psi^p = \psi_o^p \cos(\omega t) \tag{19}$$

we have

$$\psi^s = k \, \psi_o^p \cos(\omega t + \phi_k) \tag{20}$$

with

$$\phi_k \approx \sin \phi_k \approx \tan \phi_k = \frac{K}{k} \quad \text{(K negative, k positive, see Table 1) (21)}$$

Introducing Eqs. (19) and (20) into Eq. (18) yields

$$\Delta E = -\frac{2n+1}{4 \, GR} k \sin \phi_k \iint_S (\psi_o^p)^2 \, dS,$$

and with Eq. (21)

$$\Delta E = - \frac{2n+1}{4\,GR}\, K \iint\limits_{S} (\psi_o^p)^2 \; dS \tag{22}$$

For the main lunar tidal constituent M_2 we have $n=2$, and

$$\psi_o^p = \frac{3}{4}\, C\, G\, m_{\mathbb{C}}\, \frac{R^2}{D^3}\, \cos^2 \phi \tag{23}$$

with

$m_{\mathbb{C}}$: mass of the moon
D : mean distance to the moon
ϕ : geocentric latitude
C : 0.90812

Substituting Eq. (23) into Eq. (22) we get the final expression for the M_2-body tide energy dissipated during one cycle of straining

$$\Delta E = - \frac{3}{2}\, \pi\, C^2\, K\, G\, m_{\mathbb{C}}^2\, \frac{a^5}{D^6} \tag{24}$$

or alternatively

$$\Delta E = - \frac{3}{2}\, \pi C^2\, k\, G\, m_{\mathbb{C}}^2\, \frac{a^5}{D^6}\, \sin\, \phi_k \tag{25}$$

$\sin \phi_k = K/k$ is the phase shift of the secondary potential (due to the deformation) at the unstrained surface ($r = R$) with respect to the primary potential of the external forces at $r = R$. From the law of conservation of energy and momentum within the Earth-Moon-system one may easily deduce the tidal retarding torque from Eq. (24) or (25). Eq. (25) then becomes equivalent to the expression Pariiskii (1960) obtained for the tidal retarding torque, except that Pariiskii did not give an exact definition of the phase shift ϕ_k. Therefore, in Pariiskii et al. (1972) this phase shift was mixed up with the phase shift of the secondary gravity disturbance at the *deforming* surface, the latter of which was calculated by a formula of Slichter (1960). As may be seen from Section 3.3, the gravity phase shift as determined by Slichter's formula is, however, smaller by a factor of 0.6 than the phase shift of the secondary potential at $r = R$.

For the evaluation of Eq. (24) the Earth's imaginary LOVE number K is needed, which may be obtained by the method outlined in Zschau (1978a). This method was originally developed for loading tide problems, but can easily be extended to the body tide problem by merely changing one boundary condition at the Earth's surface, i.e., we must have

$$\lambda\, \text{div}\, \vec{u} + 2\mu\, \frac{\partial u}{\partial r} = 0 \qquad \text{at}\quad r = R,$$

where λ and μ are complex elastic constants. Taking account of this boundary condition we have computed the Earth's complex LOVE numbers for the Gutenberg model as tabulated in Alterman et al. (1961; see Table 1). For these computations we assumed that there are only losses of mechanical energy in shear and no losses in pure compression. From the investigation of seismic body wave Q-factors (see for instance Anderson et al., 1965; Stacey, 1977), as well as from the study of the purely compressional radial modes of free oscillation of the Earth (Slichter et al., 1966) this appears to be a very good approximation to the real behavior of the Earth. Q-values in the Earth's liquid core have been found to be extremely high, values of several thousands being

Table 1. Complex LOVE numbers (stress-free surface) and mass load coefficients (loaded surface) for the Gutenberg Earth model + LMS model for the $Q\mu$ structure of the mantle + MM8 model for the $Q\mu$ structure of the crust

Stress-free surface (body tide)

n	h	k	l	$100 \cdot H$	$100 \cdot K$	$100 \cdot L$
2	0.6079	0.3044	0.0831	-0.049	-0.032	-0.015
3	0.2869	0.0936	0.0143	-0.027	-0.011	-0.007

Loaded surface (loading tide)

n	h'	$n \cdot k'$	$n \cdot l'$	$100 \cdot H'$	$100 \cdot n \cdot K'$	$100 \cdot n \cdot L'$
0	-0.130	0	0	0	0	0
1	-0.275	0	0.117	0.015	0	0.025
2	-0.975	-0.607	0.056	0.045	0.035	0.048
3	-1.027	-0.580	0.217	0.067	0.048	0.065
4	-1.028	-0.525	0.240	0.096	0.063	0.102
5	-1.061	-0.514	0.237	0.128	0.084	0.143
6	-1.119	-0.532	0.239	0.163	0.109	0.181
7	-1.188	-0.564	0.249	0.199	0.137	0.215
8	-1.259	-0.602	0.262	0.236	0.166	0.247
9	-1.330	-0.641	0.278	0.275	0.197	0.276
10	-1.398	-0.680	0.295	0.315	0.230	0.302
11	-1.463	-0.718	0.312	0.356	0.265	0.326
12	-1.526	-0.755	0.330	0.399	0.300	0.347
13	-1.586	-0.790	0.347	0.442	0.337	0.365
14	-1.643	-0.825	0.366	0.486	0.374	0.380
15	-1.697	-0.858	0.384	0.532	0.413	0.392
16	-1.749	-0.890	0.402	0.577	0.452	0.402
17	-1.799	-0.920	0.421	0.624	0.492	0.409
18	-1.847	-0.950	0.439	0.671	0.532	0.414
19	-1.893	-0.978	0.457	0.718	0.573	0.416
20	-1.936	-1.005	0.475	0.766	0.614	0.416
21	-1.978	-1.030	0.493	0.813	0.655	0.414
22	-2.017	-1.055	0.510	0.861	0.697	0.409
23	-2.055	-1.078	0.527	0.909	0.738	0.403
24	-2.091	-1.100	0.544	0.956	0.779	0.394
25	-2.126	-1.121	0.560	1.003	0.819	0.384
30	-2.276	-1.230	0.634	1.231	1.015	0.313
40	-2.491	-1.327	0.747	1.612	1.337	0.119
50	-2.636	-1.391	0.824	1.869	1.549	-0.071
60	-2.740	-1.427	0.874	2.019	1.667	-0.219
70	-2.823	-1.448	0.908	2.092	1.718	-0.327

Table 1 continued

n	h'	n·k'	n·l'	100·H'	100·n·k'	100·n·L'
80	-2.892	-1.462	0.931	2.116	1.729	-0.401
90	-2.953	-1.472	0.946	2.111	1.715	-0.454
100	-3.010	-1.482	0.956	2.088	1.687	-0.491
120	-3.117	-1.503	0.970	2.016	1.612	-0.537
140	-3.220	-1.530	0.981	1.929	1.527	-0.558
160	-3.320	-1.563	0.993	1.836	1.439	-0.562
180	-3.419	-1.602	1.009	1.743	1.354	-0.555
200	-3.516	-1.644	1.027	1.650	1.271	-0.537
250	-3.745	-1.756	1.087	1.435	1.084	-0.469
300	-3.948	-1.867	1.157	1.251	0.929	-0.384
350	-4.121	-1.967	1.227	1.102	0.806	-0.300
400	-4.264	-2.053	1.291	0.989	0.714	-0.228
500	-4.475	-2.184	1.396	0.850	0.602	-0.129
600	-4.612	-2.271	1.469	0.789	0.553	-0.079
800	-4.760	-2.368	1.555	0.765	0.532	-0.053
1000	-4.824	-2.412	1.596	0.770	0.536	-0.054
1500	-4.866	-2.442	1.626	0.778	0.542	-0.060
2000	-4.871	-2.447	1.629	0.780	0.543	-0.060
3000	-4.872	-2.448	1.629	0.781	0.544	-0.061
5000	-4.872	-2.449	1.629	0.781	0.544	-0.061
10000	-4.873	-2.450	1.629	0.781	0.544	-0.061

typical (Quamar and Eisenberg, 1974). We therefore assumed that there are no frictional losses in the core.

In a first approach the quality factor $Q\mu$ associated with the shear component of strain was set to constant all over the Earth's mantle. The imaginary LOVE number K calculated for this model was introduced into Eq. (24) to determine the dissipation rate $\frac{1}{T} \Delta E$ (T = 12.42 h) for the M_2 body tide.

Results are given in Figure 2. It shows that if only 10 % of the astronomically observed dissipation is to be dissipated in the Earth's mantle, a constant tidal $Q\mu$ of less than 60 is needed. For these calculations the astronomically observed dissipation rate was taken to be $3.3 \cdot 10^{19}$ erg/s for the Earth-Moon-system (Morrison, 1977).

If we adopt the $Q\mu$ structure LMS as obtained from the observation of free oscillations of the Earth (Smith, 1972, see also Fig. 3), the rate of the total body tide dissipation in the mantle is

$3.19 \cdot 10^{17}$ erg/s ,

which is about 1 % of the astronomically observed dissipation rate. These results confirm the conclusion from most of the earlier approxi-

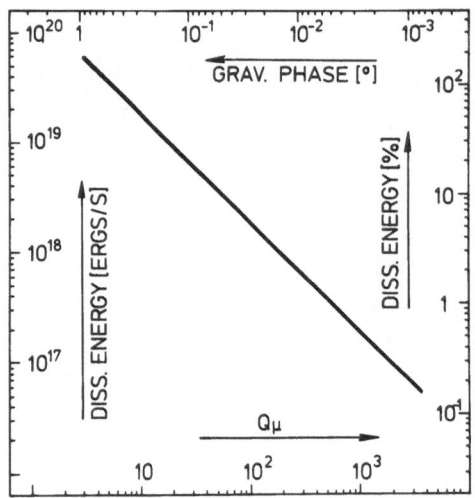

Fig. 2. M_2 body tide dissipation rate and gravity phase delay due to imperfections in the elasticity of the Earth's mantle. Results are valid for $Q\mu$ being constant within the mantle. The dissipation rate connected with losses in shear only, is given in ergs/s as well as in percent of the astronomically observed dissipation rate, here taken to be $3.3 \cdot 10^{19}$ ergs/s

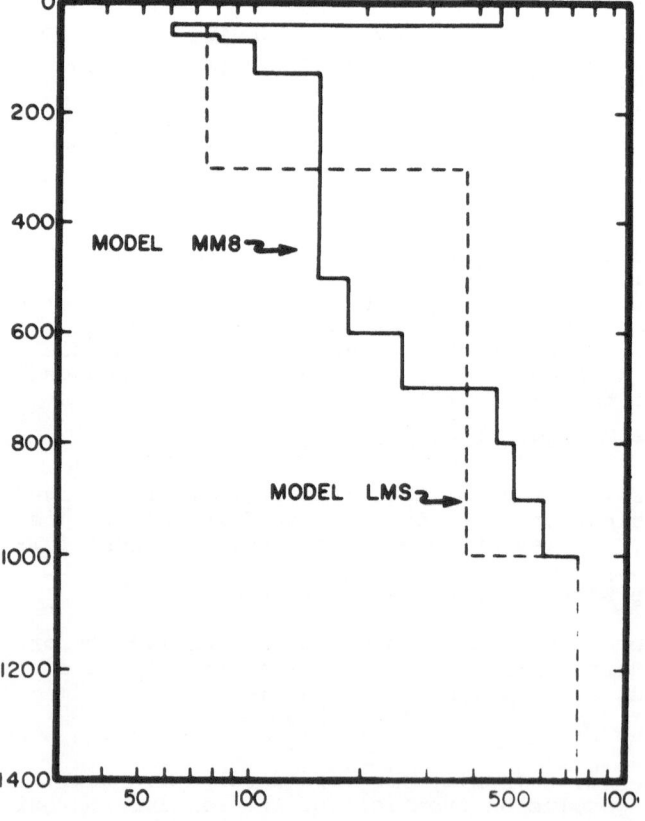

Fig. 3. $Q\mu$-values vs. depth for model MM8, used for surface-wave calculations, and for model LMS used for free-oscillation calculations (from Smith, 1972)

mative estimations of tidal dissipation in the solid Earth, which say
that body tide friction in the solid Earth probably only accounts for
a few percent of the tidal friction, the balance presumably attributed
to the seas.

2.3 Dissipation of Loading Tide Energy

Much progress has been made in calculating the present tidal dissipation in the oceans from ocean tide models. Several of these calculations gave a dissipation rate which is higher than the presently adopted
astronomical value (see Pekeris and Accad, 1969; Pariiskii et al.,
1972; Kuznetsov, 1972). However, Henderschott (1972) obtained a better
agreement with the astronomical observations when integrating the work
done over both the ocean surface and the deforming surface of the elastic solid Earth. He found a reduction of 30 % in the dissipation rate
due to the elastic deformation of the surface.

No account has yet been taken of imperfections in the Earth's elasticity when considering deformations due to the varying load of the ocean
tides. Simple two-dimensional calculations in Zschau (1978a,b), assuming a homogeneous halfspace and a sinusoidally distributed load at
the surface, suggest that, at least in the crust and in the uppermost
mantle, the potential loading tide strain energy may regionally be
much higher than the potential body tide energy (see Fig. 4). Dissipation of loading tide energy in the solid Earth might, therefore, be
comparable to the body tide dissipation.

In the following we calculate the dissipated loading tide energy due
to a variable spherical harmonic surface load $\xi = \xi_0 \cos \omega t$ of degree n
on a spherical Earth. Due to Lamb (1945; see also Henderschott, 1972)
the primary disturbing potential of such a load is

$$\psi^P = \frac{4\pi GR}{2n+1} \rho_w \xi_0 \cos \omega t . \tag{26}$$

(ξ_0 = height of the loading shell; ρ_w = density of the load, here
taken to be the density of water.)

The deformation of the Earth gives rise to a secondary potential

$$\psi^S = k^{**} \psi^P ,$$

where k^{**} is the complex mass load coefficient, $k^{**} = k'+jK'$ (k' negative, K' positive, see Table 1).

For the Earth we have $|k^{**}| \approx |k'| = -k'$, and consequently the second-
ary potential due to the deformation may be written as

$$\psi^S = \frac{4\pi GR}{2n+1} k' \rho_w \xi_0 \cos (\omega t + \phi_k) \tag{27}$$

with

$$\phi_k \approx \sin \phi_k \approx \tan \phi_k = \frac{-K'}{-k'} . \tag{28}$$

The radial displacement u may be related to the water height by the
complex mass load coefficient $h^{**} = h'+jH'$ (h' negative, H' positive,
see Table 1) using the formula $u = h^{**} \psi^P/g$.

Then similar considerations as above give

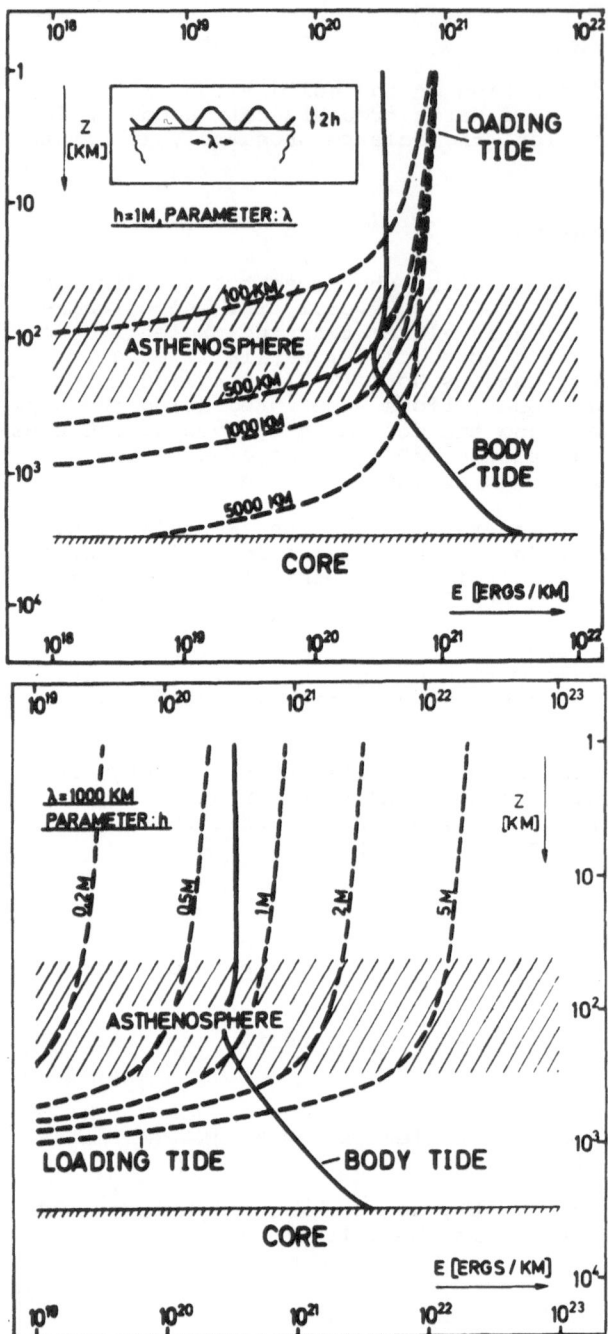

Fig. 4. Tidal shear strain energy vs. depth for marine loading tides and for the semidiurnal body tide (from Zschau, 1978a). In the loading tide case the local values for the energy density have been calculated for a two-dimensional homogeneous halfspace ($\mu = 10^{12}$ dyn/cm^2), and for a sinusoidal distribution of the load at the surface. The local values for the energy density have been multiplied by the volumes of the local 1 km thick concentric shells of the Earth to obtain the approximative values for the total energy stored within each shell. The body tide energy curve has been taken from Lagus and Anderson (1968), and has been normalized to a rigidity and an octahedral shear strain of $3.4 \cdot 10^{11}$ dyn/cm^2 and $5 \cdot 10^{-8}$, respectively, at the surface. λ, spatial wavelength of the loading; h, amplitude of the marine loading tide

$$u = \frac{4\pi GR}{2n+1} \frac{\rho w}{g} h' \xi_o \cos(\omega t + \phi_H) \tag{29}$$

with

$$\phi_H \approx \sin \phi_H \approx \tan \phi_H = \frac{-H'}{-h'} . \tag{30}$$

As in the body tide problem, the tangential stresses at $r = R$ are approximately zero, and consequently

$$\vec{\tau}_s \frac{\partial \vec{u}}{\partial t} = \tau \frac{\partial u}{\partial t} . \tag{31}$$

The normal traction at the Earth's surface consists of two terms,

$$\tau = - (\rho_w g \xi + \rho_o gu) , \tag{32}$$

the first one arising from the pressure of the water load, and the second one from the weight of the material heaped up over the surface in the unstrained state $(r = R)$. As in the body tide case, the second term does not contribute to dissipation, because

$$\oint u \frac{\partial u}{\partial t} dt = 0 .$$

Taking this into account, and introducing Eqs. (26) to (32) into Eq. (15) yields

$$\Delta E = \frac{4\pi^2 G}{2n+1} R \rho_w^2 (H'-K') \iint_S \xi_o^2 dS \tag{33}$$

for the solid Earth dissipation due to a spherical harmonic load of degree n, or alternatively

$$\Delta E = \frac{4\pi^2 G}{2n+1} R \rho_w^2 (h' \sin \phi_H - k' \sin \phi_K) \iint_S \xi_o^2 dS .$$

With $\Psi_o^P = \frac{4\pi GR}{2n+1} \rho_w \xi_o$ expression (33) takes the same form as the corresponding expression for the body tide dissipation [Eq. (22)], i.e.,

$$\Delta E = \frac{2n+1}{4 GR} (H'-K') \iint_S (\Psi_o^P)^2 dS$$

except for that in the loading tide expression there is the additional H'-term which stems from the pressure of the water on the Earth's surface.

Hendershott (1972), as well as Groten and Brennecke (1973), gave a spherical harmonic expansion of the global ocean M_2 tide. The Hendershott ocean tide model is shown in Figure 5. We have used these data to calculate the surface integral in Eq. (33) for each harmonic component of the global tide. The first column of Table 3 in Hendershott gives directly the values $\frac{1}{R} (\iint_S \xi_o^2 dS)^{1/2}$ for n = 0 to n = 25. As in the body tide problem, the Gutenberg model has been chosen to allow for the Earth's elasticity structure. Whereas the model LMS, from free oscillation data, has been taken to allow for the $Q\mu$ structure of the Earth's mantle, and the model MM8 (see Fig. 3), from surface wave data,

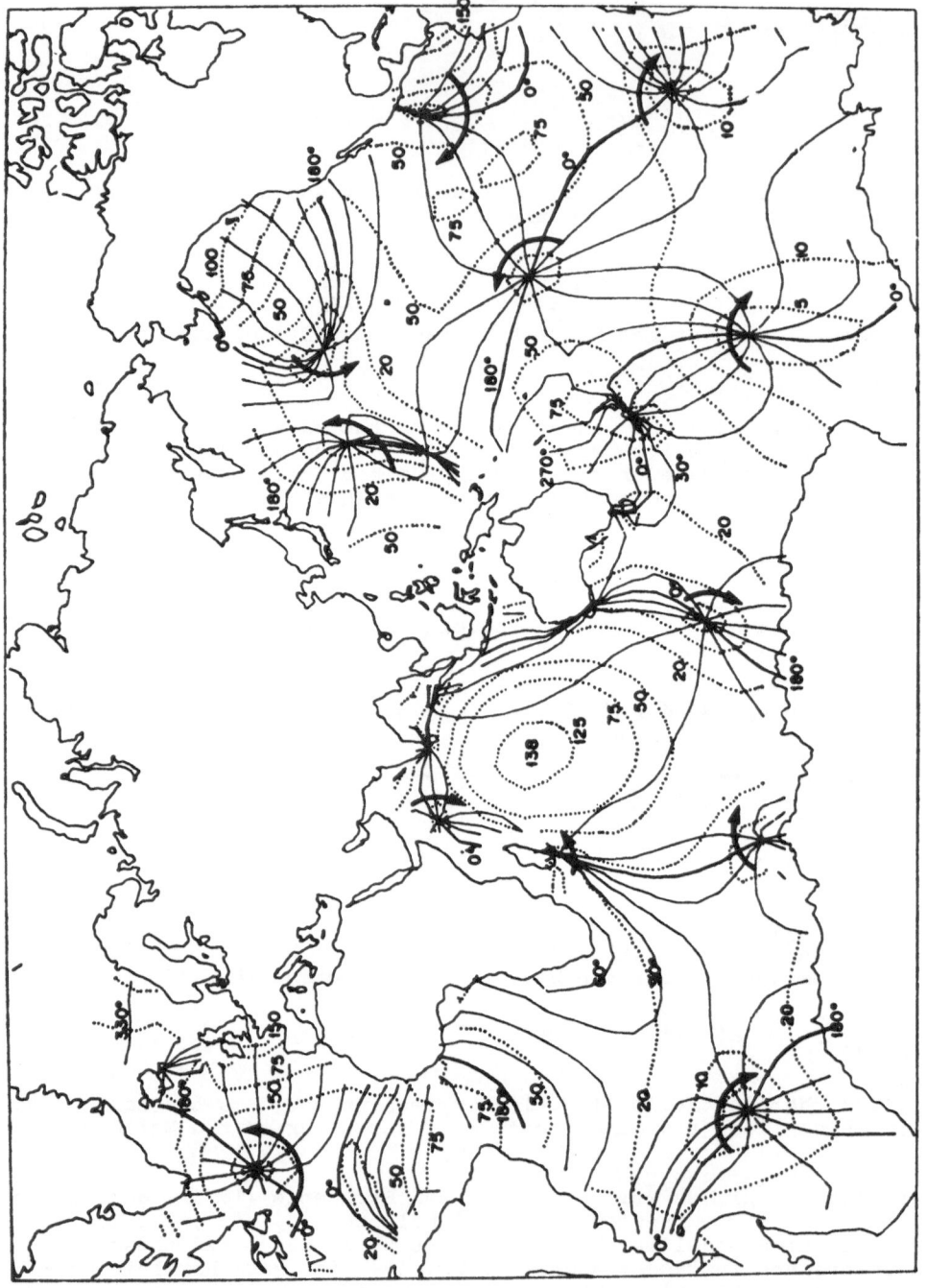

Fig. 5. Global M_2 ocean tide model used for the computation of the global M_2 loading tide dissipation within the solid Earth (from Hendershott, 1972)

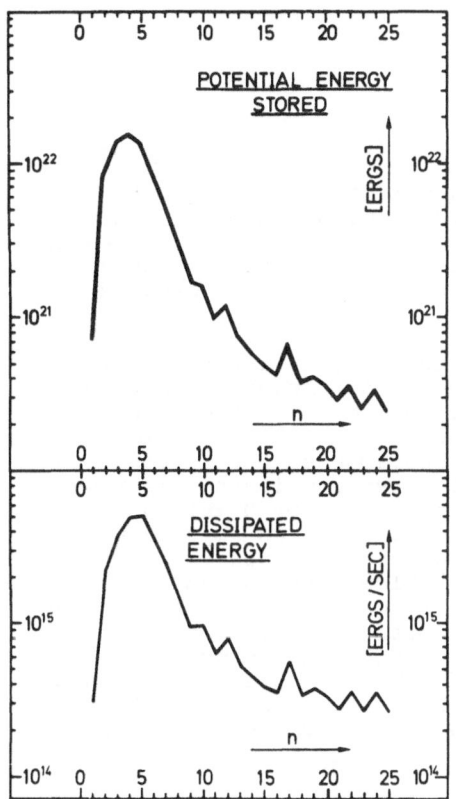

Fig. 6. M₂ loading tide shear strain
Fig. 6. M_2 loading tide shear strain energy stored and dissipated within the solid Earth. n is the degree of the spherical harmonic expansion of Henderschott's global M_2 ocean tide model (Hendershott, 1972). Results are valid for the Gutenberg Earth with the free-oscillation $Q\mu$-structure LMS for the mantle and the surface wave $Q\mu$-structure for the crust. The potential energy stored within the Earth has been obtained by introducing the numerical values for the dissipated energy (this figure) and for the loading tide $Q\mu$ (see Fig. 8) into Eq. (1)

has been chosen to allow for the high $Q\mu$ values in the crust. With H' and K' from the method outlined in Zschau (1978a; for numerical values see Table 1), and with Eq. (33) we may determine the dissipation rate $\Delta E/T$ (T = 12.42 h) for each term of the M_2 ocean tide spherical harmonic expansion. Results are plotted in Figure 6 showing that most of the energy is dissipated for n = 4 and n = 5, and that for n > 15 the dissipation rate is a rather slowly converging function of n. The total dissipation rate is merely the sum of the dissipation rates for each harmonic component. It amounts to

$$3.21 \cdot 10^{16} \text{ erg/s},$$

which is roughly 10 % of the dissipated body tide energy. This value is a minimum value because it only takes the global ocean tide into account, i.e., the low degree harmonics of the ocean tide distribution up to n = 25. The expansion does not represent the high amplitude tides in the shelf areas, which may contribute significantly to the total dissipation because the dissipated energy is proportional to the square of the tidal amplitude. Hence, we conclude that the loading tide energy dissipated in the solid Earth amounts to *at least* 10 % of the dissipated body tide energy, and therefore, may not be neglected if the body tide dissipation is considered to be important in any circumstance. However, both body tide and loading tide dissipation together most probably do not account for more than a few percent of the astronomically observed dissipation rate.

3. Global Tidal Q and Tidal Phase Shifts

3.1 Loading Tide Q Versus Body Tide Q

If the Q-factor is distributed inhomogeneously within the Earth, we must distinguish between the global Q and the local Qs of the Earth. Both of them are defined by Eq. (1). However, in the case of the global Q the *total* potential peak energy stored within the *whole* Earth as well as the *total* energy dissipated within the *whole* Earth must be inserted into Eq. (1). Whereas in the case of the local Qs the corresponding *local* energy values within the Earth have to be used in this equation.

In Zschau (1978a,b) we pointed out that there is an important difference between the storage of body tide energy and loading tide energy within the Earth: While in the body tide case most of the potential energy is stored in the lower mantle, and only a small fraction is stored in the crust and in the low Q asthenosphere, the loading tide potential energy is mainly stored in the crust and in the uppermost mantle (see also Fig. 4). On the assumption of a low Q asthenosphere in the uppermost mantle there is consequently also an important difference in the relative fraction of the dissipated energy within the Earth, when comparing the loading tide problem with the body tide problem. For the free oscillation $Q\mu$-model LMS we have found that in the loading tide problem nearly all the dissipated energy is dissipated in the low Q asthenosphere; whereas in the body tide problem 67 % of the dissipated energy is dissipated in the high Q lower mantle, and only 15 % is dissipated in the asthenosphere (see also Fig. 7). From this the global loading tide $Q\mu$ was obtained up to a factor of 6 lower than the corresponding body tide $Q\mu$.

The calculations above had been made for a two-dimensional homogeneous halfspace. We have improved these calculations for a spherical Gutenberg Earth with the $Q\mu$ structure taken from the free oscillation model LMS. As the LMS model does not allow for the high $Q\mu$ in the crust, the crustal $Q\mu$ again was taken from surface wave data (model MM8, see Fig. 3). To calculate the global tidal $Q\mu$ for a given distribution of local $Q\mu$ values within the Earth, we determine the dissipated energy for this given model as described above, and compare it with the dissipated energy for a constant $Q\mu$ distribution within the Earth's mantle. From Eq. (1) it immediately follows that

$$Q_{gl} = \frac{(\Delta E)_{const}}{(\Delta E)_{gl}} Q_{const}$$

where Q_{gl} is the global $Q\mu$ in the case of the adopted heterogeneous $Q\mu$-model, and Q_{const} is the $Q\mu$ value in the case of constant $Q\mu$ distribution. $(\Delta E)_{gl}$ and $(\Delta E)_{const}$ are the corresponding values for the energy dissipated during one cycle of straining. With Eqs. (22) and (33) the last formula may be written as

$$Q_{gl} = \frac{K_{const}}{K} Q_{const}$$

for body tides, and

$$Q_{gl} = \frac{H'_{const} - K'_{const}}{H' - K'} Q_{const}$$

for loading tides, respectively.

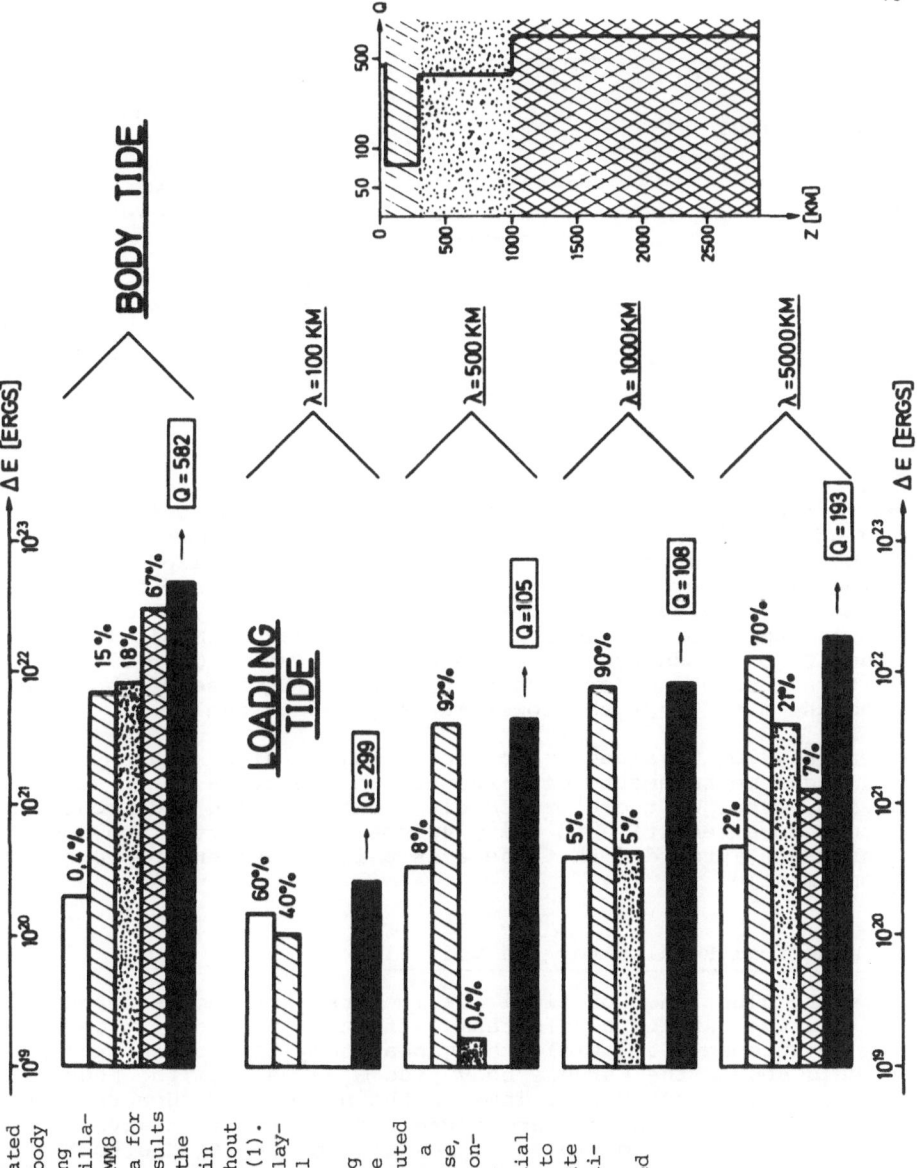

Fig. 7. Shear strain energy dissipated within the solid Earth for the M_2 body tide and for loading tides, assuming the LMS $Q\mu$-structure from free oscillation data for the mantle, and the MM8 $Q\mu$-structure from surface wave data for the crust (from Zschau, 1978a). Results have been obtained by integrating the values for the locally stored strain energy (as given in Fig. 4) throughout each layer, and making use of Eq. (1). The energy dissipated within each layer is given in percent of the total energy dissipated in the crust and mantle (black beam). In the loading tide case the absolute energy scale is true for a sinusoidally distributed ocean tide of wavelength λ, and of a 1-m amplitude. In the body tide case, the absolute scale is only conditionally true because of the arbitrary normalization of the stored potential energy as described in the legend to Figure 4. In both cases the absolute energy scale is valid for a full tidal cycle. Qs in the little boxes give the global $Q\mu$ values connected to losses in shear only

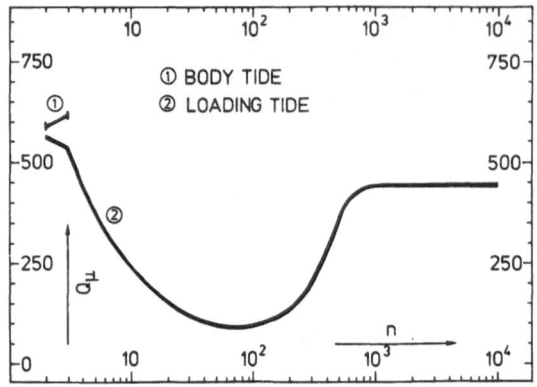

Fig. 8. Body tide Qμ and loading tide Qμ for different degrees n of the spherical harmonic expansions of the tides. Results are valid for the Gutenberg Earth with the LMS Qμ-structure from free oscillation data for the mantle, and the MM8 Qμ-structure from surface wave data for the crust. The low Qμ values for n slightly lower than 100 in the loading tide case, are due to the low Q asthenosphere in model LMS. The high Qμ values for higher n and for lower n are due to the high Qμ values in the crust and in the lower mantle, respectively

Results are plotted in Figure 8 for n = 2,3 in the body tide case, and for n up to 10^4 in the loading tide case. In contrast to the body tide case, high degree harmonics of the load are also important in the loading tide case. Figure 8 shows that the global body tide Qμ is as high as 600, whereas the global loading tide Qμ may be much lower, its value depending on the degree of the spherical harmonic loading. Due to the low Q asthenosphere it has a minimum value Qμ = 87 for n = 70, and it is increasing for higher and lower n because of the high crustal Qμ and the high Qμ of the lower mantle, respectively.

In general, low Q values are connected with "high" phase shifts of Earth tidal measurements with respect to the tides on an elastic Earth and vice versa. We therefore conclude that loading tide measurements are much more sensitive to the Q values of the Earth than body tide measurements, provided that there is a low Q asthenosphere in the uppermost mantle.

3.2   Numerical Results for the Body Tide

A complete global net of tidal gravimeters, strainmeters and tiltmeters would average out the considerable effect of the ocean tide loading on tidal measurements and yield the global body tide phase lag. From this one could deduce the Earth's body tide Q as well as the body tide energy dissipated in the Earth. This is the philosophy. However, such a global net does not exist, and ocean tide corrections have to be applied to the measurements to yield the residual phase lag due to the dissipation in the solid Earth. As the global ocean tides are not well known yet, the interpretation of the experimental residual phase lags as being due to dissipation is still very doubtful. Nevertheless, it has been pointed out by several authors that the experimental residual phase shifts in the tidal gravity variation are of the order of some tenths of a degree, and might be attributed to body tide dissipation in the solid Earth (Harrison et al., 1963; Lagus and Anderson, 1968; Pariiskii et al., 1972; Melchior et al., 1976).

We have calculated the phase shifts of the tide on the imperfectly elastic Earth with respect to that on the elastic Earth. The calculations have been carried out for the Gutenberg Earth and for different constant values of Qμ in the mantle. The following linear relationships between the phase shifts and the constant Q_μ^{-1} factors have been found (n = 2):

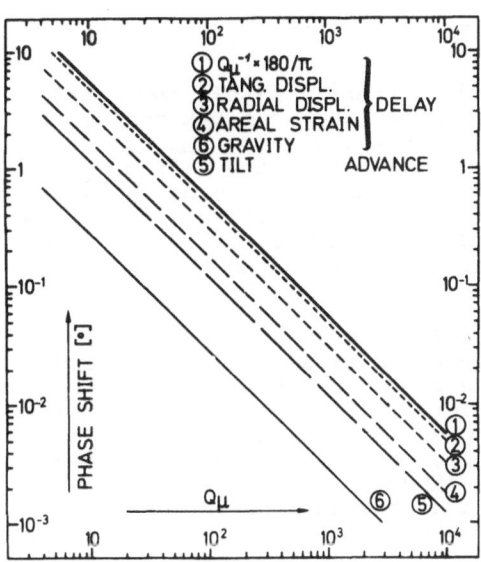

Fig. 9. Body tide phase shifts (n=2) of tangential and radial displacements, areal strain, gravity variation and tilt (including the horizontal acceleration) at the Earth's surface due to imperfections in the Earth's elasticity. "Delay" is a phase lag with respect to the elastic tide, whereas "advance" is a phase lead with respect to the elastic tide. Results are valid for $Q\mu$ being constant within the mantle

Tangential displacement: $\tan \phi = L/l = -0.897\ Q_\mu^{-1}$

Radial displacement: $\tan \phi = H/h = -0.555\ Q_\mu^{-1}$

Surface areal strain: $\tan \phi = \dfrac{H-3L}{h-3l} = -0.318\ Q_\mu^{-1}$

Gravity: $\tan \phi = \dfrac{H-\frac{3}{2}K}{1+h-\frac{3}{2}k} = -0.0508\ Q_\mu^{-1}$

Tilt: $\tan \phi = \dfrac{K-H}{1+k-h} = +0.217\ Q_\mu^{-1}$

(h,H), (k,K), and (l,L) are the complex LOVE numbers.

These results are plotted in Figure 9. It is obvious from Figure 9 that the phase shifts to be expected for reasonable $Q\mu$ values of the mantle are extremely small, especially in the case of the gravity phase shift. The experimental gravity phase shift of -1°, which had been attributed to solid Earth friction by Harrison et al. (1963) would require an absolutely unrealistic constant $Q\mu$ of less than 3 in the mantle. Furthermore, it would account for more than the astronomically observed value of the tidal dissipation (see Fig.2). If we adopt $3.06 \cdot 10^{20}$ erg/s for the total heat flux at the Earth's surface (Chapman and Pollack, 1975), the -1° phase shift would correspond to a dissipation rate which is more than 20 % of the Earth's total heat flux.

The experimental gravity phase shift of -0.8° as attributed to solid Earth friction by Lagus and Anderson (1968) is unrealistic for the same reasons.

Pariiskii et al. (1972), and Melchior et al. (1976) assume that the experimental gravity phase shift of -0.2° may be due to body tide friction in the solid Earth. Melchior et al. attribute this phase shift to a global Q of 60 which is too high for the adopted phase delay, as we will explain below. Our calculations give a constant $Q\mu$ of 14 for the same phase shift $\phi = -0.2°$ (see Fig. 9).

If we assume that $Q\mu$ is not distributed constantly within the mantle, but is distributed in the same manner as the $Q\mu$ of the LMS model, the only difference being a constant factor, then the global $Q\mu$ corresponding to the tidal gravity phase shift of $-0.2°$ takes the extremely small value of less than 3! In this case the $0.2°$ phase delay of the gravity tide would account for more than 100 % of the astronomically observed value of the dissipation rate, again. Both $Q\mu$-values, 14 in the case of constant $Q\mu$ distribution and 3 in the case of the LMS equivalent model, are unrealistic.

From Figure 9 one may see that more realistic values of $Q\mu$ (>300) as deduced from seismological data correspond to tidal gravity phase delays of only several thousandths of a degree, if a homogeneous $Q\mu$ distribution is adopted for the mantle. Furthermore, for the free oscillation $Q\mu$ model LMS, our calculations give the extremely small tidal gravity phase shift of about $0.001°$ only. We conclude that the determination of the Earth's global tidal $Q\mu$ from the tidal gravity phase delay is not unique and vice versa. The differences between possible globel $Q\mu$ factors for a given gravity phase delay, and the differences between possible gravity phase delays for a given global $Q\mu$ may be of the order of several 100 %, the exact value depending on the geometry of the $Q\mu$ distribution within the Earth.

The phase delay of tidal gravity amounts more likely to a few thousandths of a degree than a few tenths of a degree. In view of this fact we cannot see any chance of obtaining information of the mantle $Q\mu$ from body tide gravity investigations.

3.3 Theoretical Relationship between Tidal Bulge Angles, Body Tide Phase Shifts, and the Global Body Tide Q

The result presented above does not at all agree with the expected tidal gravity phase shifts published so far. We attribute this disagreement to three main reasons:

1. The determination of the tidal bulge delay angle from the tidal gravity phase shift has been done wrongly in the past.

2. The determination of the solid Earth's Q from the tidal bulge delay angle has been done wrongly in the past.

3. The relationship between the tidal gravity phase delay and the Earth's global Q is not unique, but is strongly dependent on the geometry of the Q-distribution within the Earth.

The tidal bulge delay angle is defined as the angle between the line from the Earth's center to the Moon, and the line from the Earth's center to the pole of the bulge (Fig. 10). It is sometimes determined from the tidal gravity phase shift by the formula (Slichter, 1960)

$$\alpha_g = \frac{1}{2} \frac{\delta}{\delta - 1} \phi \tag{34}$$

where δ is the gravimetric δ-factor, ϕ is the gravity phase delay, and α_g is the delay angle of the bulge. From simple geometric considerations one finds that α_g does not correspond to the bulge of displacement as usually adopted. However, it is half the phase delay of the secondary gravity disturbance at the *deforming* surface. This gravity disturbance does not inlcude the primary force of the applied external

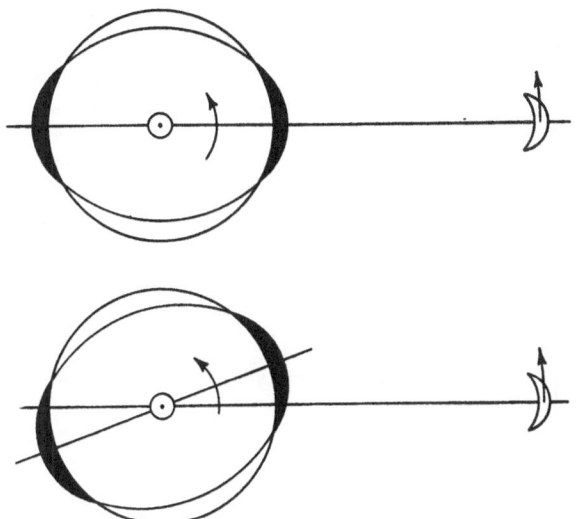

Fig. 10. The principle of tidal friction (from Munk and MacDonald, 1960). *Upper diagram* shows the tidal bulge if there is no friction. In the case of friction there is a delay in the time of high tide, and the resulting distortion of the tidal bulge leads to a deceleration in the Earth's rate of rotation, and to an acceleration in the Moon's orbital motion. The tidal bulge delay angle is defined as the angle between the line from the Earth's center to the Moon, and the line from the Earth's center to the pole of the bulge

potential. But it consists of the gravity variation which stems from the disturbed potential due to the deformation, and the gravity variation at a surface point due to the shift of this point through the Earth's gravity field when the Earth is deformed. α_g may be expressed in terms of the complex LOVE numbers:

$$\alpha_g \approx \tan \alpha_g = -\frac{1}{2} \frac{H - 3/2\,K}{h - 3/2\,k} \tag{35}$$

For simplification we will refer to this angle as the "tidal bulge delay angle of the secondary gravity at the *deforming* surface", although the designation "tidal bulge" in this context is not quite correct, because a bulge should be related to a *fixed* surface and not to a moving surface.

The tidal bulge delay angle of the Earth's radial displacement may be defined in terms of the complex LOVE numbers as

$$\alpha_H \approx \tan \alpha_n = -\frac{1}{2} \frac{H}{h} \tag{36}$$

Similarly, the tidal bulge delay angle of the secondary potential due to the deformation is

$$\alpha_K \approx \tan \alpha_K = -\frac{1}{2} \frac{K}{k} \tag{37}$$

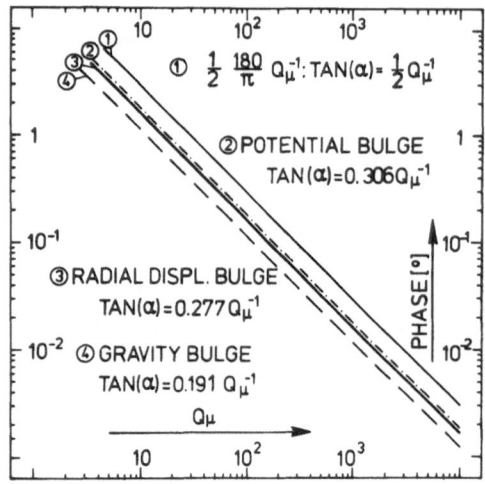

Fig. 11. Tidal bulge delay angles α (n=2) due to imperfections in the Earth's elasticity. The potential bulge is the bulge of the secondary potential (due to the deformation) at the *unstrained* surface or at any *fixed* surface outside the Earth. The radial displacement bulge is defined with respect to the unstrained surface. Whereas what is called "gravity bulge" here is related to the secondary tidal gravity at the *deforming* surface, as is obtained from the phase delay in tidal gravity measurements by using Slichter's formula. Results are valid for Qμ being constant within the mantle

The latter does *not* include the potential variation at a surface point due to the shift of this point through the Earth's gravity potential field, and, therefore, is related to a fixed surface. Numerical calculations show that the tidal bulge delay angle of the secondary potential is slightly different from the bulge delay angle of the radial displacement. Consequently, the "tidal bulge delay angle of the secondary gravity at the deforming surface", as determined from the total tidal gravity variation by using Slichter's formula, is different from both. No distinction has been made in the past between these different bulges.

The differences between these angles are graphically shown in Figure 11. Hence, the delay angle as calculated from Slichter's formula is by a factor of 0.6 - 0.7 smaller than the delay angles for the secondary potential bulge, and the bulge of the radial displacement.

The phase angle of the nonspecific bulge is currently related to the solid Earth's global Q by the relationship

$$\tan (2\alpha) = Q_{gl}^{-1} \tag{38}$$

(see for instance Munk, 1968; Munk and MacDonald, 1960). This relation originates from the fact that Q^{-1} is equal to the phase shift between stress and strain in the body under consideration (see for instance Anderson, 1967). This is definitely true if only a small volume element is considered, and in this case follows immediately from the basic definition of Q as given in Eq. (1).

However, in the case of a big planetary body we cannot understand why the body's global Q should be related to the surface value of the radial displacement phase delay or to the secondary potential phase delay in the same simple manner as in Eq. (38). MacDonald (1964; pp. 525) gave an analytic proof for Eq. (38). However, in his calculations no account has been taken of the phase delay in the total tidal potential, and no account has been taken of the hydrostatic prestress term in the energy balance. KAULA (1964) empirically found that Eq. (38) is valid. However, in his Eq. (28) he obviously assumed that the delay angle of the radial displacement bulge is the same as that of the secondary potential bulge. He also did not take into account the phase delay of the total potential.

In the following we will prove that the simple relation (38) does not hold for planetary bodies like the Earth, where selfgravitation and hydrostatic prestress cannot be neglected in the potential energy balance. This proof is given for an incompressible planet with homogeneous density.

The potential elastic energy of a deformed body may be written as

$$E = \frac{1}{2} \left(\iiint_V \rho\, \vec{u}\, \text{grad}\, \psi\, dV + \iint_S \vec{u}\, \vec{\tau}\, dS \right) \tag{39}$$

By means of Green's 1st integral transformation we have

$$\iiint_V \rho\, \vec{u}\, \text{grad}\, \psi\, dV = \iint_S \rho\, \psi\, u\, dS - \iiint_V \psi\, \rho\, \text{div}\, \vec{u}\, dV - \iiint_V \psi\, \vec{u}\, \text{grad}\, \rho\, dV$$

The second integral on the right-hand-side vanishes for an incompressible body (div $\vec{u} = 0$), and the third for homogeneous density (grad $\rho = 0$). At the Earth's unstrained surface ($r = R$) we have

$$\vec{u}\, \vec{\tau}_s \approx u\tau$$

with $\tau = -\rho_0\, gu$ being the hydrostatic prestress term (see Sect. 2.2) and $\rho_0 = \rho = $ const. It follows that

$$E = \frac{1}{2} \iint_S (\rho_0\, \psi\, u - \rho_0\, gu^2)\, dS \tag{40}$$

Neglecting small phase shifts, we may assume

$$\psi = (1 + k)\, \psi^P\,,$$
and
$$u = \frac{h}{g}\, \psi^P\,,$$

so that we have

$$E = \frac{1}{2}\, \frac{\rho_0}{g}\, h\, (1 + k - h) \iint_S (\psi^P)^2\, dS$$

If we assume the body to be strained periodically

$$(\psi^P = \psi_0^P \cos \omega t)\,,$$

the peak energy, consequently, is given by the expression

$$E^* = \frac{1}{2}\, \frac{\rho_0}{g}\, h\, (1 + k - h) \iint_S (\psi_0^P)^2\, dS \tag{41}$$

Starting with the expression (2) instead of Eq. (39), and proceeding in a smaller manner as described above in this section, we may derive the expression for the energy dissipated during one cycle of straining:

$$\Delta E = \frac{\rho_0}{g}\, \pi\, (Kh - H\, (1 + k)) \iint_S (\psi_0^P)^2\, dS \tag{42}$$

This formula gives the same numerical values for ΔE as formula (22), if both formulas are applied to an incompressible body with homogeneous density. Introducing Eqs. (41) and (42) into Eq. (1) yields the expression for the body tide global Q:

$$Q_{gl}^{-1} = \frac{Kh - H\,(1 + k)}{h\,(1 + k - h)} \tag{43}$$

It is valid for an incompressible planetary body of homogeneous density.

Q_{gl}^{-1} may also be expressed by the tidal phase shifts using the phase shift $\tan\,\phi_\psi = \frac{K}{1+k}$ of the total potential with respect to the primary potential of the external forces and the phase shift $\tan\,\phi_H = \frac{H}{h}$ of the radial displacement with respect to the primary disturbing potential:

$$Q_{gl}^{-1} = \frac{1 + k}{1 + k - h}\,(\tan\,\phi_\psi - \tan\,\phi_H)\,, \tag{44}$$

and in terms of the bulge delay angles

$$Q_{gl}^{-1} = \frac{1}{1 + k - h}\,\left\{(1 + k)\,\tan\,(2\alpha_H) - k\,\tan\,(2\alpha_K)\right\}\,. \tag{45}$$

The last expression has been derived by introducing Eqs. (36) and (37) into Eq. (43). From Eqs. (44) and (45) it is obvious that there is no such simple relation between Q_{gl}^{-1} and the delay angles of the tidal bulges as given in Eq. (38). However, if selfgravitation may be neglected, then $k \ll 1$ and (45) reduces to

$$Q_{gl}^{-1} = \frac{1}{1 - h}\,\tan\,2\,\alpha_H$$

The LOVE number h in this expression comes from the hydrostatic prestress term which is represented by the second term in the surface integral of the energy expression (40). If the body under consideration is small, then the surface acceleration g is small, and the hydrostatic prestress term may be neglected. In this case, and only in this case, the expression for Q_{gl}^{-1} reduces to the currently used simple equation

$$Q_{gl}^{-1} = \tan\,2\,\alpha_H$$

Furthermore, from Eq. (43) it is immediately clear that the global tide Q may *not* be expressed as a function of the tidal gravity phase shift $\tan\,\phi = \frac{H - 3/2\,K}{1 + h - 3/2\,k}$ alone. This is an analytic proof for the above statement, that there is no unique relationship between $\tan\,\phi$ and Q_{gl}.

3.4 Some More Numerical Examples

It has been demonstrated in Sections 3.2 and 3.3 that the errors in the determination of the body tide phase shifts, of the tidal bulge delay angles, and of the Earth's global Q may be extremely high, if one does not take into account the exact theoretical relationships between these quantities. We will give three more examples for such errors:

Let the phase delay of the body tide gravity be 0.1° with respect to the external forces. Proceeding in the conventional manner, i.e., using Slichter's formula [Eq. (34)] with δ = 1.16 and using the simple relation (38), gives the wrong global body tide Q = 79. In the case of the Gutenberg Earth, and in the case of constant Qμ within the mantle, we find the correct value to be Qμ = 28 (see Fig. 9). If we assume a low Q asthenosphere, i.e., let the real Qμ distribution within the Earth differ by only a constant factor from the free oscillation Qμ model LMS, we find that the global Qμ has to be chosen as low as 5.5 to correspond to the gravity phase delay of 0.1°. This is less than 7 % of the value Q = 79, obtained by the conventional method.

The delay angle of the secondary potential bulge has been determined from the orbits of artificial satellites to be 0.5° (Lambeck et al., 1974). Lambeck et al. relate this delay angle to a mantle Q of 60 which one obtains by using formula (38). The correct Qμ corresponding to this delay angle is 36, if the Gutenberg Earth and constant Qμ values within the mantle are adopted (see Fig. 11). For the LMS equivalent model, we have calculated the global Q which corresponds to the delay angle of 0.5° to be 34. This shows that the delay angle of the tidal potential bulge is less sensitive to the geometry of the Q distribution within the Earth than the phase delay of the tidal gravity at the deformed surface. The Q values of 34 and 36 are much lower than the lowest limit of the upper mantle seismic Q.

The usage of the exact theoretical relationship between the bulge of the secondary tidal potential and the body's global Q may also be important for other planets as for instance for Mars. From observations of the secular acceleration of the Mars satellite Phobos, Smith and Born (1976) deduced the phase angle of the secondary potential bulge due to the body tide of Mars. They related this phase angle to the global Q of Mars by the simple formula (38), and found a global Q between 50 and 150.

However, as one cannot neglect selfgravitation and the hydrostatic prestress for Mars, Eq. (38) is not applicable to the body tide of Mars, and, therefore, the Q between 50 and 150 is probably too high for Mars, provided that the observed phase angle of the tidal bulge is true.

3.5 Numerical Results for the Loading Tides

It has been pointed out in Section 3.1 that the loading tide Q may be much lower than the body tide Q, and that we, therefore, may expect larger phase shifts (attributed to imperfection of the Earth's elasticity) in the loading tide case than in the body tide case. We have computed these phase shifts between the loading tides on an Earth with imperfect elasticity and those on an elastic Earth for different realistic Earth models (see Zschau, 1978a,b). In all cases the Gutenberg model has been adopted for the elastic properties in the Earth's core and in the deeper mantle. The elastic properties of the upper mantle have also been taken from the Gutenberg model in the cases of our models Q 1 and Q 2, and from a typical continental shield model, as well as from a typical oceanic model (Harkrider, 1970), for our model CS, and OR1, OR2, respectively (see Fig. 12).

To model the mantle's imperfection of elasticity we have used viscosity depth structures and Q-structures. The viscosities below 400 km have been taken from Weertman (1970). Viscosities above 400 km have been

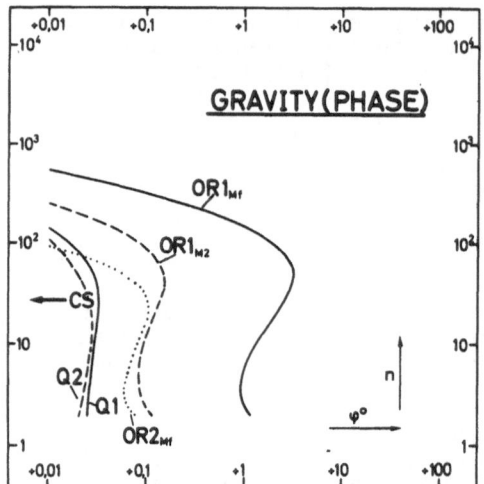 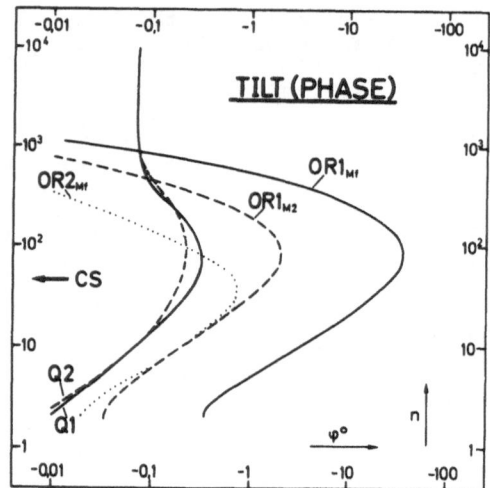

Fig. 12. Phase shifts (ϕ°) between the Earth's viscoelastic gravity-, and tilt response to harmonic loads of degree n and the theoretical elastic response (from Zschau, 1978a). The phase shifts are valid for the total effects, including the primary accelerations of the applied load. Results are presented for different Earth models (CS, continental shield model; Q1, model LMS + Gutenberg model; Q2, model MM8 + Gutenberg model; OR1, ocean ridge model 1; OR2, ocean ridge model 2), and for different frequencies of the loading (M_2, semidiurnal; M_f, fortnightly). Phase shifts are maximum for ocean ridge model 1

taken from Vetter and Meissner (1977), and have been used for our oceanic model (OR1: ocean ridge model 1) and for our continental shield model (CS). We have also used a second viscosity structure for the upper 400 km with, in general, higher viscosities (Vetter, 1978), which gives us the second ocean ridge model (OR2). In the continental shield case neither the first viscosity model nor the second has any important influence on our results. We therefore do not distinguish between the continental shield models. Model Q1 is identical with the LMS model, except that it has the crustal Q_μ values from the surface wave Q_μ model MM8 to allow for the high Q_μ in the crust. Model Q2 is equal to model MM8 (see Fig. 3).

For the viscosity models we have assumed the Maxwell law to be valid. Different from the elastic problem, the viscoelastic load response of the Earth in this case is highly dependent on the frequency of the loading. Our computations, therefore, have been carried out for the semidiurnal tide M_2, as well as for the fortnightly tide M_f. In the case of the Q models our results do not depend on the loading frequency.

From Figure 12 one may see that the loading tide gravity phase *advance* attributable to imperfection of elasticity within the Earth, amounts to several hundredths of a degree in the case of model LMS (Q1) and for low degree harmonics of the loading. These values are higher by more than one order of magnitude than the phase *delay* for the same Earth model in the body tide case (0.001°). Nevertheless they are still much too small to be measurable. This, however, is different for the ocean ridge model 1 (OR1, see Fig. 13). In this case the loading tide gravity phase advance is more than 0.1° for the semidiurnal tide M_2,

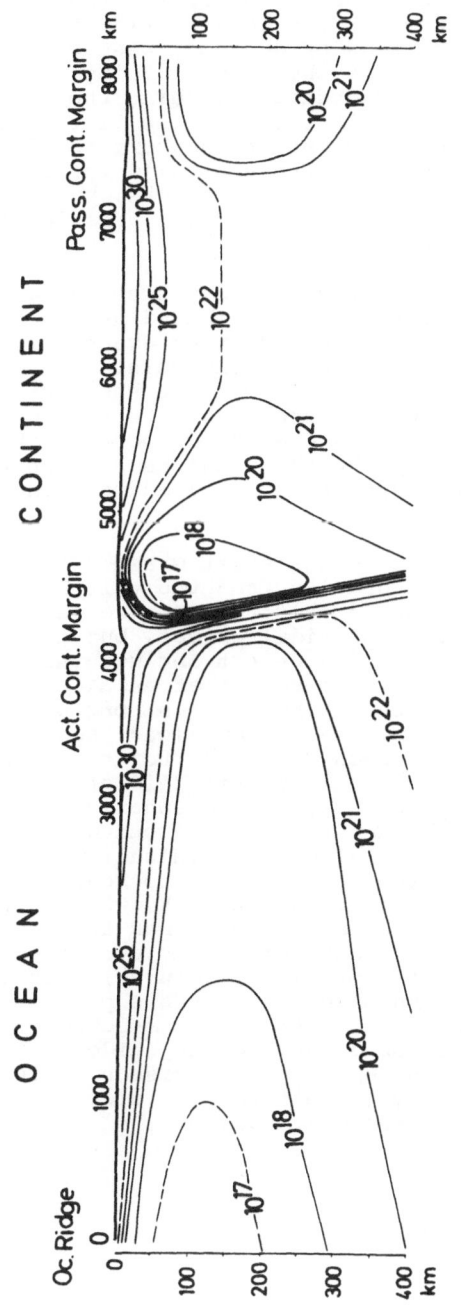

Fig. 13. Viscosity model for the upper 400 km (after Vetter and Meissner, 1977). Viscosity values are given in Poise. The viscosities take their lowest values below ocean ridges, and near active continental margins, i.e., near subduction zones

and amounts to several degrees for the fortnightly tide. The corresponding values for the loading tide tilt phase delay are more than
$1°$ and several tens of degrees, respectively. The phase shifts in
Figure 12 belong to the *total* loading tide gravity and tilt variation
at the deforming surface including the primary forces of the external
potentials. The tilt includes the tilt of the surface, as well as the
horizontal acceleration of the perturbed density field. The high values for the phase shifts are valid for loading tides which penetrate
into the low viscous asthenosphere, but do not penetrate too deep into the high viscous mantle, i.e., for loadings of harmonic degrees
slightly lower than 100.

For the determination of the viscoelastic Earth's response to loads
which are arbitrarily distributed at the Earth's surface, one needs
the frequency-dependent complex Green's functions which represent the
Earth's response to a varying point load. They are functions of the
distance from the point load. We have obtained them by summation of
the single harmonic responses with respect to the harmonic degrees.
From these functions one may deduce the phase shift Green's functions,
which are simply the phase shifts between the viscoelastic Earth's
responses to a varying point load, and the response of the corresponding elastic Earth.

In Figure 14 such phase shift Green's functions are shown for the displacements, the gravity, and the tilt in the case of ocean ridge model
1. Unlike the phase shifts in Figure 12, those in Figure 14 do not
belong to the total tidal effect, but to the secondary deformation
effect alone, which does not include the primary effect of the external forces. It may be seen that as a consequence of the low viscosity
in the asthenosphere there are strong phase delays at distances of
about 100 km from the point load: Nearly $5°$ for the tilt, $3°$ for the
gravity, and between $2°$ and $3°$ for the displacements. Phase shifts become positive several hundred kilometers away from the load. Near to
the load only the high degree spherical harmonics are important. Phases,
therefore, approach to zero. If one wants to know the maximum phase
shifts for the total effects inlcuding the primary accelerations of
the applied potential, one has to reduce the maximum gravity phase
shift in Figure 14 by about 15 %, whereas the maximum tilt phase shift
must be reduced by about 30 %. For the displacements there is no such
difference. The phase shifts may even increase, and may regionally be
stronger than the present inaccuracies in the knowledge of the ocean
tide, if one convolves the Earth's point load response with the ocean
tide to compute the integrated response of the Earth to the ocean tide
loading. This could make loading tide investigations an effective tool
for studying the upper mantle viscosity in areas with extremely low
mantle viscosity ($< 10^{19}$ Poise), i.e., near ocean ridges and near subduction zones (see Fig. 13).

4. Conclusions

For the mathematical treatment of the Earth's imperfectly elastic response to body tide forces and loading tide forces, we have introduced
the Earth's *complex* LOVE numbers, and its *complex* mass load coefficients,
respectively. We have given exact analytic expressions for the external work functions, which relate the tidal energy dissipated within
the solid Earth to the surface values of these complex characteristic
numbers. The expressions have been obtained without approximating the
real Earth by an incompressible and homogeneous one as was necessary
in former calculations.

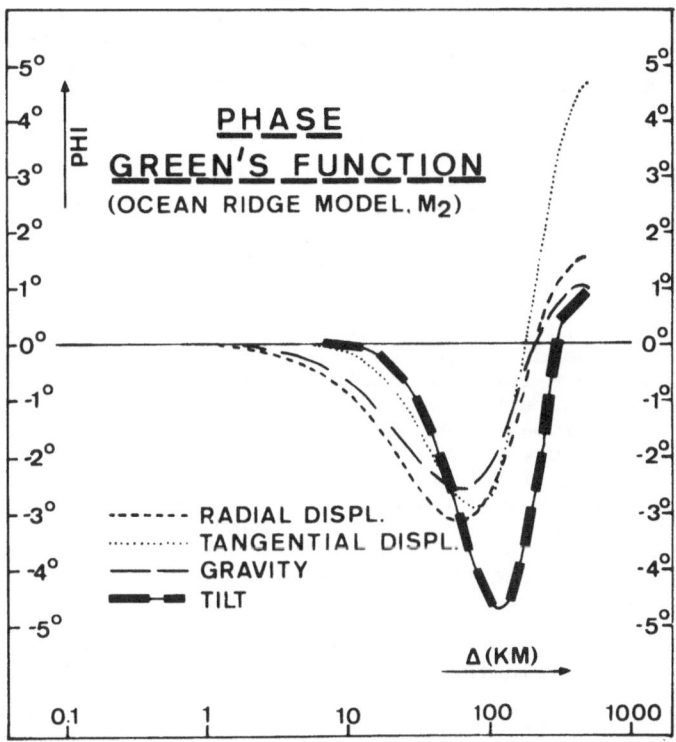

Fig. 14. M_2 phase Green's functions for ocean ridge model 1 (OR1) (from Zschau, 1978a). They represent the phase shifts between the viscoelastic Earth's responses to a varying point load, and the responses of the corresponding elastic Earth. The high phase shifts at distances of about 100 km from the load are due to the low viscous asthenosphere of ocean ridge model 1. Δ, distance from the point load; PHI, phase shift

The above expressions for the external work functions have been used to compute the rate of body tide energy dissipated in the solid Earth. The result shows that if only 10 % of the astronomically observed dissipation rate is due to solid Earth friction, a global body tide quality factor $Q\mu$ of less than 60 is needed. The $Q\mu$ model LMS, as obtained from the observation of free oscillations, yields the body tide dissipation rate of $3.19 \cdot 10^{17}$ erg/s which is about 1 % of the astronomically observed dissipation rate. Body tide friction, therefore, is considered to be of no importance for the present tidal friction.

No account had yet been taken of imperfections in the Earth's elasticity when considering deformations due to the varying ocean tide loading. Simple considerations suggest that the potential loading tide strain energy in the Earth's crust and upper mantle may regionally be much bigger than the potential body tide strain energy. If there is a low Q asthenosphere in the uppermost mantle, loading tide dissipation in the solid Earth, therefore, might be significant with respect to body tide dissipation. We have calculated the loading tide dissipation in the solid Earth by integrating over the global M_2 ocean tide as given by Hendershott (1972), adopting the $Q\mu$ structure LMS as obtained from free oscillation data for the Earth's mantle.

This has given the dissipation rate of $3.21 \cdot 10^{16}$ erg/s, which is a minimum value, because the global ocean tide model of Hendershott does not allow for the high amplitude tides in the shelf areas. Dissipation of loading tide strain energy in the solid Earth, therefore, amounts to *at least* 10 % of the dissipated body tide energy, and may not be neglected if the body tide dissipation is considered to be important in any circumstances.

It has been argued in the past that one could deduce the Earth's body tide Q, as well as the body tide energy dissipated within the Earth, from the measured phase lag between the body tide gravity variations on the real Earth's surface and those on a rigid Earth. Phase delays of several tenths of a degree up to more than 1° had been discussed in the past as being due to solid friction. Our calculations suggest that the body tide gravity phase delay due to friction within the solid Earth amounts more likely to a few thousandths of a degree than to a few tenths of a degree. For instance, the experimental phase delay of only 0.2° corresponds to a global body tide $Q\mu$ of 14, if the $Q\mu$ factor is distributed homogeneously within the mantle. It may be less than 3 for the same phase shift, if a low Q asthenosphere is adopted within the uppermost mantle. These $Q\mu$ values are not realistic, and we conclude that the measured phase lag of several tenths of a degree is due to the insufficient correction for the indirect effect of the ocean tides. There seems to be no chance at all at the moment to obtain information on the mantle Q from body tide gravity investigations.

The disagreement between our results and the relatively high phase delays expected so far may be attributed to mainly three reasons:

1. The computation of the tidal bulge delay angle from the phase delay of the tidal gravity variation by means of the currently used Slichter formula does *not* yield the delay angle of the radial displacement bulge as commonly thought. It does, however, give the angle between the Earth-Moon direction and the line to the maximum of the secondary gravity variation at the *deforming* surface, the gravity variation including the tidal variation at a surface point due to the shift of the surface through the Earth's gravity field. This angle is by a factor of 0.6 - 0.7 smaller than the delay angles of the radial displacement bulge, and the bulge of the secondary potential disturbance due to the deformation. The latter delay angle is also slightly different from the one before. There has not been made any distinction between these different tidal bulges so far.

2. The currently used basic relation $Q_{g1}^{-1} = \tan(2\alpha)$ between the tidal bulge delay angle and the global Q does not hold for planetary bodies, because — additionally to the lack of a tidal bulge specification — this relation does not allow for selfgravitation and hydrostatic prestress of the planet.

3. The relationship between the tidal gravity phase delay and the Earth's global Q is not unique, but strongly dependent on the geometry of the Q-distribution within the Earth.

If one does not take account of these three aspects, the global $Q\mu$ factor corresponding to a given tidal gravity phase delay may be computed too high by more than a factor of 14. Correspondingly the gravity phase delay computed for a given global $Q\mu$ may be too high by the same factor. If one computes the Earth's global $Q\mu$ from the tidal delay angle of the potential bulge, as for instance determined from the orbits of artifical satellites, and if one does not take into account aspect 2 for these computations, the $Q\mu$ value obtained will be too

high by about 70 %. Aspect 2 may also be important for the determination of the global Q of Mars from the secular acceleration of its satellite Phobos.

The above conclusion, that there is no chance of obtaining information
on the mantle Q from tidal gravity investigations at the moment, is
true for the body tide, but does not hold for the loading tide. Unlike
in the body case, high degree harmonics of the load are also important
in the loading tide case. We have found that the ocean loading tide of
harmonic degree of about n = 70 is strongly affected by the low Q asthenosphere of $Q\mu$ model LMS, the global loading tide $Q\mu$ (n = 70) being
nearly seven times smaller than the global body tidy $Q\mu$ (n = 2,3) for
the same model. Except for the very high harmonic degrees of the load,
the loading tide phase shifts computed for model LMS are higher by
more than one order of magnitude than those of the body tide. Near
ocean ridges and near subduction zones the loading tide phase shifts
due to friction in the solid Earth may be as high as several degrees
for the displacements as well as for gravity and tilt. From the computation of phase shift Green's functions we have found that these maximum phase shifts occur at about 80 to 100 km distance from the load,
this distance depending on the depth of the assumed low viscous asthenosphere. We conclude that, unlike body tide investigations, loading
tide investigations could become an effective tool for studying the
upper mantle viscosity in regions where we may expect viscosities lower
than 10^{19} Poise.

Acknowledgment. I thank Michael Schulze for critically reading the manuscript.

References

Alterman, Z., Jarosch, H., Pekeris, T.L.: Propagation of Raleigh Waves in the Earth,
    Geophys. J. 4, 219 (1961)
Anderson, D.L.: The anelasticity of the mantle. Geophys. J.R. Astron. Soc. 14, 135
    (1967)
Anderson, D.L.: Ben-Menahem, A., Archambeau, C.B.: Attenuation of seismic energy in
    the upper mantle. J. Geophys. Res. 70, 1441 (1965)
Biot, M.A.: Theory of stress-strain relations in anisotropic viscoelasticity and
    relaxation phenomena. J. Appl. Phys. 25(11), 1385 (1954)
Biot, M.A.: Dynamics of viscoelastic anisotropic media. In: Proceedings of the
    Second Midwestern Conference on Solid Mechanics. Res. Ser. Engeneering Experiment
    Station, Purdue University, Lafayette, Ind.: 1955, Vol. 129
Brosche, P., Sündermann, J.: On the torques due to tidal friction of the oceans
    and adjacent seas. In: Rotation of the Earth. Melchior, P., Yumi, S. (eds.).
    Dordrecht, Netherlands: D. Reidel, 1972, pp. 235-239
Chapman, D.S., Pollack, H.N.: Global heat flow: A new look. Earth Planet. Sci.
    Lett. 28, 23 (1975)
Farrell, W.E.: Deformation of the Earth by surface loads. Rev. Geophys. Space Phys.
    10, 761 (1972)
Gordon, R.B., Davis, L.A.: Velocity and attenuation of seismic waves in imperfectly
    elastic rock. J. Geophys. Res. 73, 3917 (1968)
Groten, E., Brennecke, J.: Global interaction between Earth and Sea Tides. J. Geo-
    phys. Res. 78, 8519 (1973)
Harkrider, D.G.: Surface waves in multilayered elastic media, 2, Higher mode spectra
    and spectral ratios from point sources in plane layered Earth models, Bull. Seis-
    mol. Soc. Am. 60, 1937 (1970)
Harrison, J.C., Ness, N.F., Longman, J.M., Forbes, R.F.S., Kraut, E.A., Slichter,
    L.B.: Earth-Tide observations made during the International Geophysical Year.
    J. Geophys. Res. 68, 1497 (1963)
Hendershott, M.C.: The effects of solid Earth deformation on global ocean tides.
    Geophys. J.R. Astron. Soc. 29, 389 (1972)

Kaula, W.M.: Tidal dissipation by solid friction and the resulting orbital evolution. Rev. Geophys. Space Phys. $\underline{2}$, 661 (1964)

Kuznetsov, M.V.: Calculation of the secular retardation of the Earth's rotation from up-to-date cotidal charts. Izv. Acad. Sci. USSR Phys. Solid Earth $\underline{12}$, 779 (1972)

Lagus, P.L., Anderson, D.L.: Tidal dissipation in the Earth and planets. Phys. Earth Planet. Interiors $\underline{1}$, 505 (1968)

Lamb, H.: Hydrodynamics. New York: Dover Publications Ltd., 1945

Lambeck, K., Cazenave, A., Balmino, G.: Solid Earth and ocean tides estimated from satellite orbit analyses. Rev. Geophys. Space Phys. $\underline{12}$, 421 (1974)

Lee, E.H.: Stress analysis in visco-elastic bodies. Appl. Math. $\underline{13}$, 183 (1955)

Longman, J.M.: A Green's function for determining the deformation of the Earth under surface mass loads, 1, Theory. J. Geophys. Res. $\underline{67}$, 845 (1962)

Longman, J.M.: A Green's function for determining the deformation of the Earth under surface mass loads, 2, Computations and numerical results: J. Geophys. Res. $\underline{68}$, 485 (1963)

Love, A.E.H.: A Treatise on the Mathematical Theory of Elasticity. New York: Dover Publications Ltd., 1927

MacDonald, G.J.F.: Tidal friction. Rev. Geophys. Space Phys. $\underline{2}$, 467 (1964)

Melchior, P., Kuo, J.T., Ducarme, B.: Earth tide gravity maps for Western Europe. Phys. Earth Planet. Interior $\underline{13}$, 184 (1976)

Morrison, L.V.: Tidal deceleration of the Earth's rotation deduced from astronomical observations in the period AD 1600 to the present, paper presented at the International Symposium on "Tidal Friction and Earth's Rotation", held at Bielefeld in Sept. 1977

Munk, W.H., MacDonald, G.J.F.: The Rotation of the Earth. New York: Cambridge University Press, 1960

Munk, W.H.: Once again, tidal friction. Q. J. R. Astron. Soc. $\underline{9}$, 352 (1968)

Pariiskii, N.N.: The Influence of Earth tides on the secular retardation of the Earth's rotation. Astron. J. $\underline{37}$, No.3, 543 (1960)

Pariiskii, N.N., Kuznetsov, M.V., Kuznetsova, L.V.: The effect of oceanic tides on the secular deceleration of the Earth's rotation. Izv. Acad. Sci. USSR Phys. Solid Earth $\underline{2}$, 65 (1972)

Pekeris, C.L., Accad, Y.: Solution of Laplace's equations for the M_2 tide in the world oceans. Phil. Trans. R. Soc. London, Ser. A $\underline{265}$, 413 (1969)

Pekeris, C.L., Jarosch, H.: The free oscillations of the Earth. In: Contributions in Geophysics in Honor of Beno Gutenberg. Benioff, H., Ewing, M., Howell, B., Press, F. (eds.). New York: Pergamon Press, 1958, Vol. 1, p. 171

Quamar, A., Eisenberg, A.: The damping of core waves. J. Geophys. Res. $\underline{79}$, 785 (1974)

Slichter, L.B., MacDonald, G.J.F., Caputo, M., Hager, C.L.: Comparison of spectra for spheroidal modes excited by the Chilean and Alaskan quakes. Geophys. J.R. Astron. Soc. $\underline{11}$, 256 (1966)

Slichter, L.B., Melchior, P.: Compte Rendu des Réunions de la Commission Permanente des marées terrestres à l'Assemblée Générale d'Helsinki. Marées Terr. Bull. Inf. $\underline{21}$, 369 (1960)

Smith, J.C., Born, G.H.: Secular acceleration of Phobos and Q of Mars. Icarus $\underline{27}$, 51 (1976)

Smith, S.W.: The anelasticity of the mantle. In: The Upper Mantle. Ritsema, A.R. (ed.). Tectonophysics $\underline{13}$, 601 (1972)

Stacey, F.D.: Physics of the Earth. 2nd Ed. John Wiley and Sons, 1977

Vetter, U.R.: Stresses and viscosities in the asthenosphere. J. Geophys. $\underline{44}$, 3 (1978)

Vetter, U.R., Meissner, R.O.: Creep in geodynamic processes. Tectonophysics $\underline{42}$, 37 (1977)

Weertman, J.: The creep strength of the Earth's mantle. Rev. Geophys. Space Phys. $\underline{8}$, 145 (1970)

Zschau, J.: Phase shifts of tidal see load deformations of the Earth's surface due to low viscosity layers in the interior, Proceed 8th Intern. Symp. Earth Tides, held at Bonn 1977, 1978a, in press

Zschau, J.: The influence of the Earth's viscosity on deformations by marine tidal surface loads, Proceed. Intern. Meeting on "Earth Rheology and Late Cenozoic Isostatic Movements", held at Stockholm 1977, 1978b, in press

Tidal Dissipation in the Oceans

K. Lambeck

The energy balance per unit of sea floor is

$$\frac{d}{dt}(KE) + \frac{d}{dt}(PE) + \frac{d}{dt} \cdot \nabla_s \text{ (energy flux)} = \frac{dW}{dt} + \frac{dE}{dt} \tag{1}$$

KE is the kinetic energy of the tidal currents, PE the potential energy, both per unit of surface. The third term on the left-hand side is the rate of divergence of the energy flux carried along by the tidal currents. dW/dt represents the total work done on the unit ocean surface and dE/dt represents the rate of energy dissipation. Integrating Eq. (1) over the ocean surface, the energy flux term vanishes, as there is no transport of water across the ocean-continent boundaries. The kinetic and potential energy terms also vanish if the equation is integrated over one tidal period P. Then with the definition

$$\left\langle \frac{\bar{\bar{d}}}{dt} \right\rangle = \frac{1}{P} \int_{T=T_o}^{T_o+P} \int_{\text{Oceans}} \frac{d}{dt} \, dS \, dT \tag{2}$$

$$-\left\langle \frac{\overline{dT}}{dt} \right\rangle = \left\langle \frac{\overline{dW}}{dt} \right\rangle \tag{3}$$

The rate at which energy is dissipated in the oceans equals the rate at which work is done on the ocean surface. This method of evaluating the dissipation requires a knowledge of the open ocean tide but makes no assumption about the location or mechanism of dissipation.

The rate at which work is done on the ocean consists of two parts: (1) the rate at which body forces work on the ocean

$$\rho_w \psi \, \frac{d\zeta}{dt} + \rho_w \, \nabla_s \cdot \left[\mathbf{u}(\zeta + D)\psi \right] \tag{4}$$

and (2) the rate at which the sea floor works on the ocean

$$\rho_w g \, (\zeta + D) \, \frac{d\delta}{dt} \tag{5}$$

In these expressions D is the water depth and $\mathbf{u}$ is the velocity of the tidal current. ζ is the tide elevation with respect to the sea floor and δ is the tide of the sea floor itself. The total ocean tide consists of a large number of components with different frequencies f_β. The tide ζ_β at any one frequency can be expanded into a series of spherical harmonics as

Research School of Earth Sciences, Australian National University

$$\zeta_\beta = \sum_s \zeta_{\beta,s} = \sum_s \sum_+ \bar{\Sigma} \, D^{\pm}_{\beta,st} \cos \left[2\pi f_\beta T \pm t\lambda - \varepsilon^{\pm}_{\beta \; st}\right] \cdot \rho_{st}(\sin \phi) \tag{6}$$

(Lambeck, 1977). In deriving the amplitude and phase constants in this expansion, the condition that $\zeta_\beta = 0$ on land, must be imposed.

The total potential ψ acting on the ocean, according to Eq. (4), in-cludes contributes from (1) the direct lunar or solar gravitational attraction, U_2, (2) the solid tide, $k_2 U_2$, (3) the self attraction of the ocean, and (4) the potential of the Earth's deformation under the ocean load. That is

$$\psi = (1 + k_2) \, U_\beta + \sum_s 3 \, \frac{\rho_w}{\bar{\rho}} \, \frac{(1 + k'_s)}{2s + 1} \cdot \zeta_{\beta,s} \tag{7}$$

The potential U_β raising the tide of frequency f_β is limited here to the second degree terms. ρ_w is the density of water and $\bar{\rho}$ the mean density of the Earth. With Eq. (6)

$$\psi = (1 + k_2) \, U_\beta + \sum_s \sum_t \bar{\Sigma}_+ \, \frac{4\pi GR\rho_w}{2s + 1} \, D^{\pm}_{\beta,st} \cdot \cos \left[2\pi f_\beta T \pm t\lambda - \varepsilon^{\pm}_{\beta,st}\right] \cdot$$
$$\cdot \, P_{st}(\sin \phi) \tag{8}$$

The ocean floor deformation due to the combined solid tide and ocean load becomes

$$\zeta_\beta = \frac{h_2}{g} \, U_\beta + \sum_s \sum_t \bar{\Sigma}_+ \, \frac{4\pi GR\rho_w}{2s + 1} \, \frac{h'_s}{g} \, D^{\pm}_{\beta,st} \cdot \cos \left[2\pi f_\beta T \pm t\lambda - \varepsilon^{\pm}_{\beta,st}\right] \cdot$$
$$\cdot \, P_{st}(\sin \phi) \tag{9}$$

For an elastic deformation of the sea floor $\langle d\delta/dt \rangle = 0$. Also $\zeta << D$, then with Eqs. (4) and (5)

$$\langle \bar{\bar{w}} \rangle = \rho_w \int_S \left\langle \psi \, \frac{d\zeta}{dt} \right\rangle \, dS + \rho_w g \int_S \left\langle \zeta \, \frac{d\delta}{dt} \right\rangle \, dS$$
$$= \rho_w (1 + k_2) \int \left\langle U_\beta \, \frac{d\zeta}{dt} \right\rangle \, dS + \rho_w h_2 \int_S \left\langle \zeta \, \frac{dU_\beta}{dt} \right\rangle \, dS \tag{10}$$

(Lambeck, 1977, 1978). Thus apart from modifying the actual tide, the Earth's deformation of the tidal load does not contribute to $\langle \bar{\bar{w}} \rangle$. The potential U_β can be written as

$$U_\beta(r) = \left(\frac{r}{R}\right)^2 A_\beta \, P_{2m}(\sin \phi) \cos \left[2\pi f_\beta T + m\lambda\right] \tag{11}$$

where $m = 0,1,2$ according to whether we are dealing with long-period, diurnal, or semi-diurnal tides. The amplitude factor A_β and frequency f_β are given by Lambeck (1978). Substituting Eqs. (11) and (6) into Eq. (10), and integrating over one tidal cycle and over the Earth's surface gives,

$$\langle \bar{\bar{w}} \rangle = \rho_w (1 + k_2 - h_2) \, A_\beta \, 2\pi f_\beta R^2 N^2_{2m} D^+_{\beta,2m} \cdot \sin \varepsilon^+_{\beta,2m} \tag{12}$$

with

Table 1. Summary of $\langle \overline{dE/dt} \rangle$ as deduced from tide models, the astronomical accelerations, and the phase lags in satellite orbit perturbations (Lambeck, 1978)

Tide	Models and observations	Astronomical estimates	Satellite estimates
M_2	$-3.35 \cdot 10^{19}$	$-3.06 \cdot 10^{19}$	$-3.1 \cdot 10^{19}$
N_2	-0.10		
O_1	-0.09		
S_2 (ocean)	-0.57		
T_2	-0.02		
K_1	-0.12		
P_1	-0.02		
S_2 (atmos.)	$+0.05$		
Total	-4.22 ± 0.45	-3.9 ± 0.3	-3.9 ± 0.7

$$N_{nm}^2 = \frac{4\pi \ (n+m)!}{(2n+1)(n-m)!(2-\delta_{om})}$$

The work method (3) for estimating the rate of dissipation in the oceans requires, therefore, only a knowledge of the second degree harmonics of the ocean tide expansion. The same result as Eq. (12) is obtained by evaluating dE/dt from the energy balance in the Earth-Moon system (Lambeck, 1977).

The main tide is M_2 and a number of numerical models are available that give coefficients $D_{M_2,22}^+$ and $\varepsilon_{M_2,22}^+$ that are in good agreement with each other and with those estimated from the perturbations in close Earth satellite orbits. Lambeck (1977, 1978) discusses these results further and also estimates the appropriate coefficients for the other tidal frequencies. His results are summarized in Table 1 together with recent astronomical and satellite results. The agreement is satisfactory, confirming that most of the tidal energy is dissipated in the oceans and not in the solid Earth and Moon. The nature of the dissipation mechanism, however, remains uncertain.

References

Lambeck, K.: Tidal dissipation in the oceans. Philos. Trans. R. Soc. London, A 287, 545-594 (1977)

Lambeck, K.: The Earth's variable rotation: Geophysical causes and consequences. Cambridge University Press, (in Press), 1978

The Influence of Solid Earth Deformations on Semidiurnal and Diurnal Oceanic Tides

W. Zahel

1. Introduction

Since the late nineteen sixties, tidal maps computed from various global models (Hendershott, 1973) have been added to earlier ideas of global ocean tides, which were based exclusively on observations. The use of such ocean tide models has yielded new estimates of the tidal friction in the oceans, which is supposed to be mainly responsible for secular accelerations of the Earth's rotation (Munk and MacDonald, 1960; Rochester, 1973). Brosche and Sündermann (1972) showed how to compute the effects of the tides in the sea on the Earth's rotation on the basis of tide models. It should be assumed that improved estimates of these effects depend upon the availability of ocean tide models yielding more realistic results. These models do have shortcomings, e.g., excluding stratification and interaction with the nontidal field of motion, which at present cannot be overcome. However, there are other shortcomings that can be done away with. The latter includes the neglecting of the effects of tidal loading and ocean self-attraction, which Hendershott (1972) claimed was possibly a serious shortcoming. Hendershott (1972) was the first to attempt to model the ocean tides by considering the solid Earth's deformation completely; a further approach was used recently by Gordeev et al. (1977). The approach presented in this paper is based on the inclusion of the effects of the solid Earth's deformation in the author's former 4°- ocean-tide model (Zahel, 1970; 1973), which differs considerably from those models into which Hendershott (1972) and Gordeev et al. (1977) introduced the additional effects. Reference will be made to this fact and to the widely differing results in the next Section.

The tidal deformation of the solid Earth is due primarily to the tidal body potential and to the loading shell of water caused by tidal sea-surface elevations relative to the bottom. Whereas the Earth tide due to the tidal body potential as an equilibrium tide can most easily be considered in ocean-tide models (Hendershott, 1972; Zahel, 1977)-there is a constant factor arising in the inhomogeneous terms of the tidal equations-,the loading shell of water gives rise to an essential interaction between Earth tides and ocean tides. The equations for the ocean tides dealing with this interaction appear as integrodifferential equations instead of as differential equations. The symbols involved in the presentation of modeling this interaction are listed at the end of this paper.

The solid Earth's deformations and the corresponding changes in the gravitational potential referred to above are sketched in Figure 1. There the z-coordinate points vertically upward, z = 0 defines the mean sea-surface level and z = -d(λ,ϕ), the mean sea-bottom level, both relative to the center of the Earth. The solid Earth's deformation

Institut für Meereskunde der Universität Hamburg, Heimhuder Str. 71, 2000 Hamburg 13, FR Germany

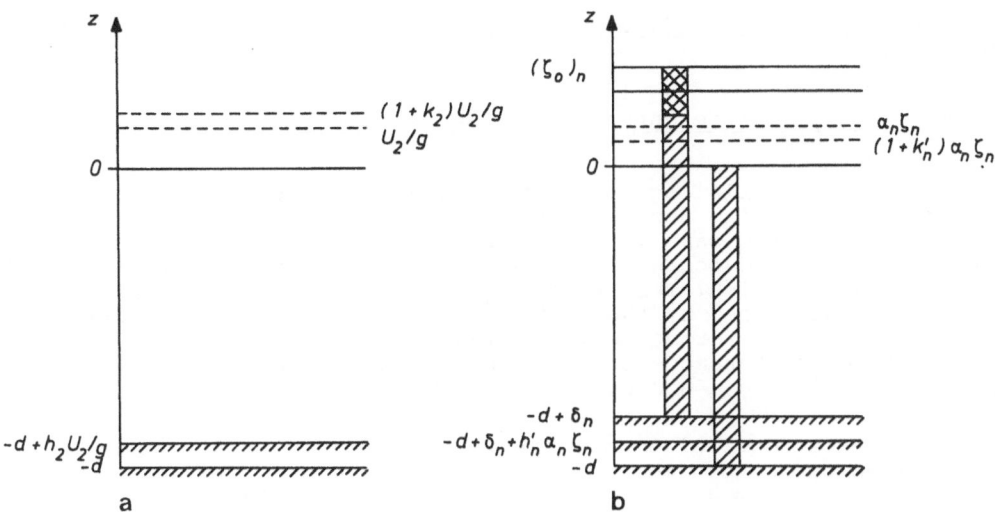

Fig. 1a and b. Effects of (a) the body potential U_2 and (b) the nth degree surface load due to $\zeta_n = (\zeta_o)_n - \delta_n$. Potentials are illustrated by an equilibrium representation (----), i.e., by a corresponding sea-surface deformation

resulting from the body potential U_m of degree m is given by an up-lift $h_m U_m/g$, while the additional gravitational potential arising from this redistribution of mass is $k_m U_m$. Thus the total potential amounts to $(1+k_m) U_m$ at which the factor $1+k_m$ allows for the attraction of the bulge by itself, the response $h_m U_m/g$ taking this self-attraction into account. The sea-surface elevation ζ relative to the moving sea-bottom is represented by spherical harmonics $\zeta = \sum_n \zeta_n$. ζ is the quantity that is observed by tide gauges and deep-sea instrument capsules. The load-ing shell defined by ζ_n gives rise to the potential $W_n = g \, \alpha_n \zeta_n$, $\alpha_n = \frac{3}{2n+1} \frac{\rho}{\rho_e}$ by ocean self-attraction. The surface load $g \rho \, \zeta_n$ yields a sol-id Earth deformation $h_n' W_n/g$ due to the weight of the additional column of water and to the opposing effect of the gravitational attraction of the solid Earth by the shell. Because of this mass redistribution, the potential due to the surface load amounts finally to $(1+k_n') W_n$. The nonloading LOVE numbers h_n, k_n as well as the loading LOVE numbers h_n', k_n' depend on the elastic properties and on the density of the ma-terial inside the Earth. While the nonloading LOVE numbers, defining the response to the body force, depend upon the Earth's global proper-ties, the loading LOVE numbers, defining the response to the ocean loads, depend essentially upon the local properties of the crust and mantle. The LOVE numbers have been determined by Longman (1963) and Farrell (1972) up to degree 40 and 10,000, respectively. For these computations the radially layered Gutenberg-Bullen Earth model was taken as a point of departure, but Farrell (1972) also computed the Green's functions, from which the LOVE numbers can be obtained, the upper 1000 km of the Earth model having been replaced by the oceanic and continental shield structures of Harkrider. The Gutenberg-Bullen Earth model and the Harkrider Earth model are tabulated in Alterman et al. (1961) and in Harkrider (1970), respectively.

2. Considering Tidal Loading and Ocean Self-Attraction in Ocean-Tide Models

2.1 Tidal Integrodifferential Equations and the Energy Equation Belonging to Them

The integrodifferential equations that arise when considering the effects of tidal loading and ocean self-attraction are, except for some transformations, the same with respect to the latter effects for the previously mentioned ocean-tide models that take into account the solid Earth's deformation. Therefore these equations will be given below as they have been used by the author when generalizing the former 4°-ocean-tide model. Reference will be made in the second part of this Section to the differences between the author's model and those of Hendershott (1972) and Gordeev et al. (1977).

Recalling the effects of a nonloading potential and an nth degree surface load by the scheme

	Nonloading potential U_m	nth degree surface load $\rho g\, \zeta_n$
Uplift of the solid Earth	$h_m U_m / g$	$h_n'\, \alpha_n\, \zeta_n$
Potential	$(1+k_m) U_m$	$(1+k_n')\, g\, \alpha_n\, \zeta_n$

it is obvious that the total vertical displacement δ of the sea-bottom and the total potential are given by

$$\delta = \delta_b + \delta_s = h_m U_m / g + \sum_n h_n'\, \alpha_n\, \zeta_n$$

and

$$V = g\bar{\zeta} = (1+k_m) U_m + \sum_n (1+k_n')\, g\, \alpha_n\, \zeta_n ,$$

respectively.

Hence

$$\bar{\zeta} - \delta = (1+k_m-h_m) U_m / g + \sum_n (1+k_n'-h_n')\, \alpha_n\, \zeta_n$$

and the vertically integrated tidal equations are:

$$\frac{\partial u}{\partial t} - 2\omega \sin\phi\, v + \frac{r}{D} (u^2+v^2)^{1/2}\, u + R_\lambda + \frac{g}{R\cos\phi} \frac{\partial \zeta}{\partial \lambda} = \frac{g}{R\cos\phi} \frac{\partial (\bar{\zeta}-\delta)}{\partial \lambda} \quad (1)$$

$$\frac{\partial v}{\partial t} + 2\omega \sin\phi\, u + \frac{r}{D} (u^2+v^2)^{1/2}\, v + R_\phi + \frac{g}{R} \frac{\partial \zeta}{\partial \phi} = \frac{g}{R} \frac{\partial (\bar{\zeta}-\delta)}{\partial \phi} \quad (2)$$

$$\frac{\partial \zeta}{\partial t} + \frac{1}{R\cos\phi} \left(\frac{\partial (Du)}{\partial \lambda} + \frac{(Dv \cos\phi)}{\partial \phi} \right) = 0 \quad (3)$$

with

$$D(\lambda,\phi,t) = d(\lambda,\phi) + \zeta(\lambda,\phi,t)$$

and

$$R_\lambda = - A_h \left(\Delta u + R^{-2}(-u(1+tg^2\phi) - 2\, \frac{tg\, \phi}{\cos\phi} \frac{\partial v}{\partial \lambda}) \right),$$

$$R_\phi = - A_h \left(\Delta v + R^{-2}(-v(1+tg^2\phi) + 2\, \frac{tg\, \phi}{\cos\phi} \frac{\partial u}{\partial \lambda}) \right).$$

Referring to the spherical harmonic decomposition of the ocean surface displacement relative to the sea bottom:

$$\zeta = \zeta_0 - \delta = \sum_n \zeta_n = \sum_n \sum_{s=0}^{n} \bar{P}_n^s(\sin\phi) \ (C_n^s \cos(s\lambda) + S_n^s \sin(s\lambda))$$

$$\begin{matrix} C_n^s \\ S_n^s \end{matrix} = \frac{1}{4\pi} \iint_B \zeta(\lambda', \phi', t) \ \bar{P}_n^s(\sin\phi') \begin{matrix} \cos(s\lambda') \\ \sin(s\lambda') \end{matrix} \ d\lambda' \ d\phi' \ \cos\phi' \ ,$$

where the normalization $\bar{P}_n^s = \sqrt{2(2n+1)\frac{(n-s)!}{(n+s)!}} \ P_n^s$, $\bar{P}_n = \sqrt{2n+1} \ P_n$ is used, one gets

$$\bar{\zeta} - \delta = \gamma_m \ U_m/g + \iint_B \zeta(\lambda', \phi', t) \ G(\lambda,\phi,\lambda',\phi') \ d\lambda' \ d\phi' \ \cos\phi',$$

where

$$\gamma_m = (1+k_m-h_m), \quad \alpha_n' = (1+k_n'-h_n') \ \alpha_n, \quad G(\lambda,\phi,\lambda',\phi') =$$

$$\frac{1}{4\pi} \sum_n \alpha_n' \sum_{s=0}^{n} \bar{P}_n^s(\sin\phi) \ \bar{P}_n^s(\sin\phi') \ \cos(s(\lambda'-\lambda)),$$

B = surface of the globe.

This expression for $\bar{\zeta} - \delta$ demonstrates how the dependent variable ζ appears in the integrodifferential equation system (1) - (3).

The energy equation belonging to Eqs. (1) - (3) will be noted subsequently, and how the additional terms contribute to it will also be shown. In establishing the tidal energy equation (Zahel, 1977), the terms due to tidal loading and ocean self-attraction contribute the following term, when considering Eq. (3) and denoting $W' = \sum_n (1+k_n')W_n$,

$$\delta_s = \sum_n h_n' \alpha_n \zeta_n, \quad Z = \sum_n \alpha_n' \zeta_n:$$

$$D \ u \ \frac{g}{R\cos\phi} \ \frac{\partial Z}{\partial \lambda} + D \ v \ \frac{g}{R} \ \frac{\partial Z}{\partial \phi} = \frac{g}{R\cos\phi} \ (\ \frac{\partial}{\partial \lambda} \ (D \ u \ Z) +$$

$$+ \frac{\partial}{\partial \phi} \ (D \ v \ \cos\phi \ Z)) + g \ \frac{\partial \zeta}{\partial t} \ Z = g \ \mathrm{div} \ (D \ u \ Z, \ D \ v \ Z) +$$

$$+ g \ \frac{\partial \zeta}{\partial t} \ Z = g \ \mathrm{div} \ (D \ u \ W'/g - D \ u \ \delta_s, \ D \ v \ W'/g - D \ v \ \delta_s) +$$

$$+ \frac{\partial \zeta}{\partial t} \ W' - g \ \frac{\partial \zeta}{\partial t} \ \delta_s = \mathrm{div} \ (D \ u \ W', \ D \ v \ W') + g \ \mathrm{div} \ (-D \ u \ \delta_s,$$

$$- D \ v \ \delta_s) + gD \ \frac{\partial \delta_s}{\partial t} + W' \ \frac{\partial \zeta}{\partial t} - gD \ \frac{\partial \delta_s}{\partial t} - g \ \delta_s \ \frac{\partial \zeta}{\partial t} \ .$$

The fifth and sixth terms contribute to the rate of change of potential energy, the second term to the rate of divergence of energy flux, and the third term to the rate of work per unit area by the moving sea bottom. Thus the completed tidal energy equation assumes the following form:

$$\frac{\partial}{\partial t} \ (1/2 \ D(u^2 + v^2)) - 1/2 \ \frac{\partial \zeta}{\partial t} \ (u^2+v^2) + r(u^2+v^2)^{3/2} + D(R_\lambda u +$$

$$+ R_\phi v) + g \ \mathrm{div} \ (D \ u \ \zeta_0, \ D \ v \ \zeta_0) + \frac{\partial}{\partial t} \ (1/2 \ g \ \zeta^2 + g(\delta_b+\delta_s)D)$$

$$= (1+k) \ \mathrm{div} \ (D \ u \ U, \ D \ v \ U) + \mathrm{div} \ (D \ u \ W', \ D \ v \ W') +$$

$$+ (1+k) \ U \ \frac{\partial \zeta}{\partial t} + W' \ \frac{\partial \zeta}{\partial t} + gD \ \frac{\partial}{\partial t} \ (\delta_b + \delta_s) \tag{4}$$

Assuming the tidal body potential to be of degree 2, subscripts in denoting the potential and the corresponding LOVE numbers have been omitted. Both tidal constituents to be referred to in this paper, i.e., M_2 and K_1, are of degree 2. Denoting the averaging over one tidal period by an overbar and spatial integration by the symbol $\langle \ \rangle$, one gets

$$\left\langle \overline{g \frac{\partial \delta_s}{\partial t} D} \right\rangle = \left\langle \overline{g \frac{\partial \delta_s}{\partial t} d} \right\rangle + \left\langle \overline{g \frac{\partial \delta_s}{\partial t} \zeta} \right\rangle = 0 + 0 = 0 \quad \text{and}$$

$$\left\langle \overline{w' \frac{\partial \zeta}{\partial t}} \right\rangle = 0 \quad \text{because of} \quad \overline{\frac{\partial \zeta_n}{\partial t} \zeta_n} = 0, \quad \left\langle \overline{(\frac{\partial \zeta}{\partial t})_m \zeta_n} \right\rangle = 0 \quad m \neq n.$$

Therefore it is obvious from Eq. (4) that, in the mean over one tidal period and integrated over the whole ocean, the energy equation does not alter when considering tidal loading and ocean self-attraction. The author's attention was first called to this circumstance by Lambeck (private communication). However, taking into account these additional effects, the tidal regime might change and thereby the rate of energy dissipation as well as the other constituents of the averaged energy equation.

2.2 Properties of Different Ocean-Tide Models

As mentioned previously, Hendershott's (1972) and Gordeev et al's. (1977) models of the ocean tides in which the effects of tidal loading and ocean self-attraction are considered are essentially different from each other and from the approach presented in this paper.

Hendershott (1972) and Gordeev et al. (1977), starting from completely linearized equations, assume simple harmonic time dependence of the dependent variables, thus obtaining an elliptical problem as far as the basic differential equations are concerned. The equations used by Hendershott (1972) are frictionless; dissipation occurs via the continental boundaries where observed surface elevations are prescribed defining inhomogeneous boundary conditions. Contrary to this, Gordeev et al. (1977) consider an impermeable coast; their model is thus independent of measurements. The dissipative terms included are linear bottom friction and eddy viscosity; the corresponding coefficients are taken to be 10^{-8} s^{-1} and 10^7 m^2 s^{-1}. Finally, the author's approach is based directly on the primitive equations given above in the generalized form (1) - (3). Thus the basic differential equations are identical with those on which the author's former ocean-tide models were based. Moreover, all characteristic features of these models are preserved, e.g., the coastal boundary is assumed to be impermeable, the no-slip condition is considered, and the coefficients of quadratic bottom friction and eddy viscosity are taken to be 0.003 and $5 \cdot 10^5$ m^2 s^{-1}, respectively.

The numerical solutions to the integrodifferential equations are obtained in all cases by performing iteration procedures. The procedure in the author's model is very different from that in the others because of the non-ellipticity of the basic differential equations. Starting from the numerical solution to the basic differential equations, Hendershott (1972) stopped his iteration procedure after the first step. Mathematical experience indicates that there is little information after the first step concerning a solution to be reached by iteration, so it must be assumed that the tidal regime obtained by Hendershott (1972) does not sufficiently reflect the effect of the solid Earth's deformation.

Moreover, in the case of the availability of a complete solution for the semi-empirical approach, the possibilities for discussing the influence of the solid Earth's deformation on the oceanic tides would be rather limited. This follows from the fact that the coastal elevations are prescribed from observations and are the same in both models, both when neglecting and when considering the additional effects. Thus the former model does not really neglect the additional effects, but includes them partially via the inhomogeneous boundary conditions. Gordeev et al. (1977) perform a complete iteration procedure. Their solutions show significant differences between both tidal regimes. However, considering the additional effects, the main features, e.g., amphidromies with their sense of rotation, quasi-nodal lines and antinodes, are not altered. Their goal was to establish the qualitative effect on the M_2-tide rather than to produce a new chart of the global tides. Indeed, the M_2-tide regimes of Gordeev et al. (1973, 1977) based on time-dependent equations and elliptic equations, respectively, neglecting the additional effects, prove to be rather different, and the former model seems to yield more realistic results.

3. Generalization of the 4°-Primitive-Equations Model

3.1 The Finite-Difference Scheme

The primitive-equations model of ocean tides, which has been generalized taking into account the interaction between the solid Earth's elastic deformations and the ocean tides, is the one presented in Zahel (1970). The same value is assigned to the eddy viscosity coefficient as in Zahel (1977); furthermore, the factor $\gamma_2 = (1+k_2-h_2) = 0.69$ is also considered as in the same paper. Thus Eqs. (1) - (3) have to be treated numerically in such a way that the original numerical scheme arises if setting $k_n' = h_n' = \alpha_n = 0$, $n = 1,2,..,N$. LOVE numbers are used up to degree $N = 50$. Their values were taken from Longman (1963) and Farrell (1972) and, as will be referred to later, both sets of values, the one based on the Gutenberg-Bullen Earth model and the one based on the Harkrider Earth model (oceanic), were considered. The combinations in which LOVE numbers appear in the tidal equations are the parameters α_n'. The maximum difference between the α_n' values of both types amounts to 7 %. Hence, just to give an impression of the magnitude of these parameters, some α_n' values referring to the Harkrider Earth model are listed below.

n	1	2	3	4	5	6	7	8	10	20	30	50
$10^{-4}\alpha_n'$	2286	1704	1359	1120	953	849	755	696	603	377	287	187

The numerical scheme applied for solving the tidal Eqs. (1) - (3) is given by:

$$D_+^t u_{i,j}^m - 2\omega \sin\phi_i \, \bar{v}_{i,j}^m + \frac{r}{\bar{D}_{i,j}^m} ((u_{i,j}^m)^2 + (\bar{v}_{i,j}^m)^2)^{1/2} u_{i,j}^m +$$

$$+ (R_\lambda)_{i,j}^m + \frac{g}{R\cos\phi_i} (D_o^\lambda \zeta_{i,j}^m \, a(\phi_i) + D_o^\lambda \zeta_{i,j}^{m+1} \, b(\phi_i)) =$$

$$\gamma_2 (K_\lambda)_{i,j}^m + \frac{g}{R\cos\phi_i} D_o^\lambda F_{i,j}^m \quad i=1,3,5,..; \quad j=1,3,5,.. \quad (5)$$

$$D_+^t v_{i,j}^m + 2\omega \sin\phi_i \bar{u}_{i,j}^m + \frac{r}{\bar{D}_{i,j}^m} ((\bar{u}_{i,j}^m)^2 + (v_{i,j}^m)^2)^{1/2} v_{i,j}^m +$$

$$+ (R_\phi)_{i,j}^m + \frac{g}{R} D_o^\phi \zeta_{i,j}^m = \gamma_2 (K_\phi)_{i,j}^m + \frac{g}{R} D_o^\phi F_{i,j}^m$$

$$i=0,2,4,6,..; \quad j=0,2,4,6,.. \tag{6}$$

$$D_+^t \zeta_{i,j}^m + \frac{1}{R\cos\phi_i} (D_o^\lambda (\bar{D}_{i,j}^m u_{i,j}^{m+1}) + D_o^\phi (\bar{D}_{i,j}^m v_{i,j}^{m+1} \cos\phi_i)) = 0$$

$$i=1,3,5,..; \quad j=0,2,4,6,.. \tag{7}$$

$$D_+^t w := \frac{w(t+\Delta t) - w(t)}{\Delta t} ; \quad D_o^\psi w := \frac{w(\psi + \Delta\psi/2) - w(\psi - \Delta\psi/2)}{\Delta\psi}$$

$$\psi = \lambda, \phi ; \quad w_{i,j}^m := w(j\Delta\lambda/2, i\Delta\phi/2, m\Delta t) \quad w = \zeta, u, v;$$

$$a(\phi_i) + b(\phi_i) = 1 \quad 0 \le a, b \le 1 ;$$

$$F_{i,j}^m := E_{i,j}^{m'} \cos(\sigma m \Delta t) + E_{i,j}^{m''} \sin(\sigma m \Delta t) ; \quad \frac{2\pi}{\sigma} = T = M \Delta t$$

$$m' := \left[\frac{m}{M}\right] M ; \quad m'' := m' + \frac{M}{4} ;$$

$$E_{i,j}^m := \sum_{k,l} \zeta_{k,l}^m G(\lambda_j, \phi_i, \lambda_l, \phi_k) \Delta\phi (\sin(\phi_k + \Delta\phi/2) + \sin(\phi_k - \Delta\phi/2)),$$

summation extends over all grid points.

Overbars denote values to be averaged on the grid points corresponding to the individual equations. With regard to the 4°-grid-point distance, implicitness is excluded, i.e., a,b are taken to be a=1 and b=0. Obviously, the term $F_{i,j}^m$ represents the tidal loading and ocean self-attraction effects.

Starting from zero values for the dependent variables, the spin-up process divides into three phases:

1. $F_{i,j}^m = 0$ (neglecting tidal loading and ocean self-attraction)

2. $F_{i,j}^m = E_{i,j}^{m'} \cos(\sigma m \Delta t) + E_{i,j}^{m''} \sin(\sigma m \Delta t)$

3. $F_{i,j}^{m+kM} = \frac{k-1}{k} F_{i,j}^{m+(k-1)M} + \frac{1}{k} (E_{i,j}^{m'} \cos(\sigma m \Delta t) + E_{i,j}^{m''} \sin(\sigma m \Delta t))$

$$m' = \left[\frac{m+kM}{M}\right] M ; \quad m'' = m' + \frac{M}{4} .$$

The first phase is finished when periodicity is obtained, and the procedure then yields the tidal regime neglecting the additional effects. Obviously, this procedure is identical with the one used in Zahel (1977). Therefore the above requirement of generalization is fulfilled. The second and third phases both belong to the difference Eqs. (5) - (7), being consistent with the tidal Eqs. (1) - (3). The third phase contributes to speeding up the spin-up process. Performing the second and the third phases, strict periodicity has been reached after nearly the same number of tidal periods, i.e., after about 40 periods, as in the case of performing the first phase. Incorporation of the integral terms of Eqs. (1) and (2) into the difference scheme was developed starting from simplified versions of the problem, for which in part analytical solutions are available. Instead of referring to the properties of the numerical method in detail, some of them will be illustrated by applying the final procedure to one of these simple cases and by comparing the results with the analytical solution.

3.2 Oceanic Tides on a Nonrotating Earth

A simple case, suitable for illustrating properties of the numerical procedure, is given by a M_2-tide wave propagating in an ocean of constant depth that completely covers a nonrotating globe. Based on the above premises and in view of the M_2-tide potential $U_2 = gcR/6 \; P_2^2(\sin\phi)$ $\cos(\sigma t+2\lambda)$, the solution is assumed to be of degree 2. With $\zeta = \zeta_2$, Eqs. (1) - (3) simplify to

$$\frac{\partial u}{\partial t} + R^* u + (1-\alpha_2') \frac{g}{R\cos\phi} \frac{\partial\zeta}{\partial\lambda} = \frac{\gamma_2}{R\cos\phi} \frac{\partial U_2}{\partial\lambda} \tag{8}$$

$$\frac{\partial v}{\partial t} + R^* v + (1-\alpha_2') \frac{g}{R} \frac{\partial\zeta}{\partial\phi} = \frac{\gamma_2}{R} \frac{\partial U_2}{\partial\phi} \tag{9}$$

$$\frac{\partial\zeta}{\partial t} + \frac{D}{R\cos\phi} \left(\frac{\partial u}{\partial\lambda} + \frac{\partial(v\cos\phi)}{\partial\phi} \right) = 0 \tag{10}$$

when introducing a linear law of bottom friction and omitting eddy viscosity, moreover. Introducing simple harmonic time dependence into Eqs. (8) - (10) and eliminating u and v yields

$$\Delta_h\hat{\zeta} + \frac{\sigma^2 - R^* i\sigma}{\beta_2 g D} \hat{\zeta} - \frac{\gamma_2}{\beta_2 g} \Delta_h\hat{U}_2 = 0 \tag{11}$$

at which $\beta_2 = 1-\alpha_2'$, $\zeta = \hat{\zeta}e^{i\sigma t}$, $\hat{U}_2 = gcR/6 \; P_2^2(\sin\phi) \; e^{2i\lambda}$.

Considering $\Delta_h\hat{\zeta} = -6 R^{-2} \hat{\zeta}$, $\Delta_h \hat{U}_2 = -6 R^{-2} \hat{U}_2$, the solution to Eq. (11) reduces to

$$\hat{\zeta} = -\frac{A + iB}{A^2 + B^2} \frac{6\gamma_2}{\beta_2 g} \hat{U}_2 \tag{12}$$

at which $A = -6 + \dfrac{\sigma^2 R^2}{\beta_2 g D}$, $B = \dfrac{R^*\sigma R^2}{\beta_2 g D}$.

Assigning to R^* and D the values 0.00003 s^{-1} and 2200 m, respectively, one finds from Eq. (12)

$$\text{Re} \; \zeta = Z \cos(\sigma t+2\lambda-\psi) \; \cos^2\phi \; , \quad \psi = \text{arctg} \frac{R^*\sigma R^2}{6\beta_2 gD-\sigma^2 R^2} \; ;$$

$$Z = 2.960 \; cm, \quad \psi = 166.2° \quad \text{if} \quad \beta_2 = 0.8167 \quad \text{and}$$

$$Z = 3.057 \; cm, \quad \psi = 165.7° \quad \text{if} \quad \beta_2 = 1 \; ,$$

i.e., neglecting the loading and self-attraction effects. The tide is indirect and far from resonance in both cases, the resonance depths being 16,667 m and 13,612 m, respectively. Therefore, the effect of the solid Earth's deformation is small, as is shown in Figure 2 by displaying the sea-surface elevations at the meridian 82° east of the Moon's position. There the sea-surface elevation is just beneath its maximum, being taken at 83.1° and at 82.85° east of the Moon's position, respectively. The differences between the numerical solutions

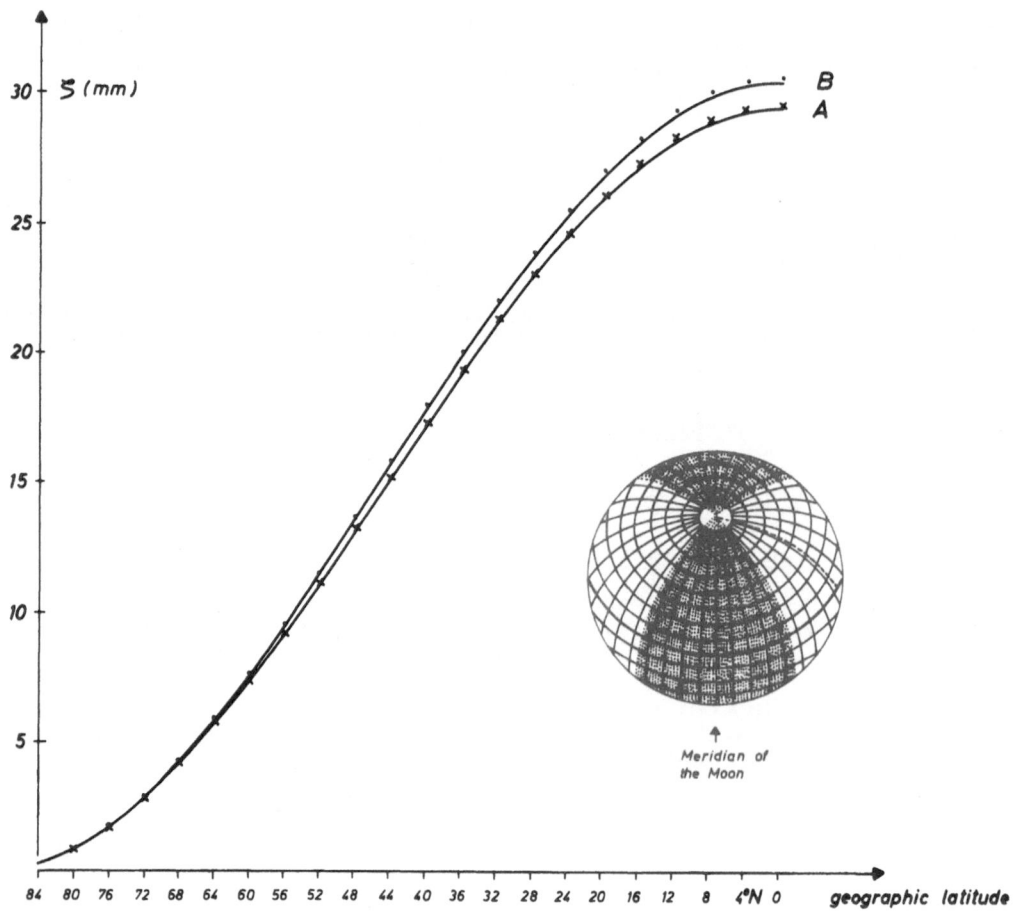

M₂ - tide in an ocean of constant depth (2200 m)
covering the whole earth, Coriolis force neglected

Surface elevation at the meridian 82° to the east from
the Moon's position

Considering tidal loading and ocean self-attraction
(A —analytical solution, × numerical solution)

Neglecting t. l. and o. s. a.
(B – analytical solution, · numerical solution)

Fig. 2. M_2-tide in an ocean of constant depth (2200 m) covering the whole Earth, Coriolis force neglected. Surface elevation at the meridian 82° east of the Moon's position. Considering tidal loading and ocean self-attraction: A, analytical solution, x, numerical solution. Neglecting t.l. and o.s.a: B, analytical solution, •, numerical solution

and the analytical solutions are small and of the same kind in both
cases. Because of the latter fact the effects of tidal loading and
ocean self-attraction can also be recognized from the numerical solutions, although in this far-from-resonance case the approximation error proves to be of the same order of magnitude as the dynamic effect.
This simple example demonstrates the necessity of a direct generalization of the numerical procedure when including the additional terms.
Furthermore in this case no significant additional approximation errors due to the discretization of the integral terms appear. At this
point, the simplifying assumption $\zeta = \zeta_2$ had not been introduced into
the numerical scheme implicitly, but the complete procedure, considering the LOVE numbers of higher degrees, was applied. Hence, in an
ill-posed scheme errors due to the contributions of higher degree,
which exist in every discrete formulation, could have spread.

4. The Computed Global M_2-Tide

As can be gathered from Figures 3 and 4, in which M_2-tide cotidal and
corange maps are presented, the main features of the computed M_2-tide
regime remain the same, considering and neglecting the effects of tidal loading and ocean self-attraction. Instead of describing the differences in detail, the discussion shall be restricted to some significant dynamic effects that can be examined in view of the observed
tide. First, there is such an effect in phase distribution on the eastern sides of the Atlantic and the Pacific Oceans, where the tide is
well observed and where it can be assumed to be dominated by resolved,
large-scale phenomena. The author's former numerical experiments concerning the M_2-tide (Zahel, 1970, 1977), in agreement with the observations, yielded an increase in phase from south to north along the
west coast of Africa. However, the computed tide was early and the
same was true for the wave progressing to the north along the west
coast of North America as well as for the wave progressing to the south
along the west coast of South America. Furthermore, the antinode at
the Galapagos shelf, well reproduced for the rest, proved to be three
lunar hours early in the computed regime. Now, what does the generalized model yield for the M_2-tide wave in these areas? As can be seen
from the global maps in Figures 3 and 4, there is a considerable phase
delay in each of the coastal areas mentioned when applying the new model. Figures 5 - 7 show for the eastern boundaries of the Atlantic Ocean,
the North Pacific, and South Pacific Ocean, respectively, observed and
computed M_2-tide phases in opposition. The latter are arranged corresponding to the positions of the tide gauges and grid points along the
coast line and clearly demonstrate the tidal wave's direction of progress. Further they show that there is a phase delay caused by considering tidal loading and ocean self-attraction. This phase delay significantly improves the agreement with the observations. There is now
good agreement between computed and observed phases in these areas.
This improvement is·most remarkable in the Pacific Ocean, the systematic error in phase having been largest there. Significant changes are
also recognized for the tidal regime south of Australia, where, in addition to coastal measurements, measurements from deep-sea instrument
capsules (Irish and Snodgrass, 1972) are available. Especially in this
area, comprising also the Kerguelen-Gaussberg Ridge and St. Paul Ridge,
agreement with observations was poor when neglecting loading and self-
attraction. Figure 8 shows a section taken from Figure 4 and demonstrates in view of observed amplitudes and phases of sea-surface elevation the considerably improved agreement there.

Finally, it shall be mentioned that the LOVE numbers obtained on the basis of the Gutenberg-Bullen Earth model were used. Moreover, in the case of the M_2-tide, the iteration process was continued replacing these LOVE numbers by those based on the Harkrider Earth model (oceanic). The tidal regime was only slightly affected by this replacement and all of the above statements remain true without any change.

Legends to Figure 7a,b and Figures 9 - 14

Fig. 3. 4°-Ocean model, M_2-tide, considering tidal loading and ocean self-attraction. Cotidal lines (——) and corange lines (----); phases in lunar hours referred to meridian passage at Greenwich,amplitudes in centimeters (10, 25, 50, 75,100, 125,150,200)

Fig. 4. 4°-Ocean model, M_2-tide. Cotidal lines (——) and corange lines (- - -); Phases in lunar hours referred to meridian passage at Greenwich,amplitudes in centimetres (10, 25, 50, 75, 100, 200)

Fig. 5. M_2-tide. Observed and computed phases at positions along the west coast of Africa. x, observed values; ● computed values considering tidal loading and ocean-self-attraction, o computed values neglecting t.l. and o.s.a. Phases in degrees referred to meridian passage at Greenwich

Fig. 6. M_2-tide. Observed and computed phases at positions along the west coast of North America. x, observed values; ●, computed values considering tidal loading and ocean self-attraction; o, computed values neglecting t.l. and o.s.a. Phases in degrees referred to meridian passage at Greenwich

Fig. 7a and b. M_2-tide. Observed and computed phases at (a) positions along the west coast of South America and (b) islands in the east part of the Pacific Ocean. Key to symbols as in Figure 6

Fig. 8. M_2-tide. Comparison of the computed tide including tidal loading and ocean self-attraction with measurements. ●, tide gauge position; 210/32, observed values phase/amplitude. Section taken from the global chart

Fig. 9. 4°-Ocean-model, K_1-tide, considering tidal loading and ocean self-attraction. Cotidal lines (——) and corange lines (- - -). Phases in degrees referred to meridian passage at Greenwich,amplitudes in centimeters (5, 10, 20, 30, 50, 75)

Fig. 10. 4°-Ocean model, K_1-tide. Symbols as in Figure 9

Fig. 11. K_1-tide. Observed and computed phases at positions along the west coast of Africa. Symbols as in Figure 6

Fig. 12. K_1-tide. Observed and computed phases at positions along the west coast of North America. Symbols as in Figure 6

Fig. 13a and b. K_1-tide. Observed and computed phases at (a) positions along the west coast of South America and (b) islands in the east part of the Pacific Ocean. Symbols as in Figure 6

Fig. 14. K_1-tide. Comparison of the computed tide including tidal loading and ocean self-attraction with measurements. ●, tide gauge position; 210/32, observed values phase/amplitude. Section taken from the global chart

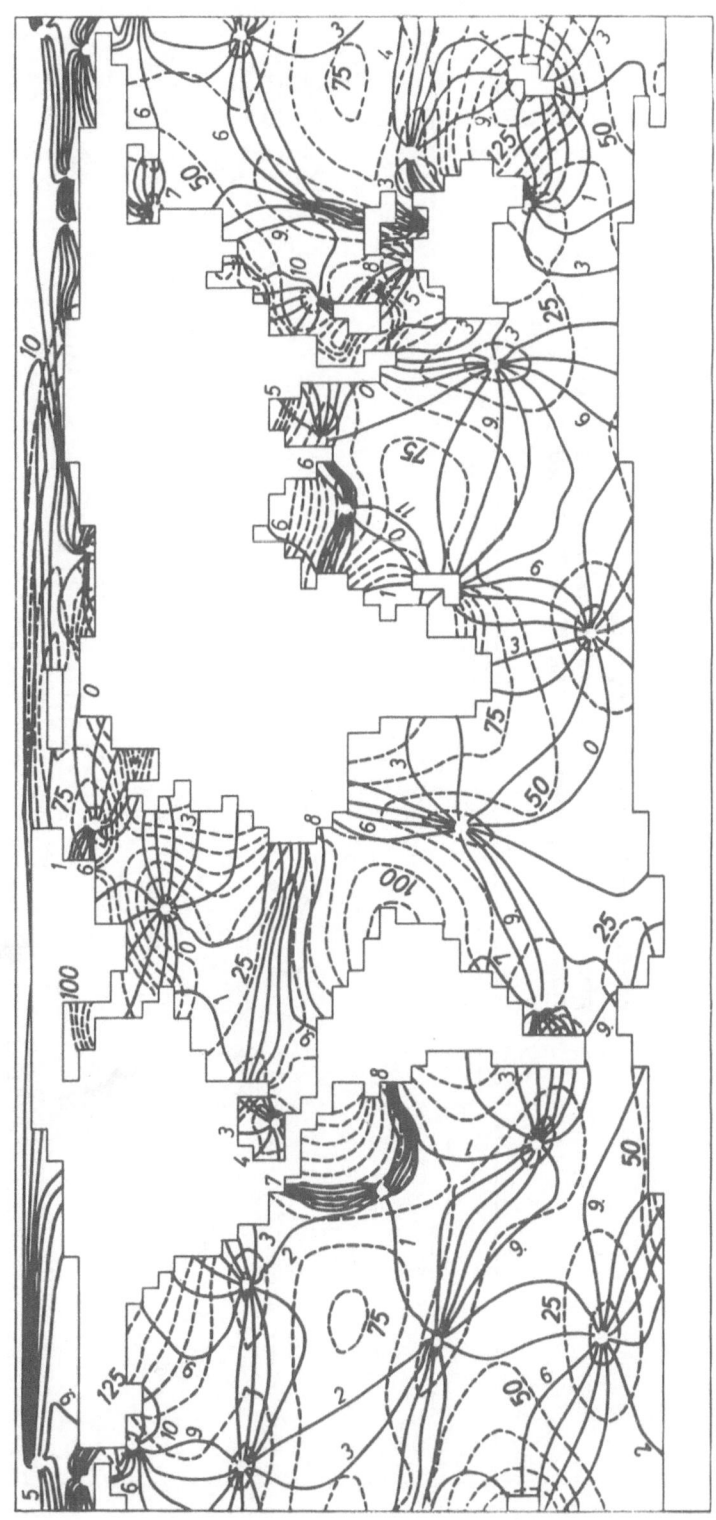

4° - OCEAN - MODEL
M₂ - Tide
considering tidal loading and ocean self-attraction

cotidal lines (——) and corange lines (- - -)
phases in lunar hours referred to meridian passage at Greenwich
amplitudes in centimetres (10, 25, 50, 75, 100, 200)

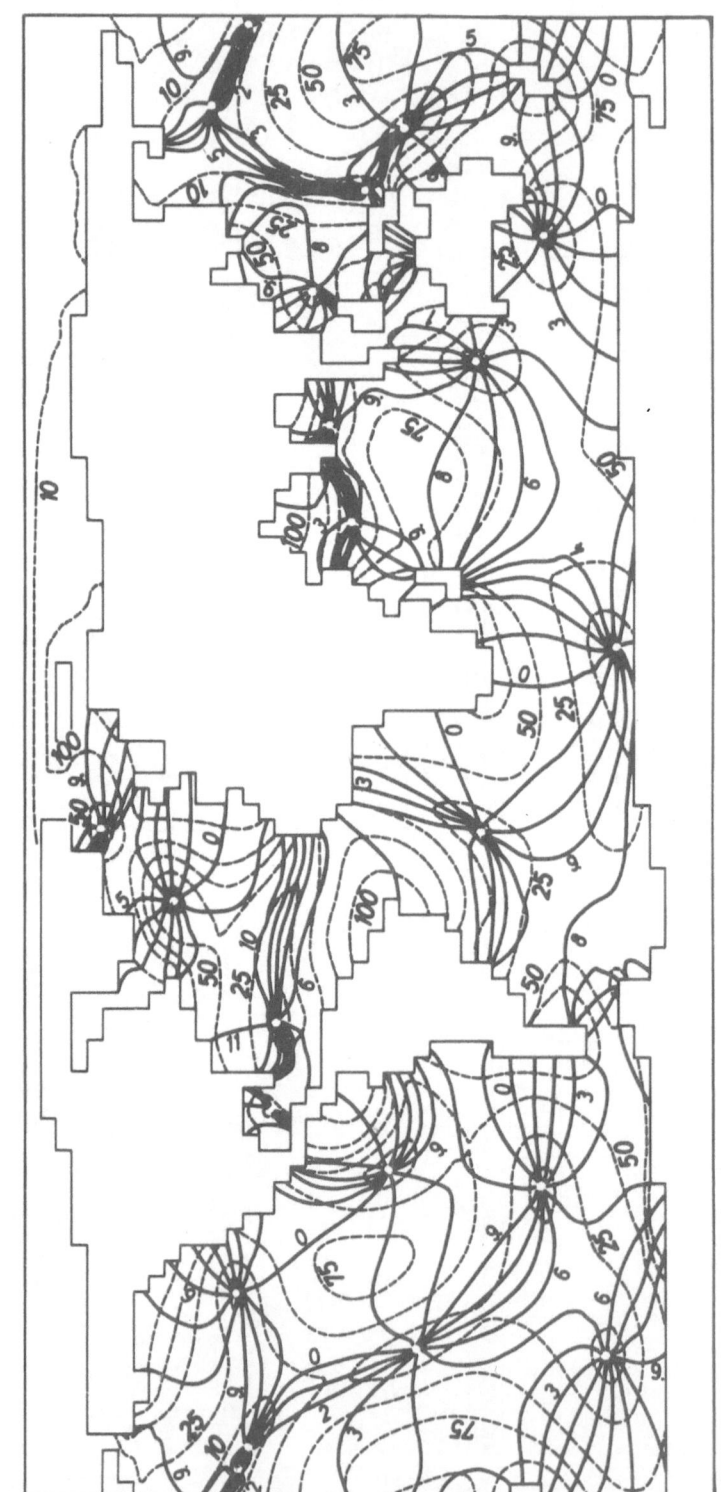

4°-OCEAN-MODEL
M_2-Tide

cotidal lines (—) and corange lines (---)

phases in lunar hours referred to meridian passage at Greenwich
amplitudes in centimetres (10, 25, 50, 75, 100, 200)

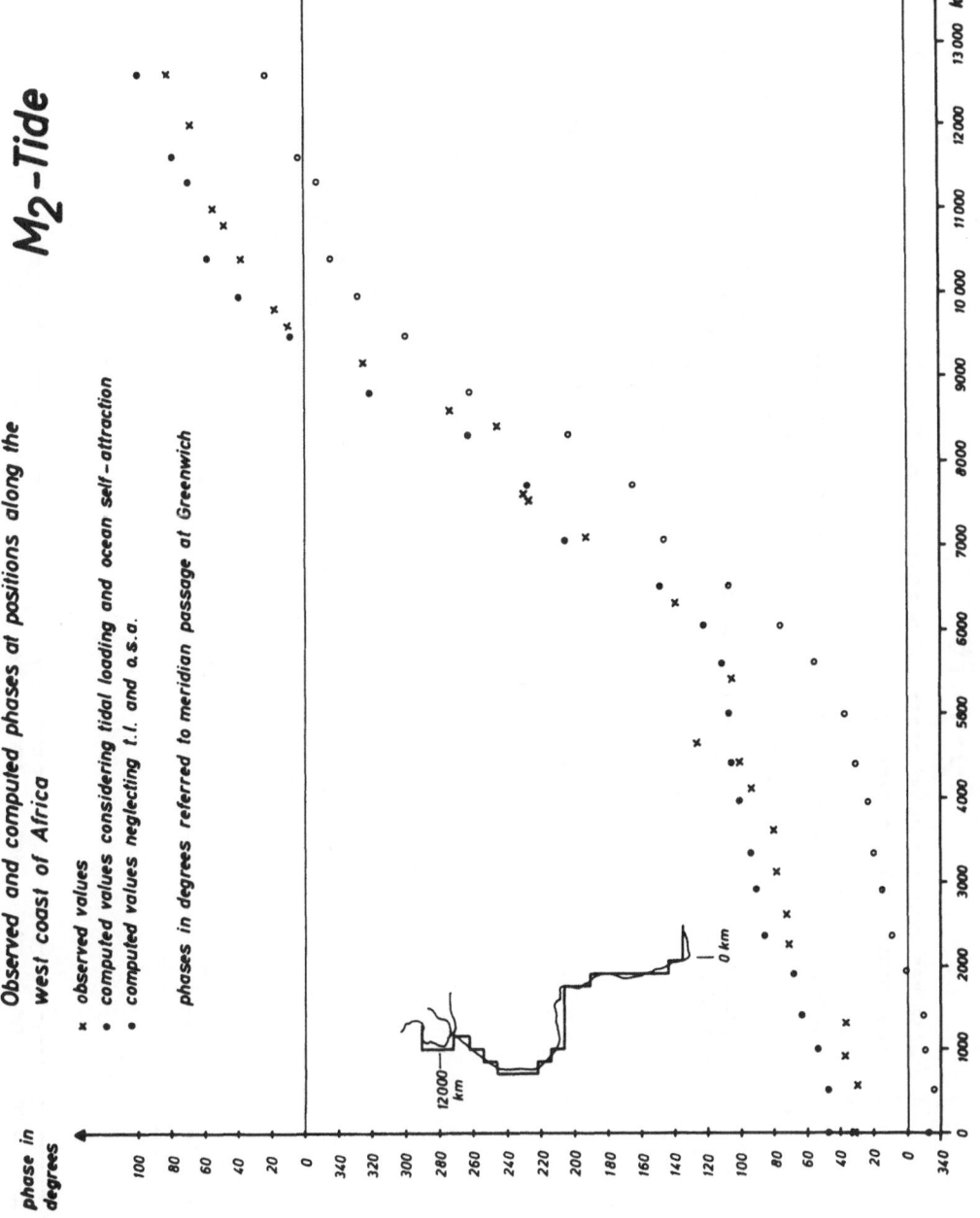

Observed and computed phases at positions along the west coast of Africa

M₂-Tide

× observed values
● computed values considering tidal loading and ocean self-attraction
○ computed values neglecting t.l. and a.s.a.

phases in degrees referred to meridian passage at Greenwich

phase in degrees

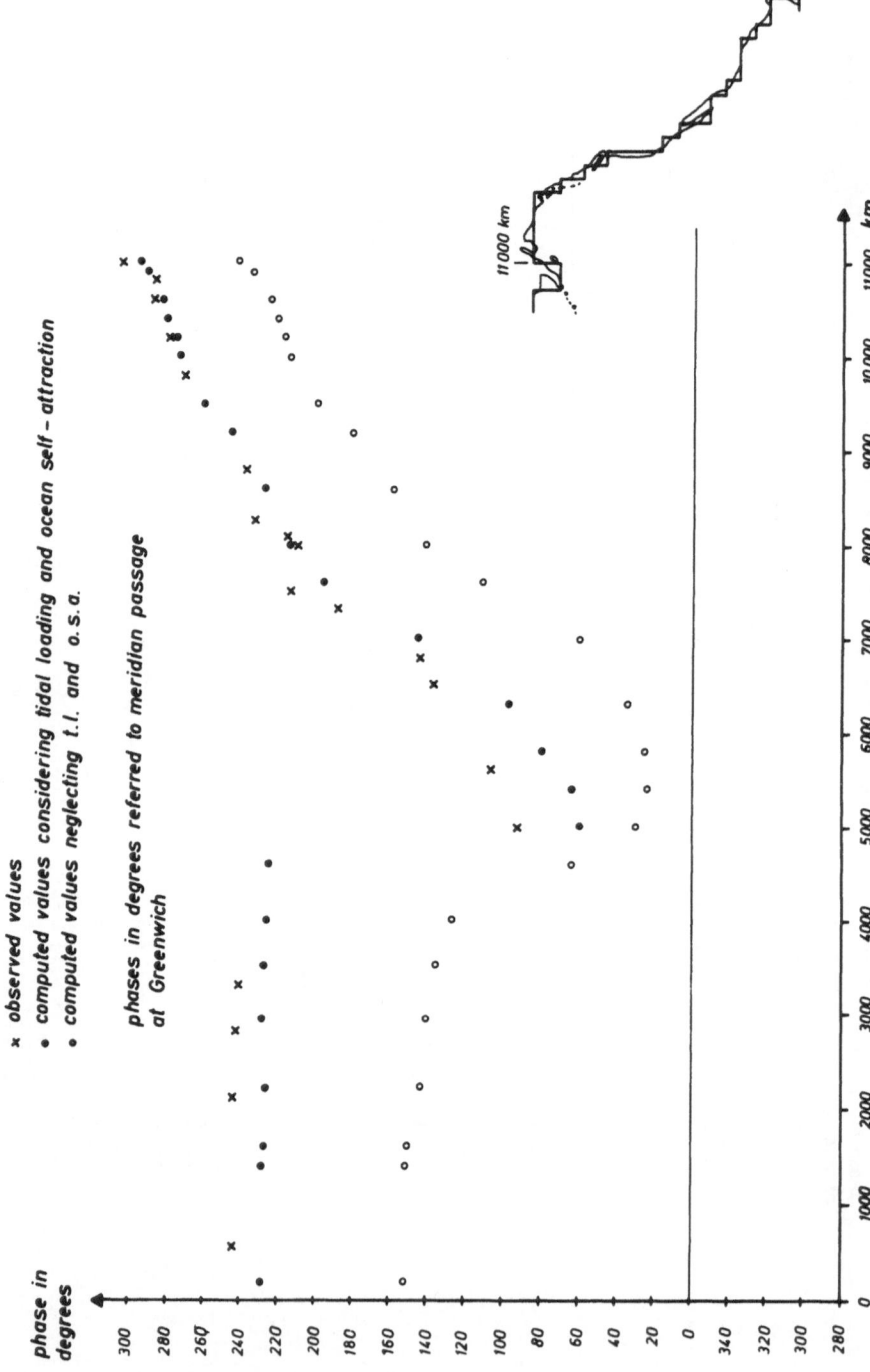

M₂-Tide

Observed and computed phases at positions along the
west coast of North America

x observed values

• computed values considering tidal loading and ocean self – attraction

• computed values neglecting t.l. and o.s.a.

phases in degrees referred to meridian passage
at Greenwich

M₂-Tide

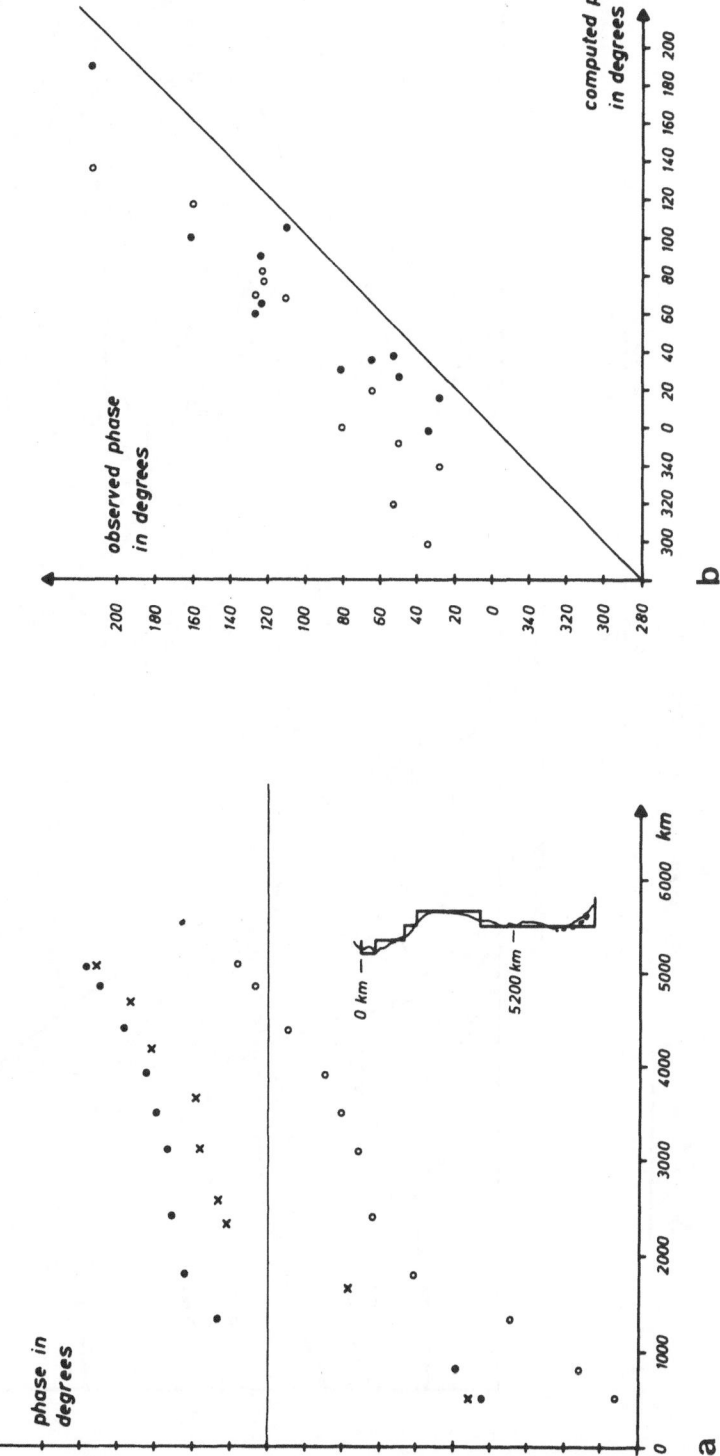

Observed and computed phases at positions along the
west coast of South America

  x *observed values*

  ● *computed values considering tidal loading and ocean
self-attraction*

  ○ *computed values neglecting t.l. and o.s.a.*

Observed and computed phases at islands
in the east part of the Pacific Ocean

  ● *considering tidal loading and ocean self-attraction*

  ○ *neglecting t.l. and o.s.a.*

phases in degrees referred to meridian passage at Greenwich

114

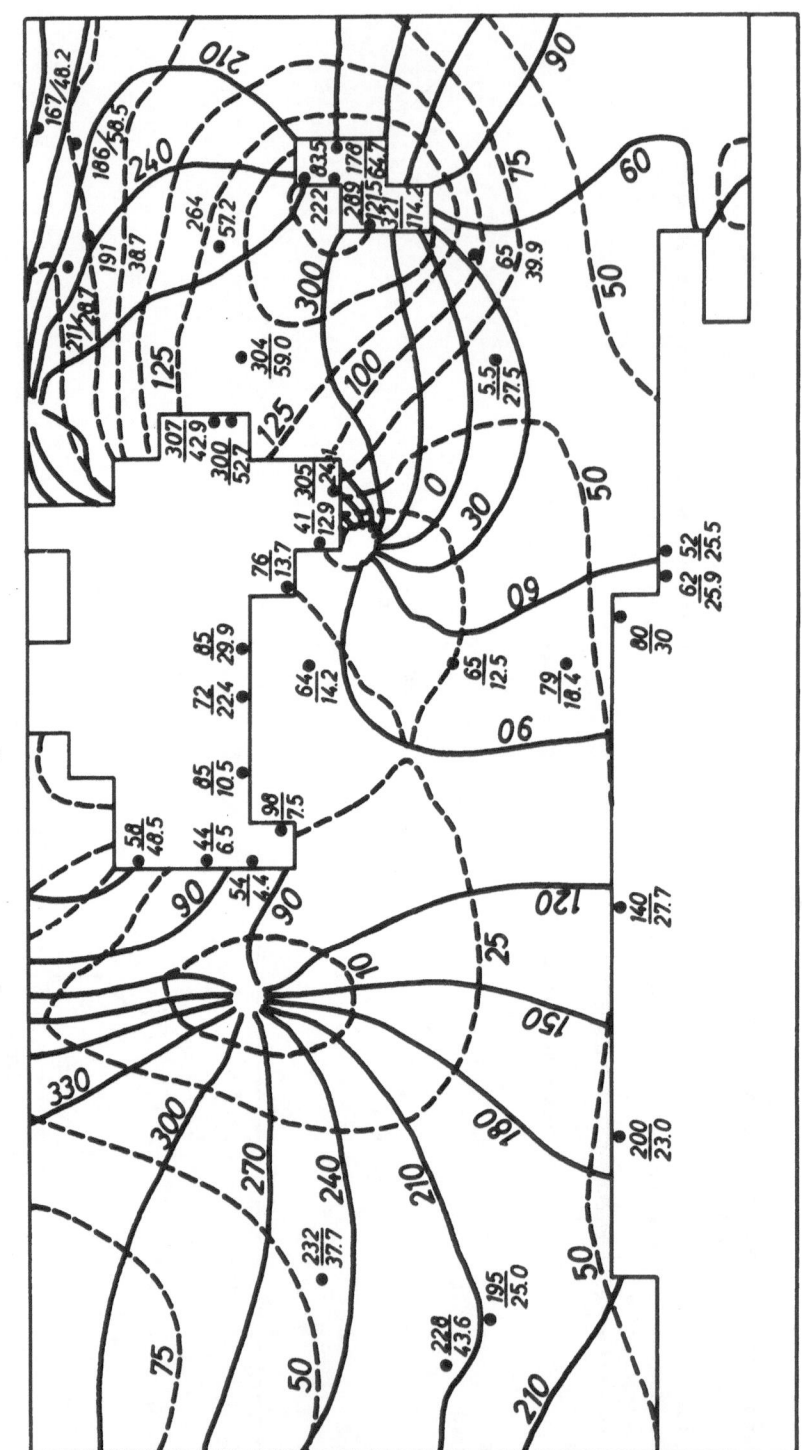

M_2 - tide

Comparison of the computed tide including tidal loading and ocean self - attraction with measurements

• tide gauge position, $\frac{210}{32}$ observed values $\frac{phase}{amplitude}$

section taken from the global chart

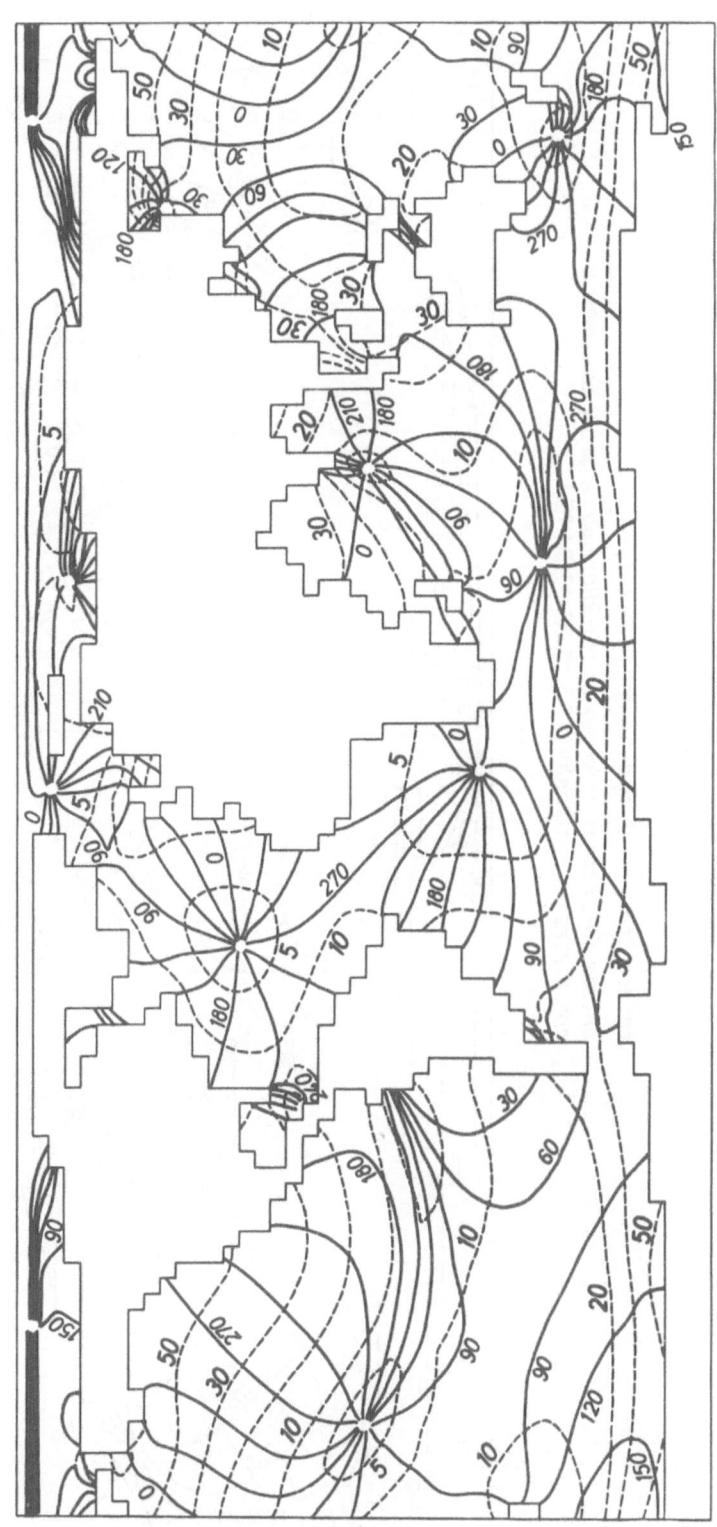

4° - OCEAN - MODEL
K_1 - Tide
considering tidal loading and ocean self-attraction

cotidal lines (———) and corange lines (- - -)
phases in degrees referred to meridian passage at Greenwich
amplitudes in centimetres (5, 10, 20, 30, 50, 75)

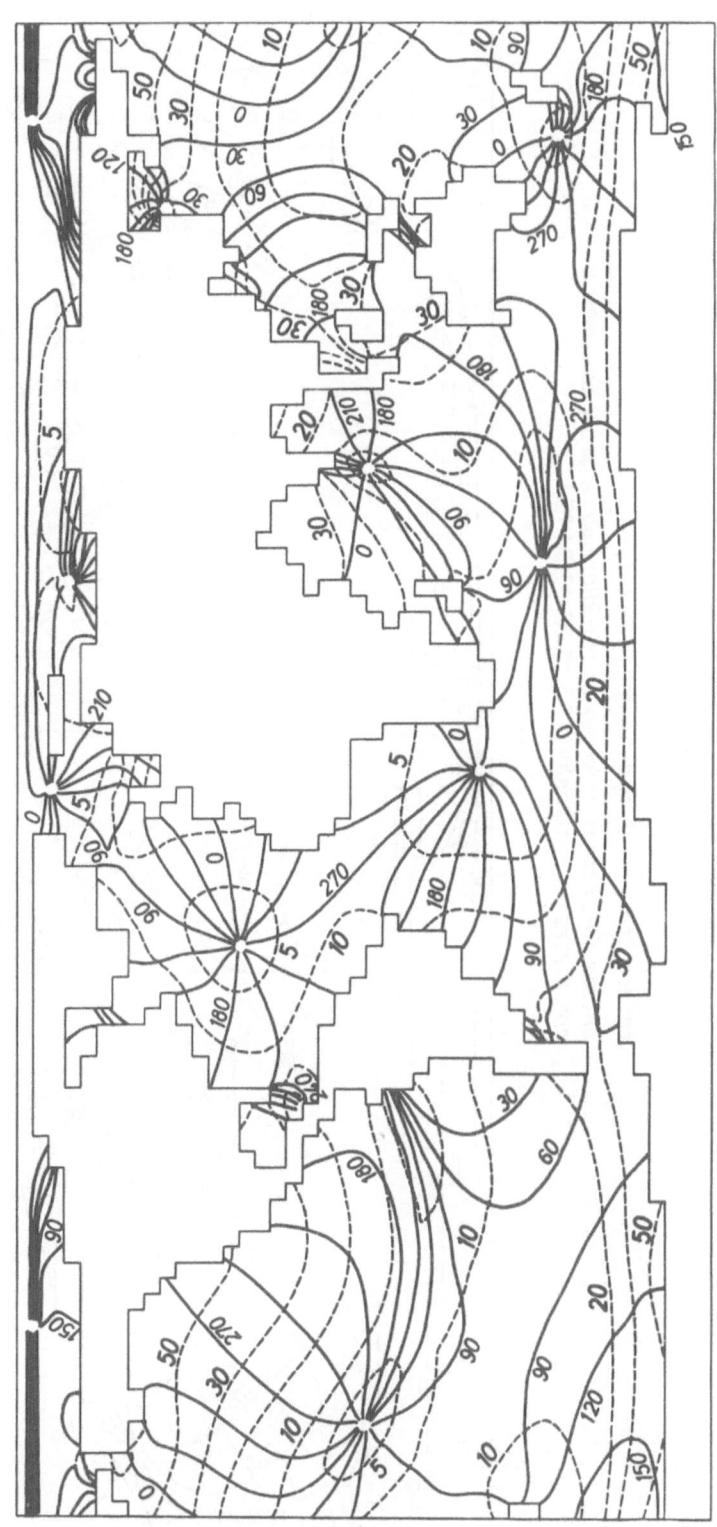

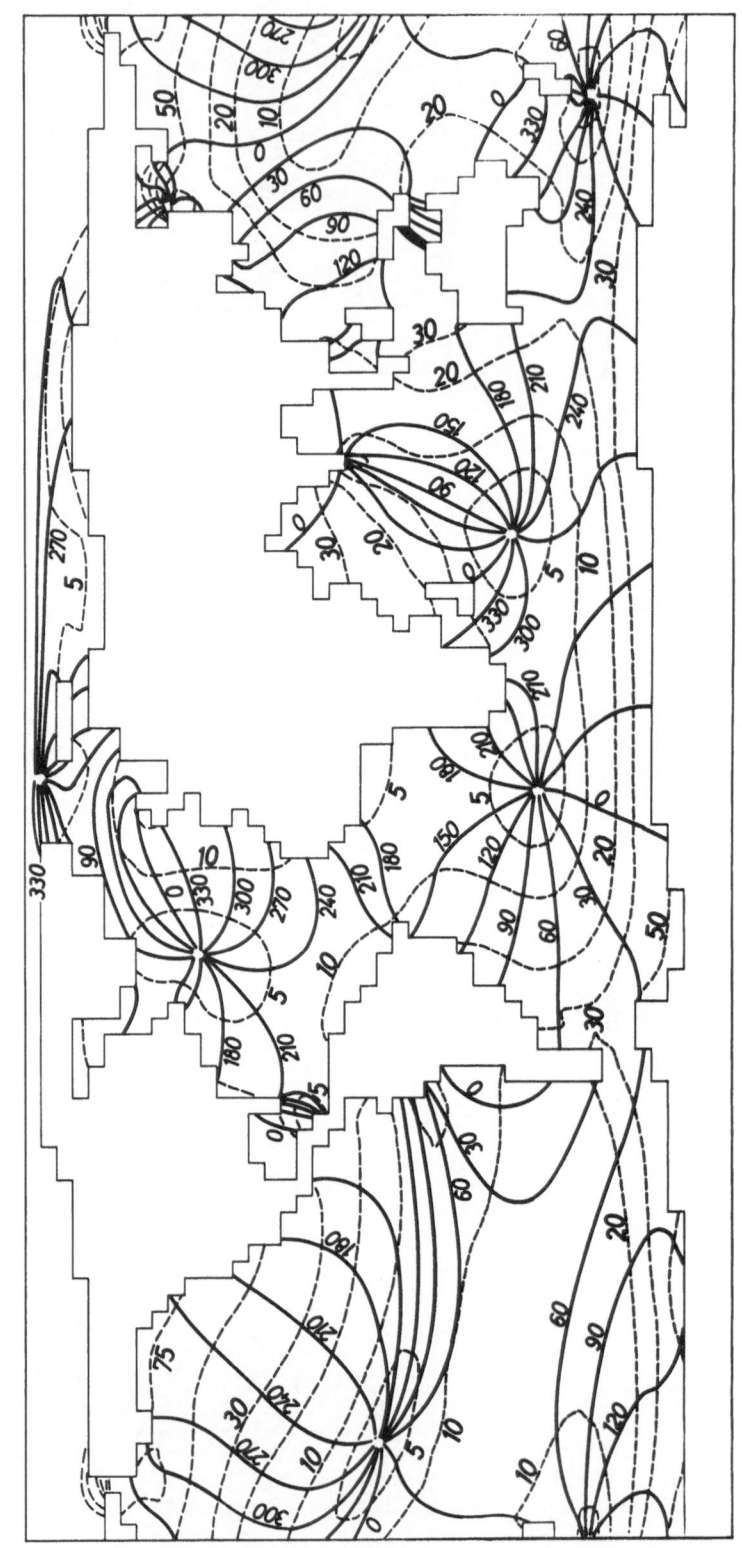

4° - OCEAN - MODEL
K₁ -Tide

cotidal lines (——) and corange lines (-----)

phases in degrees referred to meridian passage at Greenwich

amplitudes in centimetres (5,10,20,30,50,75)

K_1-Tide

Observed and computed phases at positions along the west coast of Africa

× observed values
● computed values considering tidal loading and ocean self-attraction
○ computed values neglecting t. l. and o.s.a.

phases in degrees referred to meridian passage at Greenwich

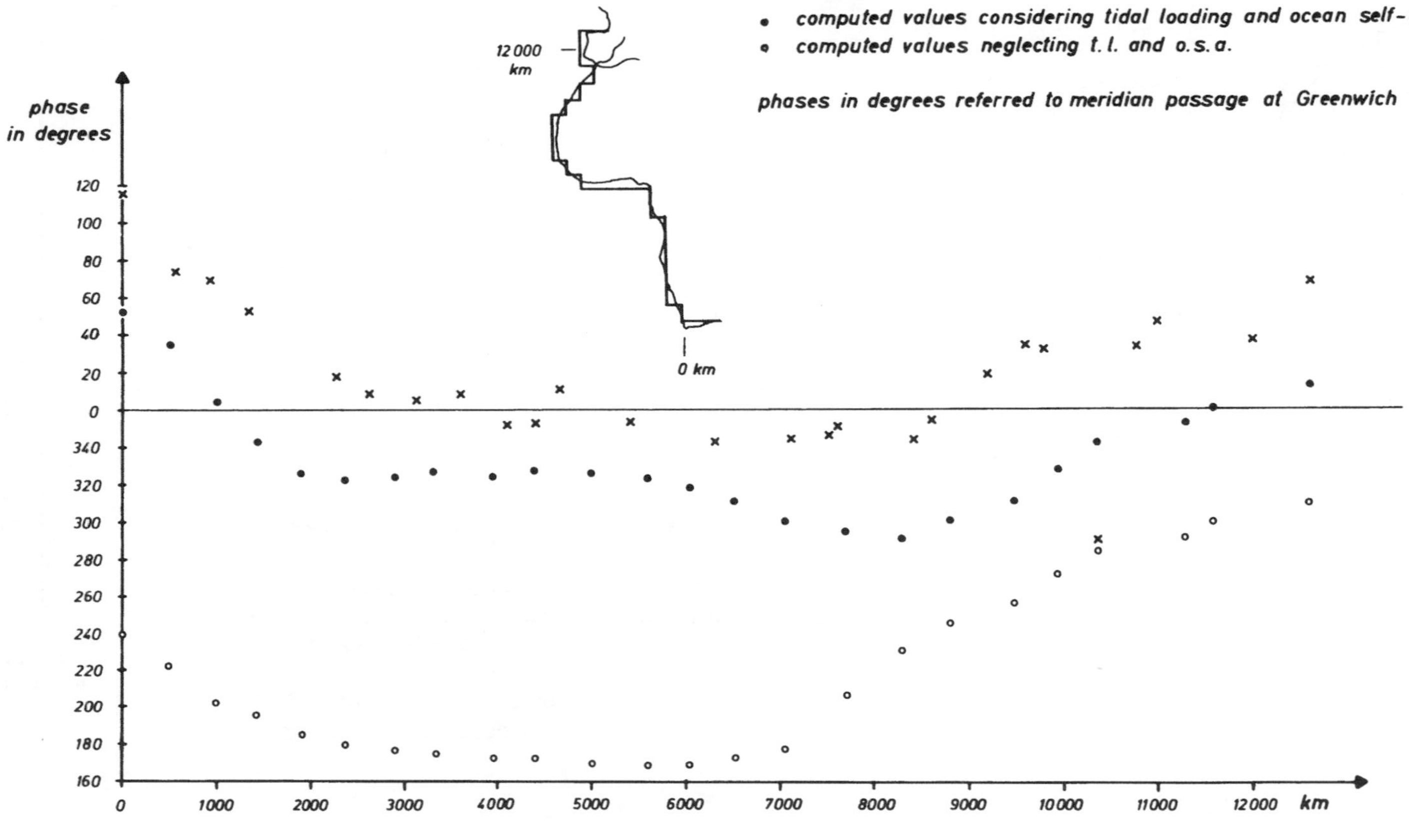

phase
in degrees

K₁-Tide

Observed and computed phases at positions along the
west coast of North America

x   *observed values*

●   *computed values considering tidal loading and ocean self-attraction*

○   *computed values neglecting t.l. and o.s.a.*

phases in degrees referred to meridian passage at Greenwich

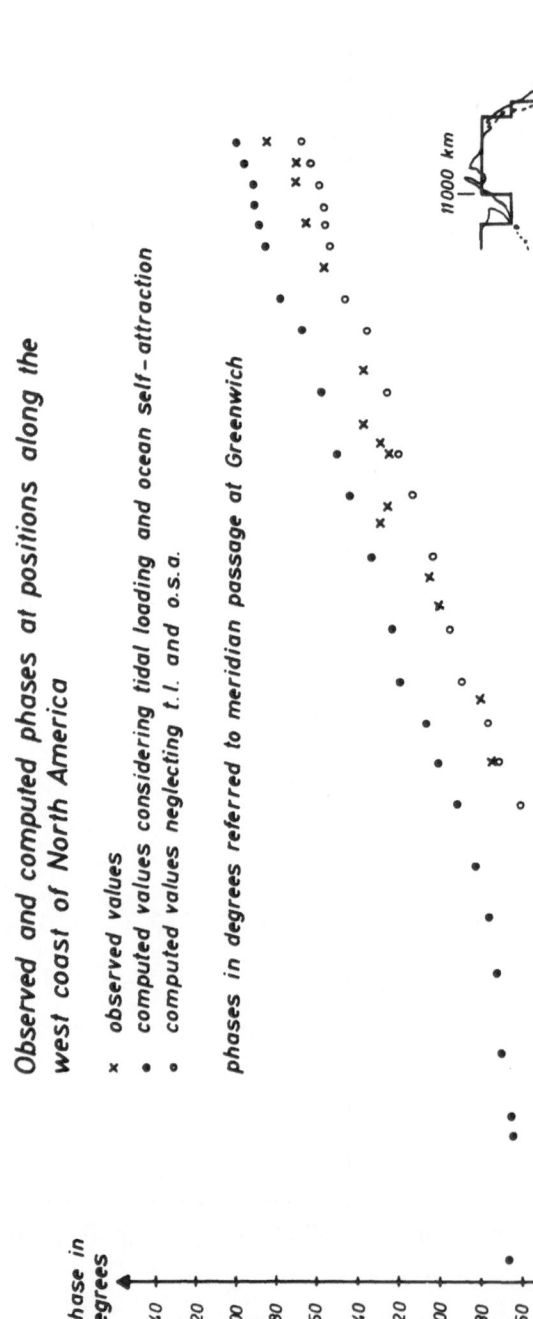
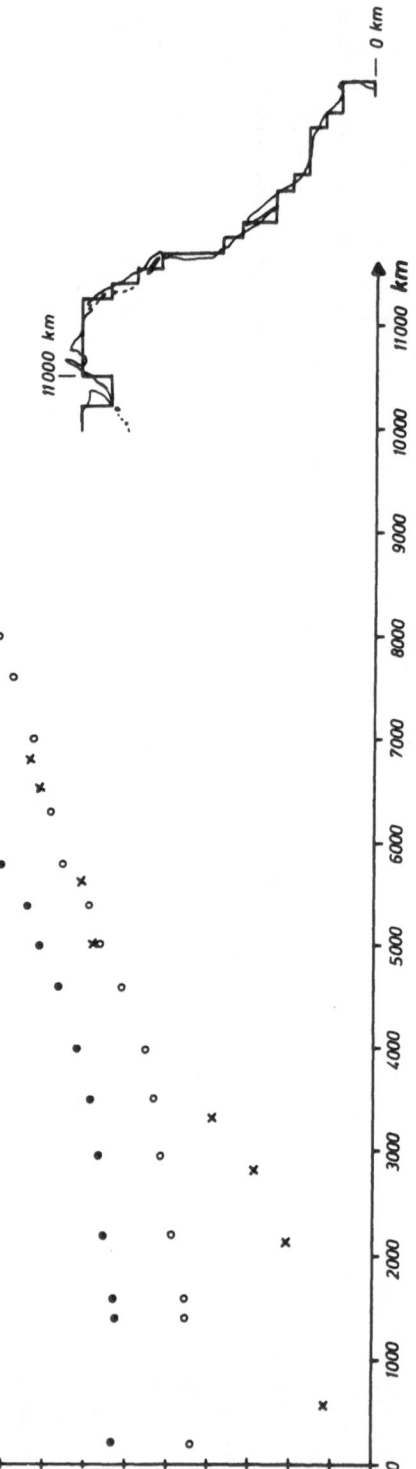

K_1-Tide

Observed and computed phases at positions along the west coast of South America

- x observed values
- • computed values considering tidal loading and ocean self-attraction
- • computed values neglecting t. l. and o.s.a.

Observed and computed phases at islands in the east part of the Pacific Ocean

- • considering tidal loading and ocean self-attraction
- ○ neglecting t. l. and o.s.a.

phases in degrees referred to meridian passage at Greenwich

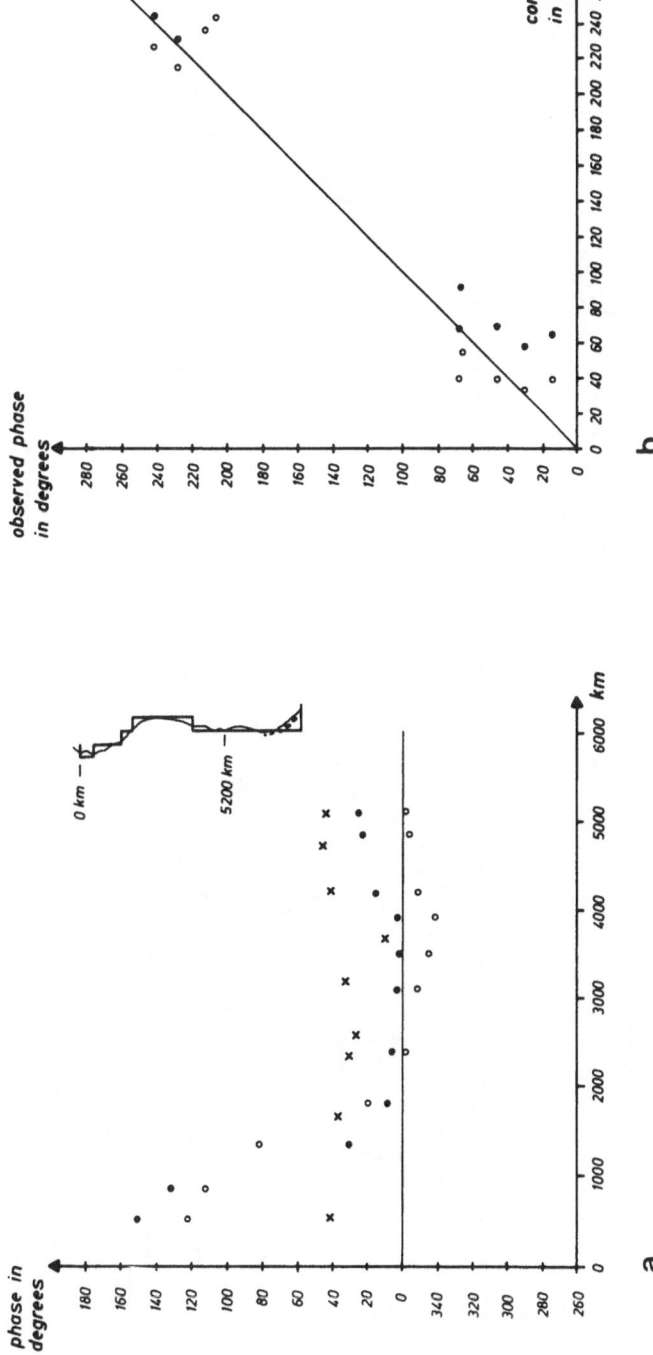

$\underline{K_1 - tide}$

Comparison of the computed tide including tidal loading
and ocean self-attraction with measurements

$\dfrac{phase}{amplitude}$

• tide gauge position, $\dfrac{210}{32}$ observed values

section taken from the global chart

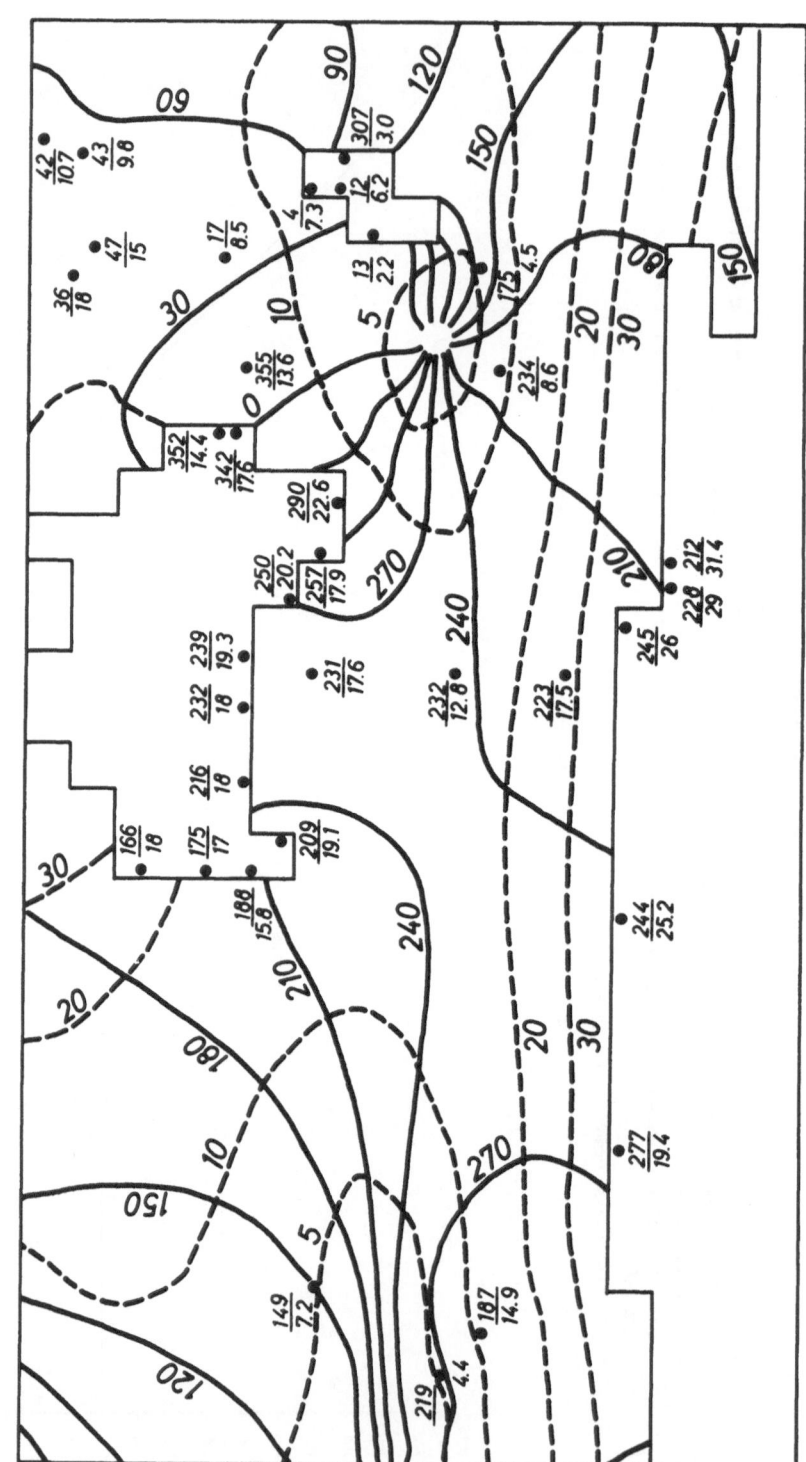

5. The Computed Global K_1-Tide

As with the M_2-tide, one recognizes from the computed K_1-tide cotidal
and corange maps, given in Figures 9 and 10, that in principle the
oscillation system remains the same when considering and neglecting
tidal loading and ocean self-attraction. Opposing the computational
results to each other and to the measurements at the same coastal areas considered in Section 4, one likewise finds significant effects,
which are of a similar kind. There is, however, a different situation
in the South Atlantic Ocean, where, taking into account loading and
self-attraction, the phase distribution becomes comparable with the
observed values. In contrast, the basic model only yielded an agreement
in phase differences among the various stations at the continental
coast and at the islands of Tristan da Cunha, St. Helena, Ascension,
and Trinidad. Figure 11 shows how the improvement in reproducing the
observed K_1-tide in the South Atlantic Ocean is reflected at the positions along the eastern boundary between Cape Agulhas and Cabo de
Finisterre.

From both Figure 9 and Figure 10, a left-hand rotating amphidromy is
recognized to dominate the K_1-tide regime in the Pacific Ocean and
hardly any difference can be detected in this system between the two
cases. However, regarding observed and computed phases at positions
along the west coast of North America in Figure 12, one finds the computed wave progressing to the north to be slightly late now, whereas
the basic model yields a phase distribution identical with the one observed there. Applying the new model, a phase delay appears at the
South American coast and thereby an improvement in reproducing the observed tide is obtained there (Fig. 13). The correlation in Figure 13
refers to islands in the eastern part of the Pacific Ocean, e.g., to
Midway, Hawaii, Easter Island, and Northern Line Islands. It obviously
reflects the situation at the American coast, which can clearly be
understood if it is accepted that the computed tidal regime is realistic in this area.

Finally, Figure 14, presenting a section taken from Figure 10, shows
that there is again a satisfactory agreement in phases as well as in
amplitudes of sea-level elevation in the area around Australia, but
in case of the K_1-tide this proves right for the results of both models.

6. Conclusions

The basic $4°$-model, in which the additional effects of tidal loading
and ocean self-attraction have been included, yielded less realistic
results than the $1°$-model (Zahel, 1977). Of course, tidal phenomena
whose resolution depends on a grid-size reduction of this degree can
only be expected to arise in a $1°$-model and this hopefully in a more
realistic manner if the influence of the solid Earth's deformation on
the ocean tide is considered. Because of the large expense in computer
time necessary for the application of a generalized $1°$-model, computations have been performed for a generalized $4°$-model as described above.
Indeed the shortcomings of the former $4°$-model results opposite to
the $1°$-model results are preserved, even when the additonal effects
of tidal loading and ocean self-attraction are considered. However,
the preceding presentation indicated that there are considerable improvements in reproducing the observed tide. These improvements could

be achieved by introducing the additional effects into a 4°-model, but
they did not appear when the grid size was reduced from 4° to 1° without considering the additional effects. Hence, in view of the comparisons performed with measurements, the few but important influences of
the solid Earth's deformation on tidal regimes appear to be realistic.
In any case, a proper numerical treatment of the primitive tidal integrodifferential equations is assumed to be available. Modifications
of parameterizations included in the equations should not alter this
fact. With regard to a better understanding of the effects in question,
numerical investigations concerning the eigenvalues and eigenfunctions
of the linear integrodifferential equations are felt to be desirable.

Acknowledgments. The author is indebted to J. Zschau for providing the LOVE numbers
based on the Harkrider Earth model (oecanic). The computations were performed on
an AEG-TELEFUNKEN computer TR 440 at the computer center of the University of Hamburg.

Glossary of Symbols

$a=a(\phi), b=b(\phi)$: Weighting factors in the discrete zonal pressure
gradient term

A_h : Coefficient of horizontal turbulent viscosity
(eddy viscosity)

$d=d(\lambda,\phi)$: Mean ocean depth

$D=D(\lambda,\phi,t)$: Actual ocean depth

D_o^λ, D_o^ϕ : Central finite-difference operators

D_+^t : Forward finite-difference operator

g : Acceleration of gravity

h_m, k_m : Nonloading LOVE numbers of degree m

h_n', k_n' : Loading LOVE numbers of degree n

c : Coefficient of the tide-generating force (M_2)

M : Number of time-steps per tidal period

$P_n^s=P_n^s(\sin\phi)$: Associated Legendre functions

$$P_n^s(\mu) = \frac{(1-\mu^2)^{s/2}}{2^n\,n!}\,\frac{d^{s+n}(\mu^2-1)^n}{d\,\mu^{s+n}}, \quad \mu=\sin\phi$$

$P_n = P_n(\sin\phi)$: Legendre polynomials $P_n=P_n^o$

$\bar{P}_n^s=\bar{P}_n^s(\sin\phi)$: Normalized associated Legendre functions

$\bar{P}_n=\bar{P}_n(\sin\phi)$: Normalized Legendre polynomials

r : Coefficient of quadratic bottom friction

R	:	Earth's radius
$R_\lambda = R_\lambda(\lambda,\phi,t)$, $R_\phi = R_\phi(\lambda,\phi,t)$	:	Eddy viscosity terms
R^*	:	Coefficient of linear bottom friction
$u=u(\lambda,\phi,t)$, $v=v(\lambda,\phi,t)$	:	Vertically averaged (from sea bottom to sea surface) horizontal current velocity components, east and north components, resp.
$U_m=U_m(\lambda,\phi,t)$	:	Tidal body potential of degree m
t	:	Time
Δt	:	Time-step
$\delta=\delta(\lambda,\phi,t)$	:	Radial displacement of the sea bottom
$\delta_b=\delta_b(\lambda,\phi,t)$	:	Radial displacement of the sea bottom due to the tidal body potential
$\delta_s=\delta_s(\lambda,\phi,t)$	:	Radial displacement of the sea bottom due to the ocean surface load
$\zeta=\zeta(\lambda,\phi,t)$	:	Sea-surface displacement relative to the moving sea bottom
$\zeta_n=\zeta_n(\lambda,\phi,t)$	:	nth spherical harmonic constituent of ζ
$\zeta_o=\zeta_o(\lambda,\phi,t)$	:	Geocentric sea-surface displacement
λ	:	Geographic longitude
$\Delta\lambda$	:	Zonal angular grid-point distance
ρ	:	Mean density of sea water
ρ_e	:	Mean density of the solid Earth
σ	:	Tidal frequency
ϕ	:	Geographic latitude
$\Delta\phi$	:	Meridional angular grid-point distance
ω	:	Earth's angular rate of rotation
Δ	:	Laplace operator
div	:	Divergence operator

References

Alterman, Z., Jarosch, H., Pekeris, C.L.: Propagation of Rayleigh waves in the Earth. Geophys. J. <u>4</u>, 219 (1961)

Brosche, P., Sündermann, J.: On the torques due to tidal friction of the oceans and adjacent seas. In: Rotation of the Earth. Melchior, P., Yumi, S. (eds.). Dordrecht, Netherlands: Reidel, D., 1972, pp. 235-239

Farrell, W.E.: Deformation of the Earth by Surface Loads. Rev. Geophys. Space Phys. 10, 761-797 (1972)

Gordeev, R.G., Kagan, B.A., Polyakov, E.V.: The Effects of Loading and Self-Attraction on Global Ocean Tides: The Model and the Results of a Numerical Experiment. J. Phys. Oceanogr. 7, 161-170 (1977)

Gordeev, R.G., Kagan, B.A., Rivkind, V. Ya.: A numerical solution of the tidal dynamics equations in the World Ocean. Dokl. Akad. Nauk SSSR 209, 340-343 (1973)

Harkrider, D.G.: Surface Waves in multilayered elastic media, 2, Higher mode spectra and spectral ratios from point sources in plane layered Earth models. Bull. Seismol. Soc. Am. 60, 1937 (1970)

Hendershott, M.C.: The Effects of Solid Earth Deformation on Global Ocean Tides. Geophys. J. R. Astron. Soc. 29, 389-402 (1972)

Hendershott, M.C.: Ocean Tides. Trans. Am. Geophys. Union 54, 76-86 (1973)

Irish, J.D., Snodgrass, F.E.: Australian-Antarctic Tides. In: Antarctic Research Series, 19, Antarctic Oceanology II: The Australian-New Zealand Sector. Hayes, D.E. (ed.), 1972

Longman, I.M.: A Green's Function for Determining the Deformation of the Earth under Surface Mass Loads. 2. Computations and Numerical Results. J. Geophys. Res. 68, 485-496 (1963)

Munk, W.H., MacDonald, G.J.F.: The Rotation of the Earth. Cambridge: University Press, 1960, p. 323

Rochester, M.G.: The Earth's Rotation. EOS, Trans. Am. Geophys. Union 54, 769-780 (1973)

Zahel, W.: Die Reproduktion gezeitenbedingter Bewegungsvorgänge im Weltozean mittels des hydrodynamisch-numerischen Verfahrens. Mitt. Inst. Meereskd. Univ. Hamburg 17, 1-50 (1970)

Zahel, W.: The Diurnal K_1-Tide in the World Ocean — A Numerical Investigation. Pure and Appl. Geophys. 109, 1819-1825 (1973)

Zahel, W.: A global hydrodynamic-numerical 1°-model of the ocean tides; the oscillation system of the M_2-tide and its distribution of energy dissipation. Ann. Geophys. t. 33, fasc. 1/2, 31-40 (1977)

Numerical Computation of Tidal Friction for Present and Ancient Oceans

J. Sündermann and P. Brosche

1. Introduction

The Earth loses rotational energy. This is evident from astronomic and paleontologic data for present as well as for ancient times. For the present state the estimates for the loss of rotational energy $\dot{E}_{rot}$ range within the interval of $4 - 7 \cdot 10^{19}$ erg/s, according to Rochester (1973), Muller and Stephenson (1975), and Kagan (1977).

This process originates from a transfer of rotational angular momentum of the Earth into orbital angular momentum of the Moon. The corresponding decelerating torque acting on the Earth can be generated principally within the solid Earth, in the ocean, and in the atmosphere. According to present knowledge the contributions of the solid Earth and the atmosphere are relatively small. They do not exceed $0.2 \cdot 10^{19}$ erg/s (Wunsch, 1975; Kagan, 1977). The key to the explanation of the momentum transfer, therefore, lies in the ocean.

As shown by Schott (1977) in a quantitative analysis of tidal energy flux in the ocean, the main dissipation takes place in the bottom friction of barotropic tides. The part of energy dissipation in the internal tides remains smaller than $0.5 \cdot 10^{19}$ erg/s. A similar value is given by Kagan (1977).

There are three approaches, in principle, for calculating the energy flux in the barotropic tides (see Munk and MacDonald, 1960):

Method 1: The energy inflow into the deep ocean from the tide generating potentials of the Moon and the Sun

Method 2: The energy dissipation by bottom friction in the shallow shelf areas

Method 3: The energy flux from the deep ocean into the shelf seas by the tidal currents

All three models have been applied in oceanographic investigations. The corresponding balances for $-\dot{E}_{rot}$ resulted in the right order of magnitude, but not more. Some representative values are given in Table 1.

Table 1. Decay of rotational energy of the Earth according to several authors

Author	Year	Method	Data	$-\dot{E}_{rot}$ $\left[10^{19} \text{ erg/s}\right]$
Pariiskii et al.	1972	Deformation of sea surface (1)	Computation	4.2
Zahel	1977	Bottom friction and horizontal viscosity (2)	Computation	3.8
Miller	1966	Tidal currents on shelf entrances (3)	Observation	1.4 – 1.7

Originally, the main cause of errors in these balances was due to the poor hydrographic material: The geographic distribution of tidal amplitudes and phases deduced from a few observations was very uncertain. All estimates that were based on field measurements suffered from this, e.g., the first calculations by Jeffreys (1920) and also those by Miller (1966).

The density of information on tidal dynamics has become considerably greater and more homogeneous by using numerical models to simulate oceanic tides. On the other hand, some new problems have been raised in this connection, which will be treated in the next Section.

The main criticism of most of the published balances, with the exception of those using "method 1", must be directed to the consideration solely in terms of energy, and specifically in a corotating frame of reference. The astronomic quantities are related to an inertial system and, consequently, they can be compared only to correspondingly computed geophysical values. All applications of "method 2", however, with the exception of those given by Brosche and Sündermann (1971, 1972, 1977), have been performed in a system moving with the rotating Earth.

Furthermore, the scalar energetic approach seems to be inadequate due to the basic physical nature of the process, namely, the transfer of the vectorial quantity angular momentum. This is obvious from the fact that the usual energy dissipation formula for the bottom friction does not depend on the geographic position and on the direction of tidal friction. On the other hand, the rotation of the Earth can be influenced only by a torque acting with respect to the axis of rotation. This torque results from the vectorial sum of partial torques caused by the contributions of different parts of the world ocean. These are indeed dependent on the direction of the acting force and on the distance from the Earth's axis, i.e., from the geographic position. The resulting torque corresponds to a change of angular momentum within the inertial coordinate system and this quantity is connected with the observed $-\dot{E}_{rot}$.

If we consider the hydrosphere in an absolute frame of reference the balance of angular momentum is given by two torques: The torque between the Moon and the water $\dot{P}_m$ and the torque between the Earth and the water $\dot{P}_e$. Since there are no microscopic sources or sinks of angular momentum in the sea, the balance contains no further terms:

$$\dot{P}_m + \dot{P}_e = 0 \tag{1}$$

If we know one quantity, the other one is uniquely determined. The change of rotational energy is given by $-\dot{E}_{rot} = - \omega\dot{P}_e$.

The energy balance includes, in contrast, besides the corresponding terms $\dot{E}_m$ and $\dot{E}_e$, additional dissipative terms $\dot{E}_b + \dot{E}_i$ describing the transition of kinetic energy into heat by bottom and inner friction, respectively. The equation reads

$$\dot{E}_m + \dot{E}_e + \dot{E}_b + \dot{E}_i = 0 \tag{2}$$

The wanted astronomically observed quantity is $\dot{E}_e$. Obviously, $\dot{E}_b$ as calculated by Jeffreys (1920) is only a lower limit for $\dot{E}_e$. It is necessary to know $\dot{E}_m$ and $\dot{E}_i$ as well, in order to obtain $\dot{E}_e$. This method requires considerably more effort than the angular momentum balance (1).

In the following the possibilities of a numerical treatment of balances (1) and (2) will be discussed.

Also the typical properties and results of the existing models will be assembled.

Since the transfer of angular momentum between Earth and Moon occurred in the geologic past as well, the corresponding balances for the paleotides are equally of great interest. In Section 4 the problems in modeling the ancient oceans will be outlined. Afterward first results for the Permian ocean, 250 million years before present (m.y. BP), are presented.

2. Computation of Tidal Friction by Hydrodynamic - Numerical Models

2.1 Basic Equations

The hydrodynamic differential equations (HDE) describing the dynamics of barotropic tides in the world ocean are obtained from the fundamental conservation laws of momentum and mass. They read in a vertically integrated form and in spherical coordinates (see Zahel, 1977):

$$\frac{\partial u}{\partial t} + \frac{u}{R \cos \phi} \frac{\partial u}{\partial \lambda} - 2\omega\sin\phi v + \frac{r}{D} |u^2 + v^2|^{1/2} u + R_\lambda = - \frac{g}{R \cos \phi} \frac{\partial}{\partial \lambda} (\zeta - \gamma \Phi)$$

$$\frac{\partial v}{\partial t} + \frac{v}{R} \frac{\partial v}{\partial \phi} + 2\omega\sin\phi u + \frac{r}{D} |u^2 + v^2|^{1/2} v + R_\phi = - \frac{g}{R} \frac{\partial}{\partial \phi} (\zeta - \gamma \Phi)$$

$$\frac{\partial \zeta}{\partial t} + \frac{1}{R \cos \phi} \left(\frac{\partial}{\partial \lambda} (D u) + \frac{\partial}{\partial \phi} (D v \cos\phi) \right) = 0 \tag{3}$$

The following notations have been used:

u, v   Components of the vertically integrated horizontal velocities in the east and north directions, respectively

ζ Water elevation = deviation from the undisturbed surface

$D = d + \zeta$ Actual water depth

d   Undisturbed water depth

g   Acceleration of the Earth

r   Friction factor (0.003)

R   Radius of the Earth

t   Time

λ, ϕ Geographic longitude and latitude

ω Angular velocity of the Earth's rotation ($0.729 \cdot 10^{-4}$ s^{-1})

$\Phi =$ $g \frac{c}{2} R \cos^2\phi \cos (\sigma t + 2\lambda)$ tidal potential of a semidiurnal tide

c   Tidal constant ($0.761 \cdot 10^{-7}$)

σ Angular velocity of the tide ($1.405 \cdot 10^{-4}$ s^{-1})

$\gamma =$ $1 + k - h$

k, h   LOVE numbers (0.302 and 0.612)

$R_\lambda = - A_h (\Delta u - \frac{1}{R^2} (u(1 + tg^2\phi) + 2 \frac{tg\phi}{\cos\phi} \frac{\partial v}{\partial \lambda}))$ lateral eddy

$R_\phi = - A_h (\Delta v - \frac{1}{R^2} (v(1 + tg^2\phi) - 2 \frac{tg\phi}{\cos\phi} \frac{\partial u}{\partial \lambda}))$ viscosity

A_h Eddy coefficient

Usually, the nonlinear convective terms in the equations of motion are neglected.

Instead of the complete Eqs. (3), frequently the Laplace tideal equations (LTE) are used. They are deduced from Eqs. (3) by linearization and neglection of any friction:

$$\frac{\partial u}{\partial t} - 2 \sin\phi \; v = - \frac{g}{R\cos\phi} \; \frac{\partial}{\partial \lambda} \; (\zeta - \gamma\Phi)$$

$$\frac{\partial v}{\partial t} + 2\omega\sin\phi \; u = - \frac{g}{R} \frac{\partial}{\partial \phi} \; (\zeta - \gamma\Phi)$$

$$\frac{\partial \zeta}{\partial t} + \frac{1}{R\cos\phi} \; (\frac{\partial}{\partial \lambda} \; (du) + \frac{\partial}{\partial \phi} \; (dv \cos\phi)) \; = \; 0 \tag{4}$$

While the set of Eqs. (3) describes the natural nonlinear behavior of the tidal motion, which may lead to deviations from the harmonic run especially in shallow water areas, Eqs. (4) admit only undamped purely sinusoidal solutions. This fact results in a characteristic simplification: The elimination of the time dependency by a harmonic approach for u, v, and ζ (see, e.g., Hendershott and Munk, 1970). This changes the type of the differential equations; thus the solution techniques for the HDE and the LTE are fundamentally different. In case (3) of the HDE, an initial-boundary-value-problem for a hyperbolic system of differential equations must be solved, whereas in case (4) of the LTE, a pure boundary-value-problem for an elliptical system is given.

The solutions we always obtain are the state quantities u, v, ζ, as functions of space (λ,ϕ), and time t. Now we can compute some of the terms appearing in the balance equations (1) and (2). Corresponding to the quadratic law for the bottom friction contained in Eqs. (3), the East-West component of the force acting on the rotating Earth by the tidal currents is given by

$$F_\lambda = r \; \rho \; Q \; |u^2 + v^2|^{1/2} \; u$$

where ρ is the density of the water and Q the area of a surface element of the world ocean.

The resulting torque with respect to the axis of rotation of the Earth is (method 2):

$$\dot{P}_z = r \; \rho \; Q \; R \; \cos\phi \; |u^2 + v^2|^{1/2} \; u \tag{5}$$

The frequently calculated quantity $\dot{E}_b$, the energy dissipation by bottom friction, is given by

$$\dot{E}_b = - r \; \rho \; Q \; |u^2 + v^2|^{3/2} \tag{6}$$

The torque $\dot{P}_z$, of course, depends on the geographic latitude and on the direction of the tidal current (in the direction of the Earth's rotation or opposite). This quantity has to be calculated for each surface element of the world ocean. Next, the several contributions must be summed up and, finally, by an integration over one tidal cycle a time-averaged mean value has to be computed. That value must be compared with the astronomically observed quantity. In contrast to that, $\dot{E}_b$ signifies only that part of tidal energy that is converted into heat and is not an equivalent description of the vectorial transfer of angular momentum.

Corresponding to method 1 described in Section 1, the torque acting on the Moon by the ocean tidal waves can be equally determined by means of the water elevations $\zeta(\lambda,\phi,t)$:

$$\dot{P}_z = - g\ c\ \rho\ Q\ R\ \cos^2 \phi\ \sin\ (\sigma t + 2\lambda)\zeta \tag{7}$$

For a consistent model of ocean tides, approaches (5) and (7), after integration over the whole world ocean and over one tidal cycle, must yield the same result. This requirement is by no means trivial. Whereas according to Eq. (5) the time-averaged value $<\dot{P}_z>$ must vanish for purely harmonic behavior of the tidal velocity, formula (7) results, in general, in this case as well, in a value different from zero due to the phase difference between the tide-generating potential and the oceanic tidal wave itself (see Pariiskii, 1972). That means that the application of the LTE leads to different values of $<\dot{P}_z>$ depending on whether Eq. (5) or (7) is used.

Only the application of the HDE causes, due to the nonlinearities, a (possibly slightly) nonharmonic run of the velocities, and hence, in the case of formula (5), also a nonvanishing mean torque.

It should be noted that Eq. (5) contains the empirical parameter r, which does not appear in the equivalent expression (7). The reason lies in the fact that the functional relation between u, v, and ζ is described by the HDE (3), which means an implicit relation between all three state quantities and r.

2.2 Analytic Considerations

The above-mentioned problem can be demonstrated for the simple case of an equatorial channel with constant depth. From Eqs. (3) then follows

$$\frac{\partial u}{\partial t} + \frac{u}{R}\frac{\partial u}{\partial \lambda} + \frac{r}{D}\ |u|u = - \frac{g}{R}\frac{\partial}{\partial \lambda}\ (\zeta - \gamma\Phi)$$

$$\frac{\partial \zeta}{\partial t} + \frac{1}{R}\frac{\partial}{\partial \lambda}\ (Du) \quad = 0$$

With the notations $\varepsilon = \frac{2u}{\sigma R}$, $p = const\ (\frac{2}{\sigma R})^2$ and $K = \frac{g\gamma}{2}\frac{\partial \Phi}{\partial \lambda}$ we obtain the ordinary differential equation

$$\frac{d\varepsilon}{dt} = \frac{K(t)\ -\frac{r}{p}(1+\varepsilon)\,|\varepsilon|\varepsilon}{\frac{\sigma R}{2}(1+\varepsilon\ -\frac{pg}{(1+\varepsilon)^2})}$$

This differential equation, in general, cannot be solved analytically. However, we can solve it numerically and obtain $\varepsilon(t)$ and $u(t)$ and, by means of the continuity equation $\zeta(t)$. Then we can calculate $<\dot{P}_z>$ according to Eqs. (5) and (7) and prove the consistency of both methods.

It becomes evident that both methods are indeed equivalent in the case of the HDE but that method 2 requires an essentially greater computational effort in order to obtain a stationary solution, particularly for deep-sea conditions. Hence, this method is not suited for practical purposes.

130

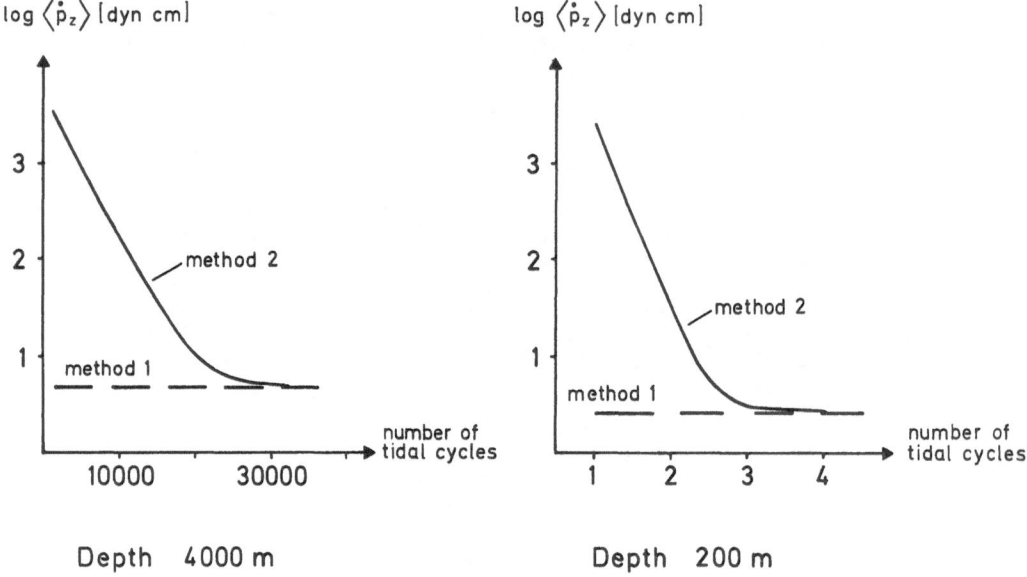

Fig. 1. The convergence of methods 1 and 2 for an equatorial channel of constant depth

2.3 Hydrodynamic-Numerical Models

Due to the complicated topography of natural seas, even the LTE cannot be solved analytically. This is true even more for the nonlinear HDE. Equations (3) and (4) are transformed, therefore, by means of several discretization techniques, to a system of difference equations to be solved algebraically. Accordingly, the natural ocean is covered by a grid net normally geographically oriented. The given geometry is approximated by representative values for each grid element, and the wanted state quantities are computed at each nodal point at certain time intervals. The whole set, containing the basic equations, initial and boundary conditions, discretization and approximation technique, is called the hydrodynamic-numerical (HN) model.

Since the first investigations made by Hansen (1952), numerous models of this type have been developed (see Hendershott and Munk, 1970). Essential features should be noted in the following.

a) Basic Equation

The LTE or the HDE are used. The fundamental differences have already been outlined.

b) Dissipation

The LTE (4) do not contain dissipative terms. A damping of tidal waves happens implicitly by prescribing the water elevations at the coast. Some authors have subsequently introduced linear friction laws into the LTE. There is still a basic difference in comparison to the HDE: Only in the HDE is the nonlinear nature of energy dissipation correctly represented. Hence, the shallow water dynamics can be adequately

Table 2. Relative importance of the nonlinear friction term related to the pressure gradient (r = 0.003)

d [m]	u [m/s]	$g \frac{\partial \zeta}{\partial \lambda}$ $[10^{-6} \text{ m/s}^2]$	$\frac{r}{d+\zeta} \|u\| u$ $[10^{-6} \text{ m/s}^2]$	Range for $\zeta = \pm 1$ m (%)
3000	0.01	10	0.0001	0.007
300	0.1	100	0.1	0.7
30	1	100	100	7

simulated only by using the HDE. The relative importance of the non-linear friction term can be seen from Table 2.

Due to the sensitivity of the momentum balance representing an equilibrium of large values with opposite signs, the deviation from harmonic behavior caused by the nonlinear friction is of central importance in applying method 2. In contrast, its influence on the integral quantity water elevation is small, which means that in the case of method 1, linear models might give useful results (see Pariiskii et al., 1972), although the basic balance Eq. (1) is violated by definition.

Besides the bottom friction, the inner friction by eddy viscosity also causes dissipation. In the numerical models of ocean tides this effect even forms the main energy sink. After Zahel (1977) in a $1°$-model of the world ocean $0.71 \cdot 10^{19}$ erg/s is obtained for $\dot{E}_b$, whereas $3.06 \cdot 10^{19}$ erg/s for $\dot{E}_i$ [see Eq. (2)].

c) Boundary Conditions

At the coastline, either observed water elevations are prescribed or a vanishing discharge is assumed: $v_n = 0$. If diffusion is considered, the order of the differential equation is raised by one and a further boundary condition is required.

d) Interaction with Solid Earth Tides

Since the work of Hendershott (1972) it is known that the oceanic tides are notably influenced by their interaction with solid Earth tides. This interaction is caused, on one hand, by the elastic deformation of the sea bottom, due to the direct action of tidal potential (leading in fact to a reduction of the effective tidal constant c). On the other hand, it originates from tidal loading and the self-attraction of water masses. The consideration of the latter (so-called dynamic interaction) requires a remarkable computational effort.

e) Grid Size

The degree of discretization depends on the available computer capacity. For the present models the grid size ranges from $5°$ to $1°$. The shelf areas that are most important for bottom friction are better resolved with increasing smallness of the grid distance. This is clearly

Table 3. Energy dissipation by bottom friction depending on the computational grid (after Zahel)

Grid size ($^{\circ}$)	$\dot{E}_b$ $[10^{19}\ erg/s]$
10	0.02
4	0.2
1	0.7

Table 4. Contribution of different tidal constituents to the global tidal torque on the Moon (after Pariiskii)

Tidal constituent	Resulting tidal torque $[10^{23}\ dyn\ cm]$	Per cent
M_2	− 5.03	88
S_2	− 0.68	12
K_1	− 0.10	2
O_1	+ 0.09	2

demonstrated in Table 3, which contains a list of values of $\dot{E}_b$ for the different models of Zahel (1970, 1977).

f) Tidal Constituents

The astronomically observed effect represents naturally the superposition of all partial constituents. Most of the HN models, however, consider only the dominating M_2-tide. The momentum balance of Pariiskii et al. (1972) using method 1 demonstrates the ratio of different partial tides (Table 4).

According to Pariiskii, only the two semidiurnal tides M_2 and S_2 are important for the momentum balance. It is interesting to note that the O_1-tide causes an acceleration of the rotating Earth.

Table 5. Some representative numerical tidal models

Author and year	Equations	Dissipation	Boundary conditions	Interaction with earth tides	Grid size ($^{\circ}$)	Tidal constituent
Bogdanow and Magarik (1967)	LTE	Coast	Coastal elevation	No	5	M_2, S_2 K_1, O_1
Pekeris and Accad (1969)	LTE	$v_n=0$ at d=1000 m	Linear bottom stress on shelves	No	1	M_2
Hendershott (1973)	LTE	Coast	Coastal elevation	Yielding, self-attraction		M_2
Kagan	LTE	$v_n=0$	Linear bottom stress, horiz. viscosity	Yielding, self-attraction	5	M_2
Zahel (1970) (1973) (1977)	HDE	$v_n=0$	Nonlin. bottom stress, horiz. viscosity	No Yielding Yielding, self-attraction	4 1 4	M_2, K_1 M_2, K_1 M_2, K_1

Table 5 gives a summary of some representative HN models of oceanic tides.

3. A Numerical Model for the Present Ocean

3.1 Requirements

According to the previous considerations, only a nonlinear HN model based on the HDE is consistent with respect to the concordance of methods 1 and 2. The only model satisfying these requirements at present is that of Zahel (1970, 1973, 1977, 1978). The following energy and momentum balances for the present world ocean are based, therefore, on the 4°-model of Zahel, initially without regard to an interaction with the tides of the solid Earth. It has been proved that this model sufficiently fulfills the fundamental conservation laws of mass, energy, and momentum.

Furthermore an extremely high degree of stationarity is required, i.e., the values of a state quantity q (λ,ϕ,t) at two subsequent identical tidal phases must differ from each other by a small number ε:

$$| \ q \ (\lambda,\phi,t + T) - q \ (\lambda,\phi,t) \ | < \varepsilon \qquad (8)$$

with T length of a tidal cycle. Deviations from this condition particularly restrict the application of the very sensitive method 2 as shown already by Figure 1. [Incidentally it is not necessary for the model to agree completely with nature, but it must satisfy the inequality (8)].

How to choose ε? In order to calculate $\dot{P}_z$ with an accuracy of two significant decimals, the difference $\Delta \dot{P}_z = \dot{P}_z$ (t + T) $- \dot{P}_z$ (t) must satisfy the condition $\Delta \dot{P}_z < 10^{21}$ dyn cm. The corresponding critical values of $\Delta \zeta$ or Δu are then found by an error analysis, separately for method 1 and 2, and depending on whether a systematic or a random error can be assumed (see Table 6).

Table 6. Maximum acceptable deviations of elevations and velocities for two subsequent tidal cycles

$\Delta \dot{P}_z < 10^{21}$ dyn cm	Method 1 $\Delta \zeta <$ (cm)	Method 2 $\Delta u <$ (cm/s)
Systematic error	0.01	0.00001
Random error	5	0.005

The accuracy required for method 1 is achieved for the 4°-model of Zahel after about 10 tidal periods (computer time: 5 min on a CDC 7600), whereas method 2 requires more than 100 tidal periods (about 1 h on a CDC 7600). For economic reasons therefore only balances on the basis of method 1 are presented.

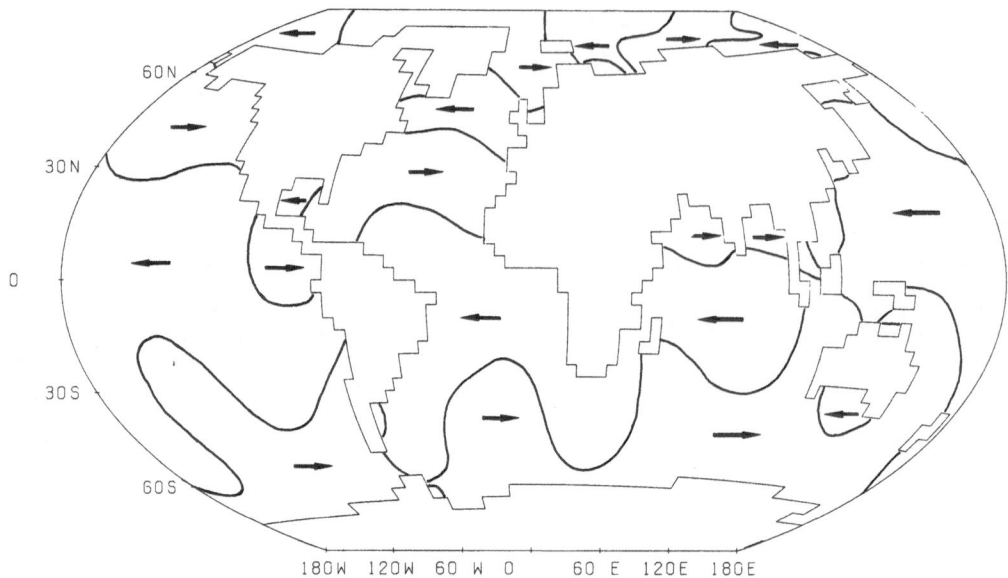

Fig. 2. Direction of tidal torques acting on the solid Earth, with respect to the axis of rotation (M_2-tide)

3.2 Balance of Angular Momentum

In the following, the results obtained under the above assumptions for a 4°-model of the world ocean and for the principal semidiuarnal lunar tide M_2 are presented and discussed. A detailed presentation of the tide itself (amplitudes, phases, stream velocities) is not the subject of this article; therefore, the reader is referred to Zahel (1970).

The balance according to Eq. (7) for a rigid Earth yields

$$- \left\langle \dot{P}_z \right\rangle = 9.6 \cdot 10^{23} \text{ dyn cm}$$

$$- \left\langle \dot{E}_{rot} \right\rangle = 7.0 \cdot 10^{19} \text{ erg/s}$$

If the elastic deformation of the sea bottom is taken into account by a reduction of the tidal constant by the factor $\gamma = 1 + k - h = 1 + 0.302 - 0.612 = 0.69$ (see Sect. 2.1), then the HN model yields

$$- \left\langle \dot{P}_z \right\rangle = 4.9 \cdot 10^{23} \text{ dyn cm}$$

$$- \left\langle \dot{E}_{rot} \right\rangle = 3.6 \cdot 10^{19} \text{ erg/s}$$

For the most recent model of Zahel (1978) (see the contribution in this volume), including the dynamic interaction of ocean and Earth tides, the corresponding numbers have not yet been computed. The consideration of this effect, as was shown by Zahel, in fact does not change the energy balance in the corotating system, but it influences the tidal phases and hence presumably also the angular momentum transfer from the Earth to the Moon.

While the last values refer to the M_2-tide alone, according to Pariiskii et al. (1972), about 12 % of this amount should be added as allowance for the S_2-tide. Then a value of $- \left\langle \dot{E}_{rot} \right\rangle = 4.0 \cdot 10^{19} \text{ erg/s}$ would be achieved, having the right sign and being close to the astronomic values of Morrison and Stephenson in this volume.

The geographic distribution of the time-averaged torques acting on the solid Earth with respect to its axis of rotation is demonstrated in Figure 2. It shows a very complicated picture: Areas of accelerating and decelerating action are distributed over the oceans. The resulting global torque is represented by the sum of these different contributions and contains finally a negative sign, which means a decelerating effect on the Earth's rotation. Since this net effect is obtained as a small difference of large numbers, a high degree of accuracy is required from the HN model.

By integrating the torque over latitude circles and by drawing these quantities as a function of geographic latitude, the full curve in Figure 3, upper picture, is obtained. It becomes evident that the decelerating torque is concentrated on a zonal belt between 10°N and 40°S. There is a zone, at about 30°N, with an accelerating torque, caused mainly by easterly directed forces in the Northern Indian Ocean. This fits qualitatively into a corresponding curve given by Pariiskii et al. (1972).

4. Numerical Model for Ancient Oceans

4.1 Necessity of Such a Model

The calculated transfer of angular momentum from the Earth to the Moon means an acceleration of the Moon in its orbit and hence a slow absentation from the Earth (as indeed presently observed). Assuming that the computed values are representative for all times, this implies very grave consequences for the history of the Earth-Moon system. Tracing back this process results in an approach of the Moon to the Earth that might indicate a critical distance of only a few Earth radii about a billion years ago (see Gerstenkorn, 1955; MacDonald, 1966). This would correspond to giant ocean tides, but there is no geologic evidence for such a dramatic situation.

The balances reported in Section 3.2, however, are only valid for the specific tidal dynamics of the present world ocean, that is, for the present length of the day and the present topographic conditions. Since these conditions are only representative of a very short period, from a geologic point of view, an extrapolation of the momentum balance into the past is very questionable. Hence a reconstruction of the history of the Earth-Moon system is only possible with knowledge of paleotides.

Due to the lack of hydrographic information, the oceanic tides of former eons can be studied presently only by means of HN models. Therefore the astronomic input quantities, such as the tide-generating potential and the angular velocity of the Earth's rotation, must be known, as well as the bathymetry of the ancient ocean. In this connection, determination of the astronomic quantities is, at least for the more recent history of the Earth, no serious problem.

4.2 Bathymetry

The modeling of ocean tides requires knowledge of the land-water distribution and of the bottom topography of the world ocean. For the more recent past of the Earth's history (until 300 m.y. ago), quite detailed maps of paleocontinents have been published within the last few years, e.g., by Dietz and Holden (1970) and Smith and Briden (1977). It is important that these presentations not only show the relative but also the absolute position of continents with respect to the geographic grid of the Earth. Within the 4°-resolution of Zahel's model, the coastline is then sufficiently determined.

The problem of a detailed paleobathymetric chart is much more difficult. A depth resolution within the meter range, as feasible for the present ocean, is certainly not obtainable for ancient times. Thus the question to the oceanographers is, to what extent can the geometry of the oceans be schematized in order to obtain balances which are still significant.

As minimum information from geology, it may be assumed that, besides the coastline, the extension of the shelf areas (up to a depth of about 1500 m) is known. Then at least a model with two representative depths can be established: One for the deep sea and one for the shelves.

In order to prove the applicability of such a two-depths model, first the tides of the present ocean are computed in this way. The representative depths for the deep and the shallow sea have been calculated by the following formula:

$$
\overline{d}_k = \frac{\sum\limits_{n,m} Q_n}{\sum\limits_{n,m} Q_n \frac{1}{d_{nm}}} \, , \qquad d_k < d_{nm} < d_{k+1} \qquad \qquad (9)
$$
$$
(k = 1, \ldots, K)
$$

Hence, the depth d_{nm} of a grid point (n,m) is substituted by a representative depth $\overline{d}_k$ of the depth interval (d_k, d_{k+1}). Formula (9) involves two weighting procedures: One according to the size of the surface element Q_n, and a second "dynamic" one reciprocal to the water depth. In that way the special importance of the shelf areas for the dynamics of ocean tides can be expressed. For $K = 2$, $d_1 = 0$ m, $d_2 = 1500$ m, $d_3 = 11,000$ m, values of $\overline{d}_1 = 241$ m (shelf seas) and $\overline{d}_2 = 3,746$ m (deep oceans) are obtained.

Taking these assumed depths, the tides of the present ocean are computed. The differences in the distribution of amplitudes and phases with respect to the ocean with natural depths turn out to be comparatively small. The influence on the angular momentum balance is most important. The following values are computed, using the model with the rigid Earth:

- $\langle \dot{P}_z \rangle = 6.2 \cdot 10^{23}$ dyn cm

- $\langle \dot{E}_{rot} \rangle = 4.5 \cdot 10^{19}$ erg/s

that means, about 65 % of the old value. The important point is that the two values are relatively close to each other. This is also confirmed by the zonal distribution of the torque, showing a run very similar to that in the case of the unschematized ocean (see the broken line in Figure 3, upper picture, p.138).

4.3 Application to the Permian Ocean

The first computation of paleotides was carried out for the Pangaea situation in the Upper Permian, 250 - 230 m.y. BP. This state was selected for two reasons: It seemed to be the oldest one with reliable paleocontinental maps, and the land-water configuration was extremely different from the present one.

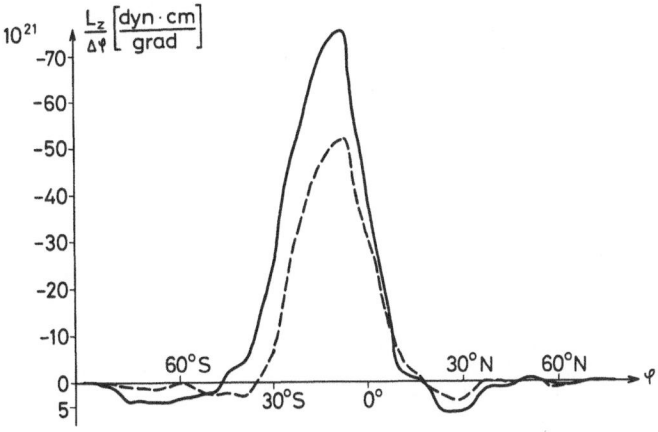

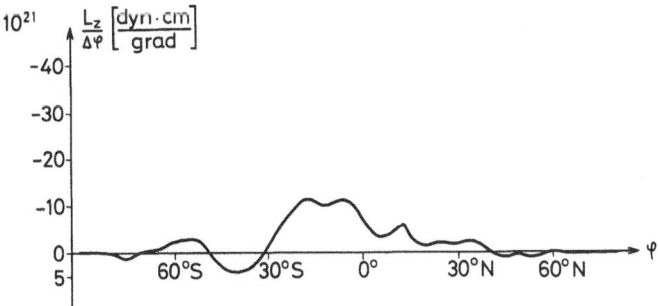

Fig. 3. The latitudinal distribution of tidal torque. *Upper picture:* The present ocean. *Full-line:* actual depths; *broken line:* two-depth model. *Lower picture:* The Pangaea Ocean, Lower Permian 280 - 270 m.y. BP

As basic material the maps of Dietz and Holden (1970) were chosen. An attempt was then made to reconstruct the true coastline by compilation of numerous geologic data. This contour was certainly different in the past since the plate boundaries have been deformed by tectonic processes. Furthermore, the approximate position of the 1,000-fathom isobath was deduced. The resulting geologic map is shown in Figure 4. A detailed description and discussion of this map is not the subject of the present paper; it will be given in another publication.

The transformation into a 4°-model of the Permian ocean with two depth levels is shown in Figure 5, p. 140.

Concerning the astronomic quantities in the Permian period, an enlargement of the angular velocity of the Earth's rotation $\Delta\omega$ by 4 % and of the tidal constant Δc by 0.16 % is assumed. This results in the input values given in Table 7, p. 140.

PALEOGEOGRAPHIC MAP OF PANGAEA IN UPPER PERMIAN
250-230 M.Y.B.P.

(CHRONOSTRATIGRAPHIC NAMES: ZECHSTEIN, THURINGIAN,
KAZANIAN - TATARIAN, GUADALUPIAN - OCHOAN,
PUNJABIAN - DZHULFIAN, BASLEOIAN - AMARASSIAN)

COMPILED BY V MAACK 1976

CONTINENTAL FIT MODIFIED AFTER R S DIETZ and J C HOLDEN 1970

LEGEND

Continental fill

Overlap areas

Land

Coast line (known, probably)

1000 - fm. isobath (probably)

Shallow sea (continental
shelf, miogeosyncline)

Deep sea (continental
rise, eugeosyncline)

① Kap Stosch
② Northern Branch
③ Western Branch
④ Boreholovka Sea
⑤ Pyrenean Basin
⑥ Lower Himalayan Basin
⑦ Carnarvon Basin
⑧ Canning Basin
⑨ Bonaparte Gulf Basin
⑩ New Guinea Basin
⑪ New Zealand, New Caledonia and
Papuan Geosyncline

Pangaea is shown on the Arbeit
projection of the globe with the
central meridian being 20°E.

PACIFIC
OCEAN

WESTERN GEOSYNCLINE

BOREAL GULF

SIBERIAN PLATFORM

ANGARA

Ural Mountains

Pre-Uralian Foredeep

RUSSIAN PLATFORM

FENNO-
SCANDIA

+30

GREENLAND

NORTH AMERICA

APPALACHIAN LANDMASS

Europe

German Basin

L A U R A S I A

+60

A N G A R A

ZECHSTEIN SEA

+60

TETHYS SEA

TETHYAN OCEAN

50

20

ARABIA

AFRICA

+30

G O N D W A N A

SOUTH AMERICA

+60

MALAGASY

INDIA

ANTARCTICA

AUSTRALIA

TASMANIA

NEW GUINEA

JAPAN

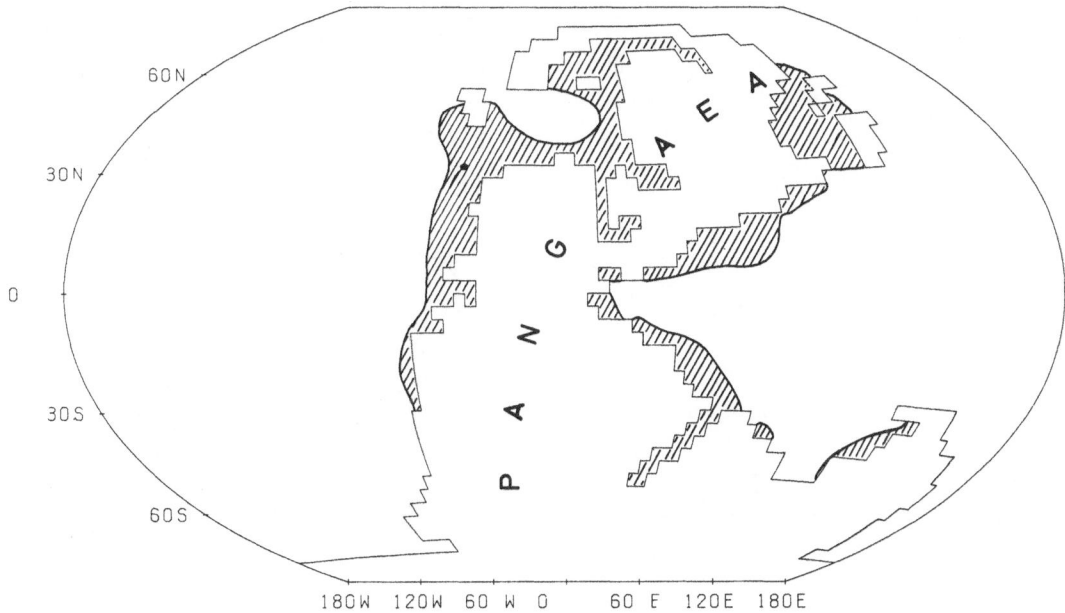

Fig: 5. 4°-model of the Upper Permian ocean. The *shadowed areas* are the shelf seas

Table 7. Astronomic input quantities for the HN models of the present and the Permian ocean

Ocean	Angular velocity ω $[1/s]$	Tidal constant c	Tidal period T $[s]$
Present	$0.729 \cdot 10^{-4}$	$0.745 \cdot 10^{-7}$	44,714
Permian	$0.758 \cdot 10^{-4}$	$0.746 \cdot 10^{-7}$	44,678

The amplitudes and phases of the M_2-tide in the Permian ocean are shown in Figures 6 and 7.

It turns out that amplitudes and phases show a very complicated geographic distribution, as they do in the present ocean. As a global phenomenon a system of four approximately symmetrically situated amphidromic points in the middle of the Permian ocean ("Pre-Pacific") may be noted. It is the origin of a broad tidal wave progressing in a westerly direction into the Tethys sea. An amplitude maximum of about 150 cm appears in the center of the ocean. It can be compared to a corresponding maximum of the present Pacific, but reaches values higher by about 50 cm. The concentration of phase isolines with numerous amphidromies on the shelves gives evidence of strong local deformation of tidal waves in the coastal zones. Summarizing, the range of amplitudes, however, does not show a significant change compared with the present situation.

◄ Fig. 4. Paleographic map of Pangaea in the Upper Permian, 250 - 230 m.y. BP (compiled by V. Maack)

140

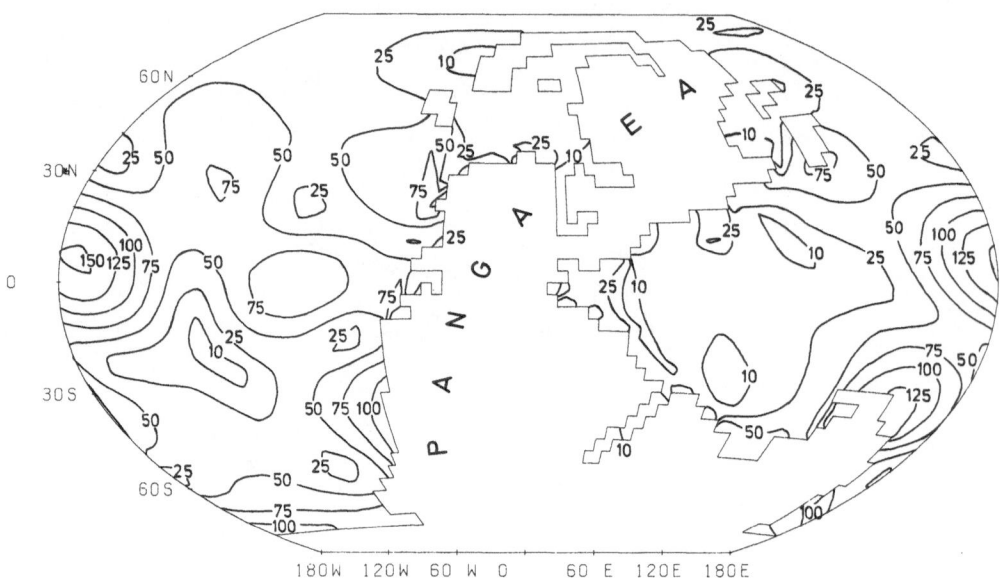

Fig. 6. Amplitudes of the M_2-tide in the Pangaea Ocean, Upper Permian, 250 - 230 m.y. BP (cm)

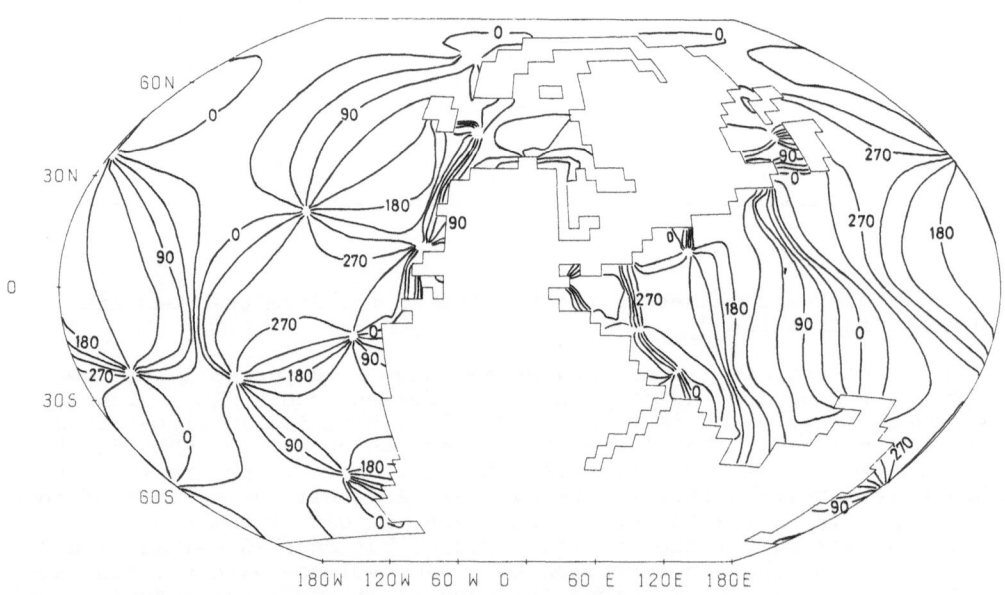

Fig. 7. Phases of the M_2-tide in the Pangaea Ocean, Upper Permian, 250 - 230 m.y. BP (degrees with respect to the Moon's transit at the 0° meridian)

The resulting momentum balance is of central importance for the present investigations. The following values for a rigid Earth are obtained for the Lower Permian Period, 250 - 270 m.y. BP (map in Brosche and Sündermann, 1977),

$$- \langle \dot{P}_z \rangle = 1.9 \cdot 10^{23} \text{ dyn cm}$$

$$- \langle \dot{E}_{rot} \rangle = 1.4 \cdot 10^{19} \text{ erg/s}$$

These numbers are remarkably smaller than those of the present ocean. This implies a significantly smaller transfer of angular momentum from the Earth to the Moon. This is equally obvious from the latitudinal distribution of the torque (see Fig. 3, lower picture).

From the model for the Upper Permian the corresponding values

$$- \langle \dot{P}_z \rangle = 4.2 \cdot 10^{23} \text{ dyn cm}$$

$$- \langle \dot{E}_{rot} \rangle = 3.0 \cdot 10^{19} \text{ erg/s}$$

were determined. A comparison with the results for the Lower Permian leads to the conclusion that moderate differences in the shape of the oceans and adjacent seas — here, e.g., the presence or absence of the Uralian Street — can have a significant bearing on the average torque.

4.4 Verification

Verification of the tides in the Permian ocean, as verification of paleotides in general, represents a serious problem. What data are available for evaluating the reliability of model results? The material is compiled in the fields of geology and paleontology, and is represented by the following quantities.

a) Length of Day

Paleontologic investigations of fossile corals, bivalves, and algae proved that the length of a day was shorter in the past (see Scrutton, 1978). From this material some information on former transfer rates of angular momentum can be deduced and, hence, indirect information on the global tidal friction. This integral quantity, which of course represents only the total tidal effect, not the special geographic distribution, seems to be at least a qualitatively useful indicator. The smaller deceleration of the Earth's rotation, as computed for the Permian period, is, in principle, in agreement with the conclusions of Scrutton (1978).

b) Partial Tides

It has been shown by Panella (1976) that even different tidal cycles representing the superposition of different partial tides can be identified from the growth rhythmus of bivalves. Thus, the type of a tide and also the relative amplitudes of different tidal constituents can be reconstructed. The fossils, however, originate generally from the coastal zone where the tidal phenomena are determined mainly by the local topographic situation. The global tide system calculated in the $4°$-model is by no means representative of the coastal zones. Thus, an evaluation of its quality by means of local paleontologic data is an improper way of verification. The grid net of the HN model should be considerably refined in the shelf area (down to 50 km grid distance), in order to use these data for comparison.

c) Tidal Range

In the geologic-paleontologic literature there are some estimations on paleotidal ranges in certain parts of the world ocean. The values going back to the Precambrian appear to be very similar to the present ones (see Klein, 1971). For comparison with the computations the same restrictions as for the partial tides are valid.

d) Stream Marks

Directions and velocities of tidal streams can be deduced from certain sedimentation patterns in the ocean that have been conserved throughout the past (Tarling, 1971). As long as this information is only of local importance, a direct comparison with computations seems to be questionable.

Summarizing, it can be stated that the possibilities of verification of paleotidal models by means of geologic-paleontologic data are very restricted at present.

However, because of the scientific weight of the problem, all possible ways should be considered. In any case, a grid resolution better than that within a 4° net is required.

5. Further Activities

The above considerations led to the conclusion that the model of the present ocean as well as that of ancient oceans must be improved and extended.

In order to resolve in a better way the important shelf areas, a model is under development that implies a successive refinement from a 4° basic grid up to 1/2° in the shallow areas. Furthermore, according to the findings of Zahel (1978), the dynamic interaction with the Earth tides will be included in all models.

As further typical situations, the tides of an Ice Age, of the Cretaceous ocean (Upper Cambrian, 75 - 70 m.y. BP) and, as soon as feasible, of a Precambrian period will be investigated. The results of Piper (1978) seem to render this latter model possible in the near future. A systematic study of a whole class of schematic land-water distributions and its tendential influences on the momentum balances would also be of interest.

Additionally, the completed HN models permit, without any big effort, the investigation of related geophysical problems such as the wind-induced circulation in the ocean, the magnitude of tidal forces on different continental plates, and the path of the Earth's barycenter within a tidal cycle. Equally, an expanding Earth could be modeled.

Acknowledgments. The numerical computations were carried out by J. Krohn. Geologic assistance was given by V. Maack who also compiled the map of the Permian ocean.

References

Bogdanov, K.T., Magarik, V.: Numerical solutions to the problem of distribution of semidiurnal tides M_2 and S_2 in the world ocean (transl.). Dokl. Adad. Nauk. SSSR 172, 1315-1317 (1967)

Brosche, P., Sündermann, J.: Die Gezeiten des Meeres und die Rotation der Erde. Pure Appl. Geophys. 86, 95-117 (1971)

Brosche, P., Sündermann, J.: On the torques due to tidal friction of the oceans and adjacent seas. In: Rotation of the Earth. Melchior, P., Yumi, S. (eds.). Dordrecht: D. Reidel, 1972, pp. 235-239

Brosche, P., Sündermann, J.: Effects of oceanic tides on the rotation of the Earth. In: Scientific Applications of Lunar Laser Ranging. Mulholland, J.D. (ed.). Dordrecht-Boston: D. Reidel, 1977, pp. 133-141

Dietz, R.S., Holden, J.C.: Reconstruction of Pangaea: Breakup and dispersion of continents, Permian to present. J. Geophys. Res. 75, 4939-4956 (1970)

Gerstenkorn, H.: Über Gezeitenreibung beim Zweikörperproblem. Z. Astrophys. 36, 245-274 (1955)

Hansen, W.: Gezeiten und Gezeitenströme der halbtägigen Hauptmondtide M_2 in der Nordsee. Dtsch. Hydrogr. Z., Ergänzungsheft 1, 1-46 (1952)

Hendershott, M.C.: The effects of solid Earth deformation on global ocean tides. Geophys. R. Astron. Soc. 29, 389-403 (1972)

Hendershott, M.C.: Ocean Tides, EOS, Trans. Am. Geophys. Union. 54, 76-86 (1973)

Hendershott, M.C., Munk, W.H.: Tides. Ann. Rev. Fluid Mech. 2, 205-224 (1970)

Jeffreys, H.: Tidal friction in shallow seas. Philos. Trans. R. Soc. London, Series A, 221, 239-264 (1920)

Kagan, B.A.: Global Interaction of Ocean and Earth Tides (in Russian). Leningrad: Gidrometeoizdat, 1977, p. 48

Klein, G.: A sedimentary model for determining paleotidal range. Geol. Soc. Am. Bull. 82, 539-546 (1971)

MacDonald, G.J.F.: Origin of the Moon: dynamical considerations. In: The Earth-Moon System. Marsden, B.G., Cameron, A.G.W. (eds.). New York: Plenum Press, 1966, pp. 165-209

Miller, G.R.: The flux of tidal energy out of the deep oceans. Journ. Geophys. Res. 71, 2485-2489 (1966)

Morrison, L.V.: Tidal deceleration of the Earth's rotation deduced from astronomical observations in the period A.D. 1600 to the present. In: Tidal Friction and the Earth's Rotation. Brosche, P., Sündermann, J. (eds.). Berlin-Heidelberg-New York: Springer, 1978, pp. 22-27

Muller, P.M., Stephenson, F.R.: The acceleration of the Earth and Moon from early astronomical observations. In: Growth Rhythms and History of the Earth's Rotation. Rosenberg, G.D., Runcorn, S.K. (eds.). London: J. Wiley, 1975, pp. 459-534

Munk, W.H., MacDonald, G.J.F.: The Rotation of the Earth. Cambridge: Cambridge University Press, 1960, p. 323

Panella, G.: Tidal Growth Patterns in recent and fossil mollusc bivalve shells: a tool for the reconstruction of paleotides. Naturwissenschaften 63, 539-543 (1976)

Pariiskii, N.N., Kuznetsov, M.V., Kusnetsova, L.V.: On the influence of ocean tides on the secular deceleration of the Earth's rotation (in Russian). Fiz. Zemli 2, 3-12 (1972)

Pekeris, C.L., Accad, Y.: Solution of Laplace's equation for the M_2-tide in the world oceans. Philos. Trans. R. Soc. Lond., Series A, 265, 413-436 (1969)

Piper, J.D.A.: Geological and geophysical evidence relating to continental growth and dynamics and the hydrosphere in Precambrian Times: a review and analysis. In: Tidal Friction and the Earth's Rotation. Brosche, P., Sündermann, J. (eds.). Berlin-Heidelberg-New York: Springer, 1978, pp. 197-241

Rochester, M.G.: The Earth's rotation. EOS, Trans. Am. Geophys. Union 54, 769-780 (1973)

Schott, F.: On the energetics of baroclinic tides in the North Atlantic. Ann. Geophys. 33, 41-62 (1977)

Scrutton, C.T.: Periodic growth features in fossil organisms and the length of the day and month. In: Tidal Friction and the Earth's Rotation. Brosche, P., Sünder-mann, J. (eds.). Berlin-Heidelberg-New York: Springer, 1978, pp. 154-196

Smith, A.G., Briden, J.C.: Mesozoic and Cenozoic Paleo-Continental Maps. Cambridge: Cambridge University Press, 1977, p. 64

Stephenson, F.R.: Pre-telescopic astronomical observations. In: Tidal Friction and the Earth's Rotation. Brosche, P., Sündermann, J. (eds.). Berlin-Heidelberg-New York: Springer, 1978, pp. 5-21

Tarling, D.H.: Principles and Applications of Paleo-Magnetism. London: Chapman and Hall, 1971, p. 164

Wunsch, C.: Internal tides in the ocean. Rev. Geophys. Space Phys. $\underline{13}$, 167-182 (1975)

Zahel, W.: Die Reproduktion gezeitenbedingter Bewegungsvorgänge im Weltozean mittels des hydrodynamisch-numerischen Verfahrens. Mitt. Inst. Meereskd. Univ. $\underline{17}$, 1-50 (1970)

Zahel, W.: The diurnal K_1-tide in the world ocean — a numerical investigation. Pure Appl. Geophys. $\underline{109}$, 1819-1825 (1973)

Zahel, W.: A global hydrodynamical-numerical 1°-model of the ocean tides; the oscillation system of the M_2-tide and its distribution of energy dissipation. Ann. Geophys. $\underline{33}$, 31-40 (1977)

Zahel, W.: The Influence of Solid Earth Deformations on Semi-Diurnal and Diurnal Oceanic Tides. In: Tidal Friction and the Earth's Rotation. Brosche, P., Sünder-mann, J. (eds.). Berlin-Heidelberg-New York: Springer, 1978, pp. 98-124

The Earth's Palaeorotation

K. Lambeck

Scrutton (this Volume) and Lambeck (1978a) have independently reviewed the bivalve, coral, and stromatolite evidence for changes in the length of day and length of month. It is thought that the growth rhythms of these organic structures reflect three astronomical cycles:

1. The number N_1 of solar days per year

2. The number N_2 of solar days per synodic month, and

3. The number N_3 of synodic months per year.

These three quantities relate to the mean motions of the Sun ($n_\odot$) and Moon ($n_\mathcal{C}$) and to the rotation of the Earth (Ω) by

$$N_1 = \frac{\Omega}{n_\odot} - 1 ,$$

$$N_2 = \frac{\Omega - n_\odot}{n_\mathcal{C} - n_\odot} ,$$

$$N_3 = [n_\mathcal{C} - n_\odot]/n_\odot .$$

Only Ω and $n_\mathcal{C}$ are thought to have varied by significant amounts during the last 10^9 years so that the rates of change of N_1, N_2, and N_3 provide estimates of $\dot{\Omega}$ and $\dot{n}_\mathcal{C}$. That is

$$\frac{dN_1}{dt} = \frac{\dot{\Omega}}{n_\odot}$$

$$\frac{dN_2}{dt} = \frac{\dot{\Omega}}{n_\mathcal{C} - n_\odot} - \frac{\Omega - n_\odot}{(n_\mathcal{C} - n_\odot)^2} \cdot \dot{n}_\mathcal{C}$$

$$\frac{dN_3}{dt} = \frac{\dot{n}_\mathcal{C}(t)}{n_\odot}$$

Figure 1 illustrates the change in N_i with time on the assumption that tidal friction is the only phenomenon responsible for $\dot{\Omega}$ and $\dot{n}_\mathcal{C}$. The solar contribution has been assumed to remain constant except insofar as the Earth's polar moment of inertia depends on Ω. Dissipation mechanisms have been assumed to be constant. At present there is no reasonable alternative to these assumptions and we consider this as a hypothesis against which the palaeontological results have to be tested. The integrations have been carried out by assuming that the equivalent lag angle of the semi-diurnal tide is about twice that of the diurnal tide, as is suggested by the present-day situation. The present value for the semi-diurnal tide's equivalent lag is about $6°$ (Lambeck, 1978a).

Research School of Earth Sciences, Australian National University

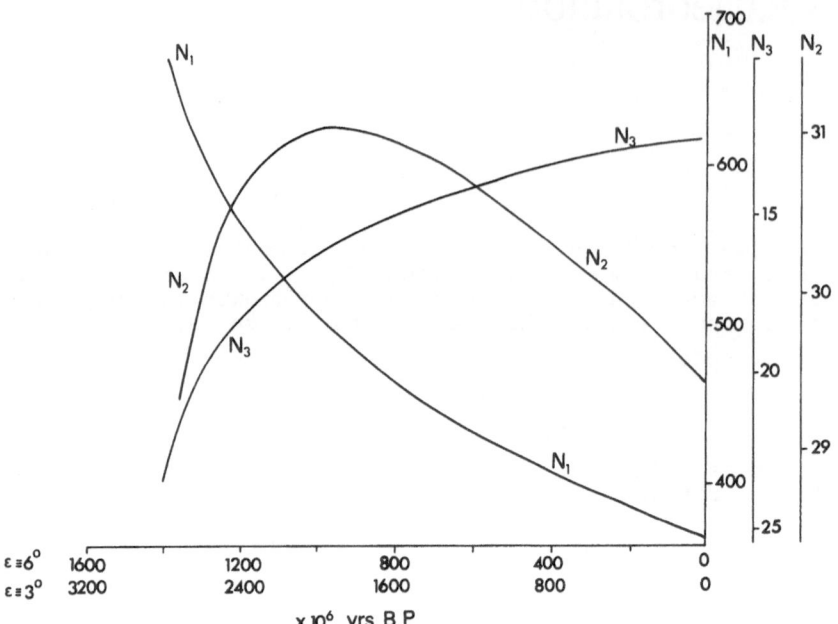

Fig. 1. Number of solar days per year (N_1), number of solar days per synodic month (N_2), number of synodic months per year (N_3), as a function of years before the present (B.P.). The two time scales correspond to two equivalent phase lags, $6°$ equal to the present value and $3°$

If the interpretation of the palaeontological data is valid and the above assumptions are reasonable, the observed N_i should fall somewhere on these curves and determine the average equivalent phase lag and the accelerations.

The data used in this study are summarized in Tables 1 to 5. The reasons for the choice and for the adopted standard deviations are discussed in Lambeck (1978a). In solving for the accelerations, the simplest approach is to assume that $\Omega(t)$ and $n_{\mathbb{C}}(t)$ vary linearly with time, that the accelerations are constant since over the last $400 \cdot 10^6$ years the Earth-Moon distance was only a few percent less than it is now. The data also suggest that there may be significant systematic errors in the counts, and we should write

$$\Omega(t) = \Omega(t_o) + \Delta\Omega + \dot{\Omega}t$$

$$n_{\mathbb{C}}(t) = n_{\mathbb{C}}(t_o) + \Delta n_{\mathbb{C}} + \dot{n}_{\mathbb{C}}t . \qquad (1)$$

Then the condition equation are

Table 1. Summary of the number of growth increments per seasonal annulation in fossil corals, based on Wells (1970) and Scrutton (1970)

Epochs	Geologic time M.Y.	No. of counts	No. of samples	Range of counts	Adopted standard deviation
Wells (1970)					
Upper Carboniferous	-300	385	2	380 - 390	4.2
Lower Carboniferous	-320	398	1		6
Middle Devonian	-370	398	12	385 - 405	1.7
Middle Silurian	-420	400	1		6
Upper Ordovician	-440	412	1		6
Modern	0	360	Several	?	4
Scrutton (1970)					
Middle Devonian	-370	401	1		

Table 2. Summary of the number of growth increments per lesser annulation in fossil corals

Epoch	Geologic time M.Y.	No. of counts	No. of samples	Range of counts	Adopted standard deviation
Scrutton (1970)					
Middle Devonian	-370	30.66	10	27 - 35	0.5
Johnson and Nudds (1975)					
Lower Carboniferous	-330	30.2	at least 6	?	0.4

$$\frac{1}{n_\odot} \begin{pmatrix} 1 & 0 & t & 0 \\ 1 & -\beta & t & -\beta t \\ 0 & 1 & 0 & t \end{pmatrix} \begin{pmatrix} \Delta\Omega \\ \Delta n_{(\!(} \\ \dot{\Omega} \\ \dot{n}_{(\!(} \end{pmatrix} = \begin{pmatrix} N_1 + 1 - \Omega(t_o)/n_\odot \\ \dfrac{\Omega(t_o) - n_\odot}{n_\odot} \quad (N_2+1) - \dfrac{\Omega(t_o)}{n_\odot} \\ N_3 + 1 - n_{(\!(}(t_o)n_\odot \end{pmatrix} \quad (2)$$

with

$$\beta = [\Omega(t_o) - n_\odot]/[n_{(\!(}(t_o) - n_\odot] .$$

The observations are weighted according to $(1/\text{adopted standard deviation})^2$.

1. The Coral Data

Only estimates of N_1 and N_2 are available (Tables 1 and 2). Also, a meaningful separation of the four unknowns in Eqs. (1) is not possible and we impose the condition that, for the present epoch t_o,

$$N_2(t_o) = 29.5 \pm 0.05 . \tag{3}$$

Then

$$\Omega(t) = (366.2 - 5.13 - 0.100t) n_\odot$$
$$\pm 3.93 \pm 0.012$$
$$n_\mathbb{C}(t) = (13.35 - 0.16 - 0.0024t) n_\odot$$
$$\pm 0.23 \pm 0.0005 \tag{4}$$

where t is in units of 10^6 years and negative when referred to the past. The accelerations are

$$\dot{\Omega} = (-6.3 - 0.7) 10^{-22} \text{ rad s}^{-2}$$
$$\dot{n}_\mathbb{C} = (-1.5 - 0.4) 10^{-23} \text{ rad s}^{-2} \tag{5}$$

and

$$\dot{\Omega}/\dot{n}_\mathbb{C} \simeq 42.0 \pm 11.1 .$$

2. The Bivalve Data

For three epochs, N_1, N_2, and N_3 are available while for other epochs only one or two of the N_i are available. Solving Eqs. (2) for all of Pannella's bivalve data (Tables 3, 4 and 5, pp. 150/151), without the constraint (3), yields

$$\Omega(t) = (366.2 - 4.8 - 0.095t) n_\odot$$
$$\pm 1.8 \pm 0.009$$
$$n_\mathbb{C}(t) = (13.35 - 0.04 - 0.021t) n_\odot$$
$$0.09 \pm 0.0005 \quad . \tag{6}$$

Thus

$$\dot{\Omega} = -(5.9 \pm 0.6) 10^{-22} \text{ rad s}^{-2}$$
$$\dot{n}_\mathbb{C} = -(1.3 \pm 0.3) 10^{-23} \text{ rad s}^{-3} \tag{7}$$

and

$$\frac{\dot{\Omega}}{\dot{n}_\mathbb{C}} \simeq (46 \pm 8) \tag{8}$$

These accelerations agree well with both the coral data and with the modern results. This can be taken as a measure of the reliability of the palaeontological data and of its interpretation, or it may be that both sets of results are biased towards a priori "expected" results. I object to the removal of the sentence in the script. However, until more data become available we will consider the results as significant, yet treat them with some caution. Special caution in interpreting the bivalve data is necessary in that the growth rhythms are now known to follow more complex patterns than initially appreciated (see Pannella, 1975; Scrutton, this Volume).

Table 3. Summary of number of growth increments per seasonal annulation in fossil bivalves. Data are from Pannella (1972)

Period	Geologic time 10^6 years	Increments per annulation	No. of specimens	No. of annulation	Range of counts	Adopted standard deviation
Recent	0	359	9	32	353 – 366	2.1
U. Cretaceous	-70	375	3	16	371 – 379	3.5
M. Triassic	-220	372	3	7	365 – 375	3.5
U. Carboniferous	-290	383	3	11	380 – 389	3.5
L. Carboniferous	-340	398	2	9	397 – 399	4.2
M. Devonian	-360	406	1	6		6

Table 4. Summary of the number of growth increments per lesser annulation in fossil bivalves according to Pannella (1972) and Berry and Barker (1975). The former are believed to be estimates of N_2, the latter of $N_2/2$

Period	Geologic time 10^6 years	Increments per lesser annulation	No. of specimens	No. of lesser annulation	Range of counts	Adopted standard deviation
Pannella (1972)						
Recent	0	29.2	7	186	29.0-29.6	0.4
U. Tertiary	-14	29.4	5	197	29.2-29.8	0.5
M. Tertiary	-38	29.8	2	40	29.6-29.9	0.8
L. Tertiary	-54	29.6	3	141	29.4-30.0	0.6
U. Cretaceous	-70	29.9	7	159	29.6-30.2	0.4
M. Triassic	-220	29.7	3	77	29.4-30.0	0.6
U. Carboniferous	-300	30.2	4	59	29.9-30.7	0.6
U. Devonian	-350	30.4	4	168	30.2-30.5	0.6
U. Ordovician	-445	30.3	2	84	29.8-30.7	0.8
Berry and Barker (1975)						
Pleistocene	1	14.75	19	115	14.2-15.0	0.14
Pliocene	4	14.83	9	40	14.3-15.0	0.16
Oligocene	30	14.82	10	49	14.7-15.0	0.09
Eocene	48	14.87	3	14	14.6-15.0	0.27
Palaeocene	60	14.82	6	20	14.7-15.0	0.17
Cretaceous	100	14.88	17	1189	14.7-15.0	0.12
Jurassic	160	14.90	10	12	14.7-15.3	0.18
Triassic	230	14.91	10	17	14.8-15.0	0.07
Carboniferous	310	15.09	17	29	14.7-15.5	0.17
Devonian	370	15.25	1	8	---	0.60

Table 5. Summary of the number of lesser annulations per seasonal annulation in fossil bivalves. From Pannella (1972)

Period	Geologic time 10 years	Lesser annulations per annulation	No. of specimens	Range of counts	Adopted standard deviation
Recent	0	12.3	7	12.0-12.7	0.17
Cretaceous	-70	12.6	3	12.5-12.8	0.26
M. Triassic	-220	12.6	3	12.5-12.7	0.26

Berry and Barker (1975) found that, in their specimens, there is a pronounced clustering of the growth lines in groups of about fifteen, and they suggest that this provides a measure of $N_2/2$. Their data are insufficient to separate $\dot{\Omega}$ and $\dot{n}_{\mathbb{C}}$ but the rate of change of N_2 with time appears reasonably well established. These data indicate marginally smaller accelerations than do Pannella's data.

3. Stromatolite Data

The bivalve and coral fossils provide estimates of the paleorotation for only some 10 % of the age of the Earth since the fossil records do not extend back much beyond the Ordovician. Prior to the Paleozoic, the only potential source of information on past astronomical cycles lies in the stromatolite formations that extent back to the Cryptozoic. Pannella (1972) investigated the growth patterns in fossil stromatolites back to the Archean, but only small digitate forms, from the Biwabik-Gunflint formation dated at about $2 \cdot 10^9$ years provide what may be significant results. Pannella's conclusions are at variance with those drawn by Mohr (1975), also based on the Biwabik formation. Pannella suggests that there may have been 39 days per synodic month $2 \cdot 10^9$ years ago but, as seen from Figure 1, it is unlikely that such a high value could ever have been attained. Mohr's results suggest 26 days per synodic month, which would fall on the rapidly descending part of the N_2 curve. This is equally unsatisfactory. Pannella's (1976) more recent discussion of the stromatolite evidence is much more cautious. Much further work will be required before we can have much confidence in the stromatolite values.

4. Combined Data

The nature of the systematic errors in the N_i are such that $\Delta\Omega$ and $\Delta n_{\mathbb{C}}$ are likely to differ for the two data sets [cf. solutions (4) and (6)]. Thus in a combined bivalve-coral solution, we must solve for the two sets of biases. Constraint (3) is not imposed. The accelerations are

$$\dot{\Omega} = -(5.28 - 0.23)10^{-22} \text{ s}^{-2}$$

$$\dot{n}_{\mathbb{C}} = -(1.23 - 0.16)10^{-23} \text{ s}^{-2}$$

(9)

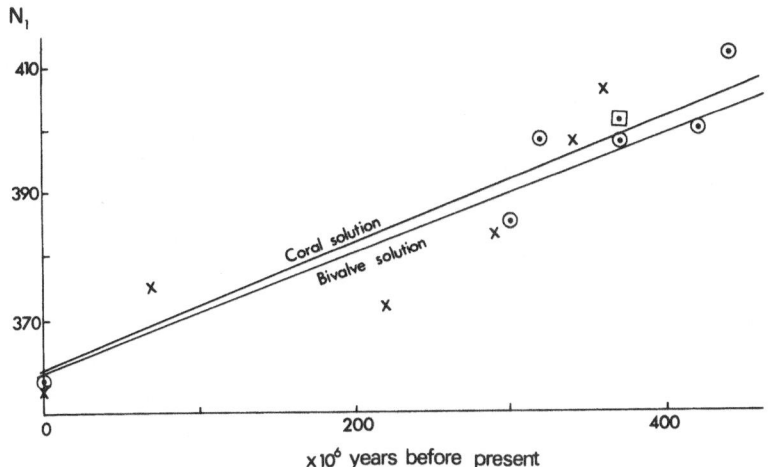

Fig. 2. Estimates of N_1 according to Wells (⊙), Scrutton (□), and Pannella (x). One line represents the coral solution (4), the other the bivalve solution (5)

with

$$\dot{\Omega}/n_{\mathbb{C}} \simeq 43 \pm 5 . \tag{10}$$

The introduction of Berry and Barker's results (Table 4) give slightly smaller accelerations. The recent telescope and eclipse observations indicate the following relation between the tidal acceleration of the Earth $\dot{\Omega}_T$ and $\dot{n}$ of

$$\dot{\Omega}_T - 51.4\dot{n}_{\mathbb{C}} = 0 \pm 5.1\dot{n}_{\mathbb{C}} \tag{11}$$

(Lambeck, 1978a). Comparing this with the observed ratio (10) indicates that there have been no significant non-tidal accelerations during at least the previous $400 \cdot 10^6$ years. Thus we can also solve for the accelerations, using both coral and bivalve data, by imposing the condition (11). Now

$$\dot{\Omega} = -(5.2 \pm 0.5)10^{-22} \; s^{-2}$$

$$\dot{n}_{\mathbb{C}} = -(1.0 \pm 0.1)10^{-23} \; s^{-2} \tag{12}$$

The precision estimates are only as good as the validity of the underlying assumptions made about the growth rhythms. Yet the values are in surprisingly good agreement with the modern astronomical values (Lambeck, 1978a)

$$\dot{\Omega} = -(5.5 \pm 0.5)10^{-22} \; s^{-2}$$

$$\dot{n} = -(1.35 \pm 0.10)10^{-23} \; s^{-2} .$$

Figures 2 and 3 compare the observed N_i with those computed from the palaeo-accelerations (6) and (9). The residuals show no obvious systematic trends, indicating that the above simple analysis is probably

152

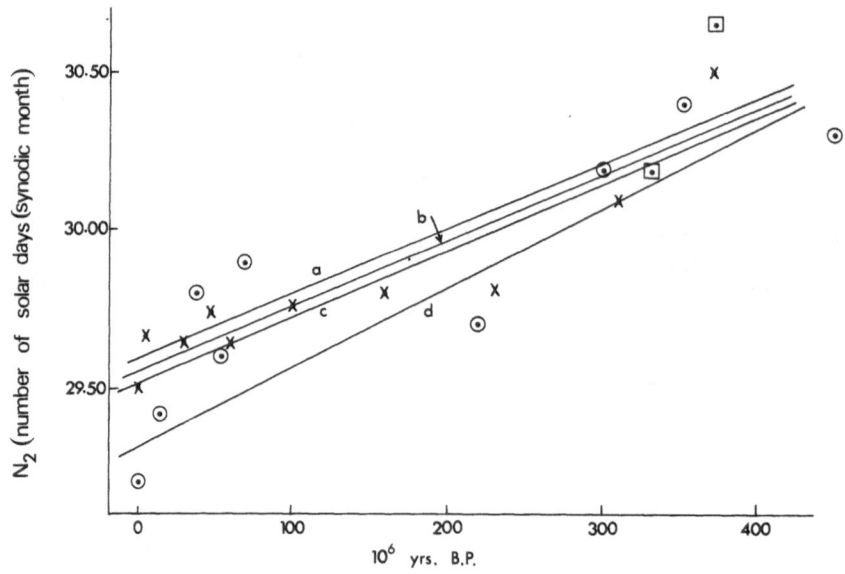

<u>Fig. 3.</u> Estimates of N_2 according to Pannella (⊙), Berry and Barker (x) and the coral data (▣). The four lines represent $N_2(t)$ based on (a) solution (4) of coral data, (b) an unweighted best fit to Berry and Barker's observed data, (c) an unweighted best fit to Pannella's observed N_2 data, and (d) solution (5) from all of Pannella's bivalve data

adequate. The geological implications and consequences of those accelerations on the Earth's rotation and lunar orbit are discussed in Lambeck (1978a,b).

References

Berry, W.B.N., Barker, R.M.: Growth increments in fossil and modern bivalves. In: Growth Rhythms and the History of the Earth's Rotation. Rosenberg, G.D., Runcorn, S.K. (eds.). London: John Wiley, 1975, pp. 9-24

Johnson, G.A.L., Nudds, J.R.: Carboniferous coral geochronometers. In: Growth Rhythms and the History of the Earth's Rotation. Rosenberg,G.D., Runcorn, S.K. (eds.). London: John Wiley, 1975, pp. 27-41

Lambeck, K.: The Earth's variable rotation: Geophysical Causes and Consequences. Cambridge University Press, in Press, 1978a

Lambeck, K.: The history of the Earth's rotation. In: The Earth: Its Origin, Structure and Evolution. McElhinny, M.W. (ed.). London: Academic Press, 1978b

Mohr, R.E.: Measured periodicities of the Biwabik stromatolites and their geophysical significance. In: Growth Rhythms and the History of the Earth's Rotation. Rosenberg, G.D., Runcorn, S.K. (eds.). London: John Wiley, 1975, pp. 43-55

Pannella, G.: Paleontological evidence on the Earth's rotational history since Early Precambrian. Astrophys. Space Sci. <u>16</u>, 212-237 (1972)

Pannella, G.: Paleontological clocks and the history of the Earth's rotation. In: Growth Rhythms and the History of the Earth's Rotation. Rosenberg, G.D.,Runcorn, S.K. (eds.). London: John Wiley, 1975, pp. 253-284

Pannella, G.: Geophysical inferences from stromatolite lamination. In: Stromatolites. Walter, M.R. (ed.). Amsterdam: Elsevier, 1976, pp. 673-685

Scrutton, C.T.: Evidence for a monthly periodicity in the growth of some corals. In: Palaeogeophysics. Runcorn, S.K. (ed.). London: Academic Press, 1970, pp. 11-16

Wells, J.W.: Problems of annual and daily growth rings in corals. In: Palaeogeophysics. Runcorn, S.K. (ed.). London: Academic Press, 1970, pp. 3-9

Periodic Growth Features in Fossil Organisms and the Length of the Day and Month

C. T. Scrutton

1. Introduction

It has long been known that animals and plants reflect environmental rhythms in their tissues. Periodic growth features, such as the annual rings of trees, sheep's horns, and sea shells, have been used both for calculating the ages of organisms and their growth rates.

The present phase of interest in this field follows from the realisation that in some cases the influence of more than one environmental rhythm can be recognised in the same skeletal structure. Although Wells (1937) referred to such a combination of growth features in the Devonian coral *Heliophyllum halli* in passing, it was not until his now famous contribution of 1963 that the potential of the interrelationships thus made available was more fully and directly exploited. Well's original idea was that the numerical relationship between circadian and annual growth features in fossil corals could be used as an independent check of absolute age based on radioactive isotopes. These latter were plotted against days per year calculated on the basis of estimates of the rate of dissipation of the Earth's rotational energy by tidal friction. In fact, of course, the subsequent history of investigation has very much switched the emphasis. Radiometric estimates of geological age are accepted with little comment and the palaeontological data are seen as an important factor in assessing the history of the Earth's rotation.

Since Wells (1963) first provided data on days per year in the geological past from corals, a range of other figures for days per year, days per lunar month, and lunar months per year derived from several groups of organisms have been published. Additional data derived from corals have come mainly from Wells (1970), Scrutton (1965, 1970), Mazzullo (1971), and Johnson and Nudds (1975). Data from bivalves have been published by Pannella and MacClintock (1968), Pannella et al. (1968), Pannella (1972b), and Berry and Barker (1968, 1975), from brachiopods by Mazzullo (1971), from cephalopods by Pannella and MacClintock (1968), and from stromatolites by McGugan (1967), Pannella et al. (1968), Pannella (1972a,b, 1976a), and Mohr (1975). In addition a number of reviews of the growth increment work and its geophysical implications have been published (Scrutton and Hipkin, 1973; Clark, 1974b; Pannella, 1975) as well as a book combining much useful biological, and astronomical research in this field (Rosenberg and Runcorn, 1975).

The great interest generated in the palaeontological data through their potential to contribute to an understanding of the Earth-Moon system has stimulated biological research into the genesis of growth increment records. The quality of these data needs careful critical assessment. It is therefore of fundamental importance that the way in which

Department of Geology, University of Newcastle upon Tyne

organisms react to environmental stimuli is understood. Such work must
be done first on living representatives of the groups providing the
data and then the reliability with which these results can be extrapolated to distant relatives in the geological past must be assessed.
It is already clear, for example, that the complexity of the bivalve
growth record was not appreciated when the currently available data
were accumulated.

This review begins, therefore, by summarising current knowledge of how
growth increments form in various organisms, how well established is
their temporal significance, and how faithfully they record elapsed
time. It will be seen that there is still a great deal to be accomplished in this area. The methods by which the data are recorded and
their statistical treatment are discussed, followed by a review of the
data published so far, which is reproduced here in tabular and graphical form. The data are of varying degrees of reliability and an attempt
has been made to assess this qualitatively. Finally the implications
of the data for the history of the Earth's rotation rate and the Earth-
Moon system are summarised.

2. Biological Considerations

Theoretically, any organism with preservable hard parts with a continually additive mode of growth is potentially a source of information on environmental rhythms. In practice, different organisms respond with different degrees of sensitivity to these stimuli depending
on their habitat, rate of growth, and detailed skeletal structure among
other factors. The groups of organisms that have proved of interest so
far are all marine, or at least subaqueous, invertebrates, or algal
structures. Among invertebrates, corals (Phylum Cnidaria), bivalves,
cephalopods (Phylum Mollusca), and brachiopods (Phylum Brachiopoda)
have provided data, with the bulk coming from the first two groups.
Cephalopods, including particularly ammonite aptychi and belemnites
(Pannella, 1972b, p. 226), may prove to be more useful with further
study but the two latter have not yielded data so far. No other invertebrate groups at the moment show signs of proving useful although
outside possibilities include stromatoporoids, some bryozoa, calcareous
worm tubes, and, over short time periods, echinoderm plates and arthro-
pod exuviae. The other major field of interest is in stromatolites,
laminated pads and columns of sediment formed by the activities of
algae and bacteria. Because of their long geological history, extending
back as far as 3000 m.y. compared with the maximum range of only some
550 m.y. of the other groups, there is a very great incentive to ex-
tract data from stromatolites if at all possible. Finally, vertebrates
have not yet shown great potential to contribute to this field. Some.
vertebrate structures, such as fish otoliths (Pannella, 1971), show
clearly developed growth periodicities but they have yet to contribute
data from the geological record.

In this review, discussion will concentrate on the three principle
sources of data so far: the corals, bivalves, and stromatolites.

2.1 Growth Increments in Corals

The coral skeleton (see Fig. 2a) consists of a framework of vertical
(septa) and horizontal (dissepiments and tabulae) plates, the whole
surrounded in some cases by an external wall, the epitheca of solitary

Fig. 1a-c. Growth increments in rugose corals. (a) Annulation and bands; scale 1 cm; (b) bands; scale 1 cm; (c) enlargement of part of (b) to show growth ridges in four complete bands; scale 2 mm. Both specimens *Heliophyllum halli*: (a) Middle Devonian, Hamilton Group; New York, U.S.A.: (b) and (c) Middle Devonian, Traverse Group, Michigan, U.S.A.

corals and individual corallites in a colony, or the holotheca of the colony as a whole. So far data have been derived only from growth features on the surface of the epitheca (Fig. 1). Whereas an epitheca is a more or less universal feature of Palaeozoic corals (the Tabulata and Rugosa) may genera of Mesozoic to Recent corals (the Scleractinia) do not develop this structure. Efforts to utilise growth periodicities in septa have not been particularly successful so far. Although incremental growth can be detected (Jell and Hill, 1974; Sorauf and Jell, 1977), its significance, susceptibility to diagenetic change, and potential for yielding data of astronomical interest have yet to be established.

An annual periodicity in coral growth has long been proposed (Whitfield, 1898; Ma, 1933, 1934a,b). This was linked to seasonal variation in water temperature from the beginning, and Ma (1934a,b) assembled considerable circumstantial evidence in support of this link. The cycle consists of alternations in density of internal tissues, which, according to Ma, are well marked in corals growing further from the equator where there is a distinct seasonality in water temperature. Closer to the equator, the density differences become less distinct and eventually disappear (Ma, 1934a). Recent work, however, has suggested a more complex and confusing picture of density banding in hermatypic corals. Some observations support a close link between growth rates and water temperature (Buddemeier and Kinzie, 1975; Weber et al., 1975) but Stearn et al. (1977) report that linear growth rates of Barbadian corals have been shown to vary little with the seasons, and Dodge and Vaisnys (1975)

demonstrate a correlation of above-average annual band width and be-
low-average maximum air temperature (as an indicator of water temperature) in coral growth over a 60-year period in Bermuda. Furthermore,
Buddemeier et al. (1974) and Buddemeier and Kinzie (1975) note that
density banding is still present where temperature shows little variation over the year. They regard light levels as important in these
cases, the denser units of growth corresponding to lowered light levels. Such a correlation was denied by Weber et al. (1975), who regard
temperature as the entraining factor in density banding, but supported
by Stearn et al. (1977), who criticised some of Weber et al.'s (1975)
data as too generalised, and suggested more indirectly by Dodge and
Vaisnys (1975). Finally the time of the year at which the denser component of the annual band forms has been variably reported as in the
winter (Dodge and Thompson, 1974; Buddemeier and Kinzie, 1975), in the
summer (Weber et al., 1975a; Weber et al., 1975b) and in the autumn
(Stearn et al., 1977).

These apparently conflicting results need resolution. To some extent
they may simply reflect the fact that corals react to different stimuli
in different environmental conditions. It seems, however, that some of
the techniques and data used have not been sufficiently precise and
that there is a need for detailed records of local environmental variables to be plotted against high resolution analysis of the density
banding. All authorities agree, however, that the skeletal density
banding, whatever its cause, is an annual rhythm.

Ma (1937) related these internal density changes to regular circumferential swellings, or annulations, of the epitheca (Fig. 1). It is
the expression of the annual cycle as annulations that has been used
to derive data from corals (Wells, 1963; Scrutton, 1965). Only a relatively small proportion of fossil corals show unambiguous annulations,
however, and some corals, presumably those that grew in more uniform
environments, do not show these structures at all.

Whenever coral epithecae are well preserved, very fine circumferential
growth ridges are clearly developed parallel to the growing margin of
the epitheca (Fig. 1). Wells (1963) first proposed that these growth
ridges represent a circardian growth rhythm. He showed that *Manicina
areolata*, a recent epithecate coral, developed about 360 growth ridges
in the space of a year's growth and that experimental evidence on hermatypic corals suggested a strong diurnal control of calcification
(Goreau, 1959; Goreau and Goreau, 1959). Later Barnes (1972) showed
experimentally how the growth ridges develop in some living hermatypic
corals, including *Manicina areolata*. Their form is controlled by the shape
of a lappet cavity formed where the edge zone of the polyp meets the
lip of the epithecal wall (Fig. 2). Upoward growth of the epitheca is
accommodated by refolding of the lappet cavity at intervals, each refolding resulting in a reorientation of the cavity and leading to zig-
zag upwards growth. The corrugated leading edge of the epitheca is
gradually thickened and smoothed over on the inside so that the final
expression of this process is a series of ridges on the outer surface
only of the epitheca. The periodicity of the growth ridges therefore
is a reflection of the periodicity of the refolding of the lappet cavity. In Barnes' experimental corals he concluded that a daily cycle
of expansion and contraction of the polyp was responsible for the reorientation of the lappet cavity but he pointed out that other stimuli
could be involved.

Is it coincidental that the size of the lappet cavities in these corals
exactly matches calcification over a 12-h period? In other corals with
slower or faster growth rates could refolding be stimulated by filling
of the cavity on a longer or shorter time scale? These possibilities

158

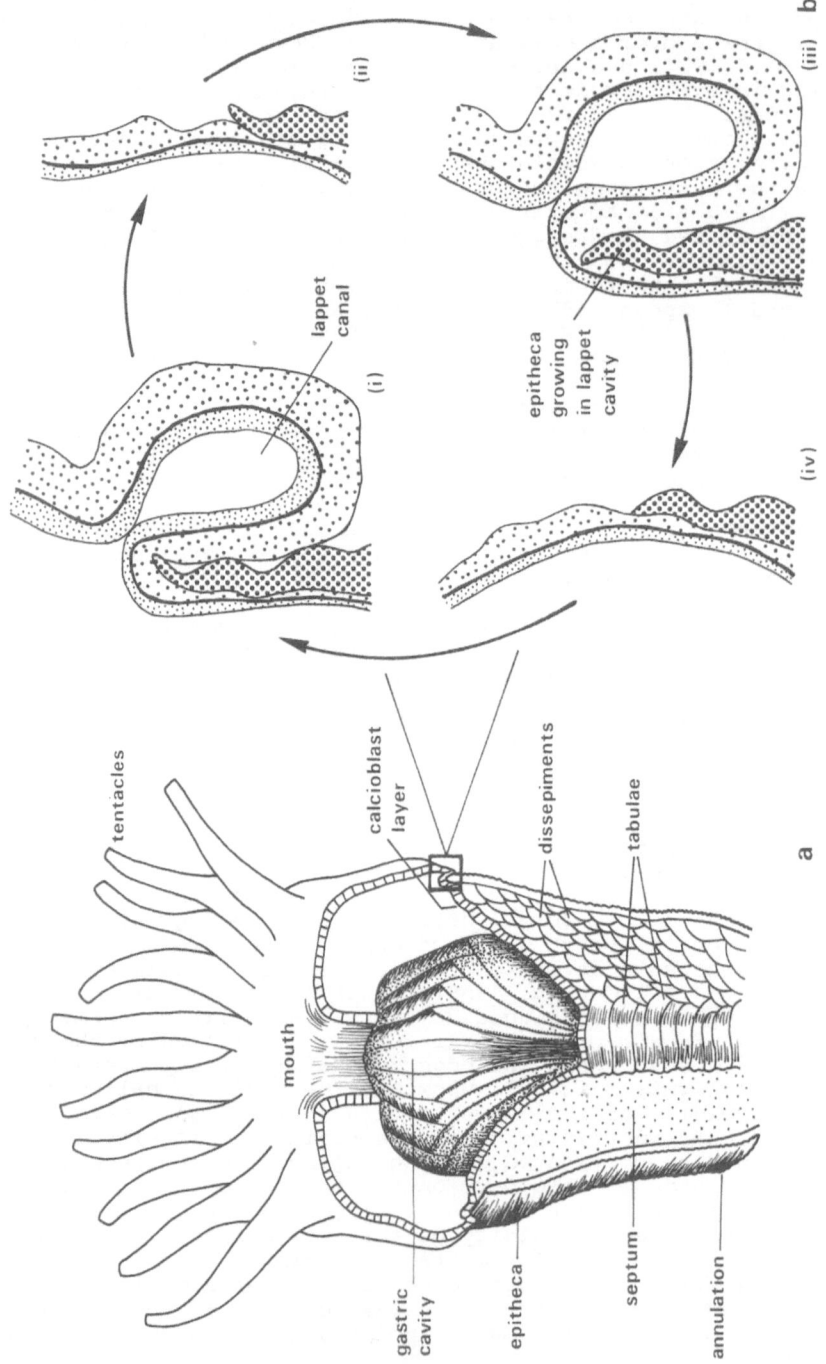

Fig. 2a and b. Formation of growth ridges in rugose corals. (*a*) Hypothetical section of a rugose coral showing relationship of polyp to skeleton. (*b*) Cycle of refolding of lappet cavity to give zig-zag upward growth of epitheca. Figure 2b is based on growth in the extant scleractinian coral *Manicina areolata* in which (i) 10.00 h, (ii) dusk, (iii) 23.00 h, (iv) dawn (based on Barnes, 1972, Fig. 7)

are of particular interest in view of the presence of growth ridges
of the same order of size on the epithecae of ahermatypic corals.
Hermatypic corals are distinguished by the presence of symbiotic algae (zooxanthellae) in their tissues. The strong diurnal cycle of activity in the algae influences their coral hosts, reinforcing sensitivity to diurnal environmental rhythms. Ahermatypic corals on the
other hand do not possess symbiotic algae and are not limited to warmer,
near-surface waters. They live at depths of up to several thousand metres, in environments where conditions in terms of the total darkness
and temperature are uniform. Yet corals from these depths still show
growth ridges on their epithecae (Stubbs, 1966, Fig. 2). Wells (1970,
p. 7) suggested several possible mechanisms to account for this, such
as daily fluctuations in nutrient supply, tidal flux, magnetic forces,
or the operation of a biological clock inherited from a distant an-
cestor inhabiting shallower waters. It could be, however, that the mechanism is purely mechanical: a function of growth rate and lappet cavity
size. As Barnes (1972, p. 347) noted, ahermatypic corals are thought
to grow much more slowly than hermatypes. At the moment, however, there
does not seem to be much direct evidence on growth rates in these corals.

In fossil corals, the presence or absence of zooxanthellae is impossible to prove directly and difficult to assess indirectly the further
one goes back in time. In the Palaeozoic, with coral groups not directly related to the Scleractinia (I do not regard the Rugosa as directly
ancestral to the Scleractinia), the problem is more acute. The Rugosa,
the group that has provided the data, show no convincing signs of an
algal symbiosis. On the other hand the particular corals studied by
Wells (1963), Scrutton (1965), Mazzullo (1971), and Johnson and Nudds
(1975) are all interpreted as living in fairly shallow but not intertidal waters, presumably well within the photic zone. These corals
could have been strongly influenced by the daily cycle in illumination
directly, with or without the participation of symbiotic algae.

In both living and fossil corals the surface of the epitheca is easily
worn and specimens on which growth ridges can be clearly seen are fairly rare. In some particularly well-preserved material of the Devonian
rugose coral *Eridophyllum archiaci*, however, Hipkin (1972) observed considerable variety in growth ridge development from 100 µm to 10 µm in
size and including sequences in which ridges graded from uniformly
spaced, to paired, to double-crested, to paired, back to uniformly
spaced in a manner strongly suggestive of the interference of a tidal
cycle with a solar daily cycle of growth. A complete cycle of transition occupied about 35 ridges (presumably with double-crested ridges
counting singly; Hipkin noted that counting was ambiguous). Unfortunately Barnes' (1972) study did not test tidal influence on growth-ridge
formation, although he mentioned complexities in ridge morphology in
naturally occurring modern corals. Other workers on corals have not
commented particularly on the morphology of the growth ridges, so it
is not known if, or to what extent, the phenomenon observed by Hipkin
may affect published data. Certainly this is an area requiring further
study.

Some but not all epithecate corals exhibit a grouping of fine growth
ridges into units termed bands by Scrutton (1965), defined by changes
in the diameter of the epithecal sheath and interpreted as reflecting
a lunar monthly periodicity (Wells, 1937; Scrutton, 1965) (Fig. 1).
More recently, Buddemeier (1974) and Buddemeier and Kinzie (1975) reported a fine structure of density band pairs within the annual density banding of the internal skeletal elements of some recent corals.
This they also regard as reflecting the lunar cycle. In both cases,
the interpretation of the temporal significance of the pattern is based

on its numerological relationship to another order of growth increment. Although living corals are known to be influenced by the lunar cycle there is no direct evidence to show what the link might be. Scrutton (1965, 1970) argued that a lunar breeding periodicity might be the most likely factor, with either monthly periods of gametogenesis resulting in reduced calcification or, alternatively, contractions of the polyp at planulation affecting the lappet morphology (Barnes, 1972, p. 348).

A distinction in the development of the epithecal and internal monthly units should be noted. Although the epithecal bands vary in thickness, no reference to any pattern in this variation has been made so far; indeed the variation in the Devonian and Carboniferous material appears to be relatively slight, and records are regarded as essentially complete (Scrutton, 1965; Johnson and Nudds, 1975). Buddemeier and Kinzie (1975), however, noted that in their corals, all of which had fairly clear annual density banding, the lunar monthly record was nearly always incomplete. The fine banding was not visible in the narrow, denser bands of winter growth (when the growth rate dropped below about 0.5 mm/month in *Porites lobata*, which in their material had an average annual growth rate up to 13.00 mm/year). The most obvious explanation is that the fossil corals grew in areas of more even year round environmental conditions. Nevertheless, it would be interesting to see if the internal growth density patterns of Buddemeier and Kinzie could be matched in epithecal patterns. It is not impossible, although rather unlikely, that the two different monthly patterns result from different stimuli.

In conclusion, then, corals may show three levels of growth periodicity: daily, lunar monthly, and annual. The daily increment, present on all unworn coral epithecae, is probably a reflection of the solar day but may show interference in response to the tidal cycle. In some corals the fine growth ridges may not be diurnal at all. The lunar monthly and annual cycles may be detected in epithecae or internal density banding, but are not always present. The synodical month is recognised numerologically rather than experimentally whereas the annual periodicity appears to be well established, if of uncertain environmental control. There is little evidence, however, to suggest how precise the growth-ridge record may be over these longer periods. In addition to these regularly recurrent features, random events ranging from storm effects, damage to the edge of the epitheca perhaps by attempted predation, to gravitational reorientation of unstable coralla can all leave their record in skeletal growth. The collection of data from corals is restricted by the fact that the supposed diurnal growth periodicity can only be readily studied on the epithecal surface so far, and this surface is too worn in most corals for details of the growth ridges or indeed any ridges at all to be seen.

2.2 Growth Increments in Bivalves

Bivalve molluscs, as their name implies, consist of two valves, hinged along one edge, which in most cases can close to completely seal the internal cavity containing the viscera. Although the external valve surfaces show growth increments, as has been recognised for centuries, these increments are developed even more clearly in the internal microstructure of the valves of many species (Fig. 3). These can be studied in section, and although susceptible to recrystallisation and diagenetic replacement, simple abrasion of the shell surface is not the problem as it is with corals. The history of growth increment studies particularly in bivalves has been reviewed by Clark (1974b).

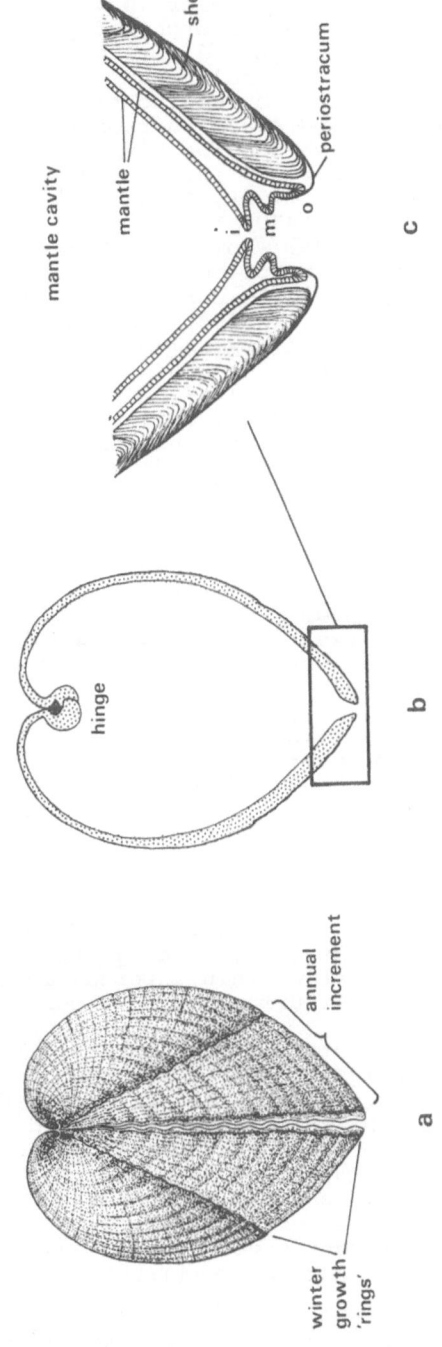

Fig. 3a-c. Growth increment formation in the extant bivalve *Cerastoderma edule*. (*a*) Anterior view of shell. (*b*) Dorso-ventral section at right angles to plane of opening. (*c*) Edges of left and right valves showing position of mantle during shell secretion (*i* - inner, *m* - middle, *o* - outer mantle lobes)

Although annual growth bands had long been known and even daily growth increments demonstrated earlier (Davenport, 1938), Barker (1964) was the first to describe a hierarchy of growth increments in sections of bivalve shells. He recognised five orders of increment groupings, a subdaily tidal rhythm, a solar rhythm, a fortnightly tidal cycle, a half yearly disturbance by equinoctial tides and storms, and an annual temperature-controlled cycle. Barker's work was followed by a considerable volume of observations on bivalves, much of it aimed at testing experimentally his largely inferential conclusions (Pannella and Mac-Clintock, 1968; Clark, 1968, 1974a, 1975; House and Farrow, 1968; Rhoads and Pannella, 1970; Farrow, 1971, 1972; Dolman, 1975; Evans, 1972, 1975; Hall, 1975; Thompson, 1975; Whyte, 1975). At the same time data on days per month (principally) and year in the past based on analyses of bivalve shells were published by Pannella and MacClintock (1968), Pannella et al. (1968), Pannella (1972b), and Berry and Barker (1968, 1975). Much more experimental evidence and data exist on bivalves than on corals.

A clear understanding of the daily increment and its substructure is obviously fundamental to the interpretation of the growth record and the accumulation of meaningful data from it. The case of the daily increment in bivalves may well be an indication of the complexities yet to be understood in other groups less well studied. Pannella and MacClintock's (1968) work appeared essentially to confirm Barker's interpretation. Their experiments with *Mercenaria mercenaria* suggested a very accurate record of daily growth increments in the shell (Fig. 4). Specimens notched, returned to a mid-intertidal environment and killed after 368 days (3 specimens) and 723 days (10 specimens), showed between 360 to 370 increments (in all 3) and 720 to 725 (in several of the 10), respectively.

Shell growth takes place when the valves gape and the mantle margin is able to extend to cover the accreting surface (see Fig. 3). Nocturnal calcification has been reported by Pannella and MacClintock (1968) in *Mercenaria mercenaria* and by Whyte (1975) in *Cerastoderma edule*. External or internal stimuli causing the mantle to withdraw, usually accompanied by closure of the valves, causes a surface of discontinuity in the deposition of calcium carbonate. The discontinuity is emphasised by a thin organic layer, corresponding to periods of non-deposition, and its thickness and definition will tend to reflect the duration of the period of shell closure. Each daily growth-increment in *Mercenaria mercenaria* shows a number of thin internal layers reflecting opening and closing of the shell during the 24-h period.

Pannella and MacClintock (1968) distinguished two basic types of daily increment, a simple increment, bounded by sharp surfaces and usually showing two intergrading layers of different colour and thickness, and a complex increment in which a pronounced internal surface subdivides the increment into two parts, each showing the internal layering of a simple increment. Increment thickness is extremely variable under different conditions, up to a maximum of about 100 µm. They also showed that the form of the increment was very variable both within the same species from different environments and from species to species. Experiments by House and Farrow (1968) and Farrow (1971, 1972) on *Cer-*

Fig. 4. Growth record of *Mercenaria mercenaria* from 1964 to 16 Aug 1967 when the ➤ shell was killed (k). Transplanted from a subtidal (u) to an intertidal environment (i) and notched (n) on 24 Aug 1965. Shows strong bidaily increments at b and synodical month patterns (s). Reproduced by kind permission of Dr. Giorgio Pannella and Dr. Copeland MacClintock from *Journal of Paleontology*, 42 (1968), memoir 2, plate 2

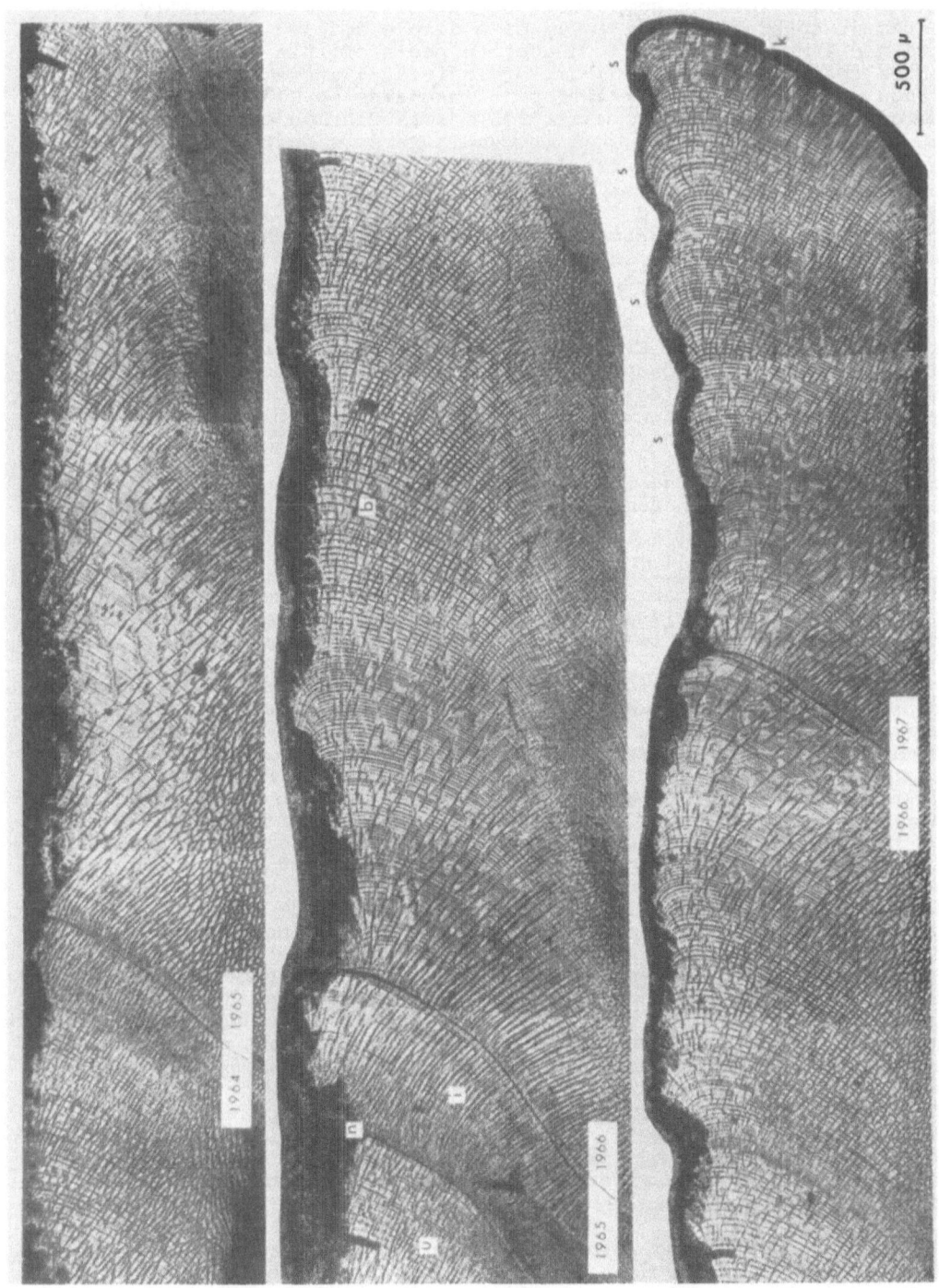

astoderma edule further supported the identification of a daily growth increment consisting apparently of a simple doublet of a thin white band and a broad grey band in acetate peels of this species. They presented their data as histograms of individual increment thickness, a method used in a slightly different form also by Clark (1968) and favoured for analysis by Dolman (1975). Both Clark's experimental work on *Pecten*, in which he counted growth increments on the external surface of the shell, and Farrow's results, suggested a tendency for some individuals to record less than the elapsed number of days. Rhoads and Pannella (1970) provided further data on the variation in growth increment size and appearance in the same and different species from a variety of environments but using fundamentally the same interpretation of the daily increment as in previous work.

An important advance in the interpretation of the daily increment, however, resulted from Evan's (1972, 1975) work, particularly on *Clinocardium nuttalli*. He showed convincingly that this species had a growth increment series that faithfully recorded the effects of the tidal cyle at its site of growth but with no apparent influence of the solar day (Fig. 5). Sharp increment bounding lines corresponded to exposure at low tide, experienced diurnally during spring tides but semi-diurnally during neaps. The increment pattern is thus an alternation of simple and complex increment types but each increment recording not a solar but a lunar day.

It has since been realised that for intertidal and immediately subtidal bivalves, increments are likely to reflect environmental stimuli with both a solar day and lunar day periodicity. Pannella (1975, p. 267) now regards growth patterns in intertidal *Mercenaria mercenaria* to be the result of the "interplay of solar, circadian and lunar rhythms". The apparent close correlation of the increment record with the solar day in earlier experiments was fortuitous. He also suggests that House and Farrow (1968) and Farrow (1971, 1972) consistently counted tidal (12 h 25 min) increments as daily and that this misinterpretation could account for the missing daily increments reported by Farrow. Pannella (1975, p. 269) introduced the term "switch zone" to refer to the complex change in increment form produced by the interference of solar and tidal stimuli in different species and tidal settings (Fig. 6). He illustrates (1975, Fig. 8) a number of idealised patterns which certainly underline the complexities and variation of the record, although they would be easier to understand with more comprehensive explanations of their formation. As yet, no similar complexity has been reported in the external growth ridges of species of pectinids by Clark (1968, 1974a, 1975) except that *Argopecten gibbus* may under "some conditions" form two lines per day (1975, p. 110). Most of his detailed work, however, has been on sublittoral species and conducted in aquaria. Under these artificial conditions, the growth ridges have a solar daily periodicity. For an accurate interpretation of the temporal significance of growth increments in bivalves, therefore, a continuous series including several fortnightly tidal patterns needs careful analysis to determine if possible, the relative roles of illumination and tidal flux in their formation. Although their sensitivity in this respect is of great value palaeoenvironmentally, the extraction of data from the record becomes much more complex.

The recognition of longer-term growth rhythms in bivalves also poses difficulties and there are several aspects still requiring further research and clarification. Intertidal and immediately subtidal bivalves show a clear cyclic growth rhythm reflecting the periods of neap and

165

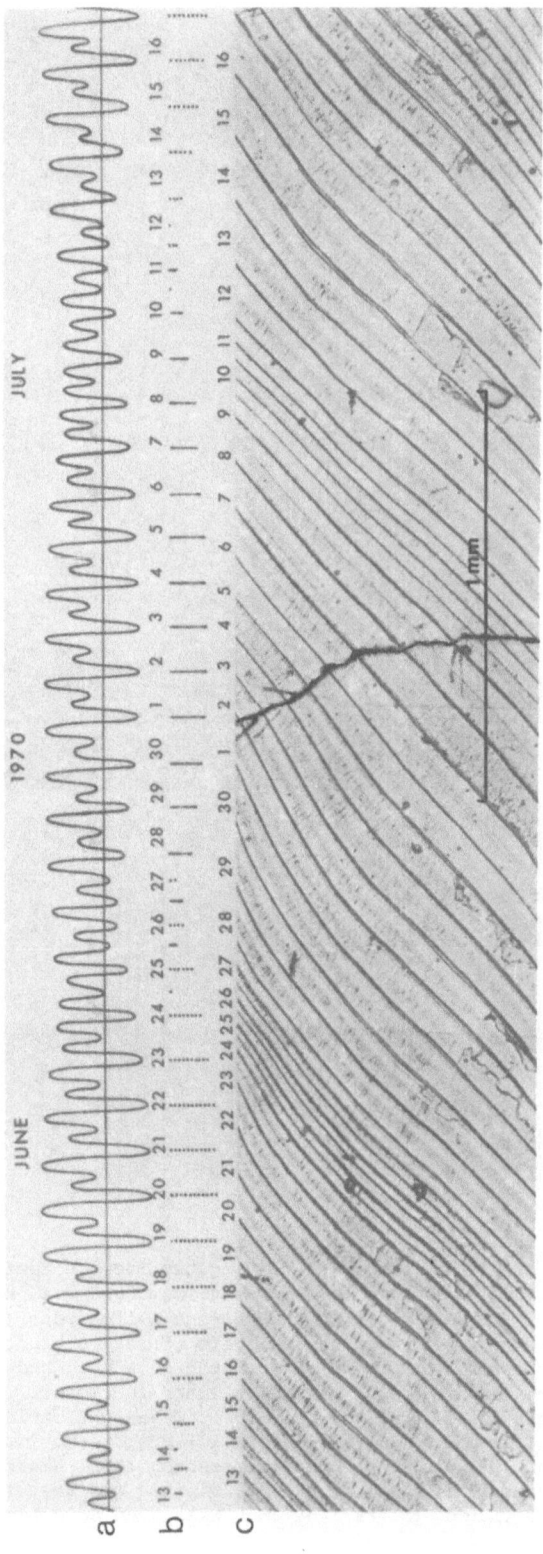

Fig. 5a-c. Growth record of *Clinocardium nuttalli*, Oregon coast, U.S.A. (*a*) Tide curve for Empire, Oregon, 13 June to 16 July 1970: Dates at noon position for each day. Line indicates approximate life position of *C. nuttalli*. (*b*) Predicted time and exent of exposure at low tide: Alternate low tides shown by dotted and solid lines. (*c*) Shell deposited during this period. Dates located at presumed noon positions relative to shell growth each day. Reproduced by kind permission of Dr. John W. Evans and that publishers from *Science*, 176 (28 April 1972), Figure 1. Copyright © 1972 by American Association for the Advancement of Science

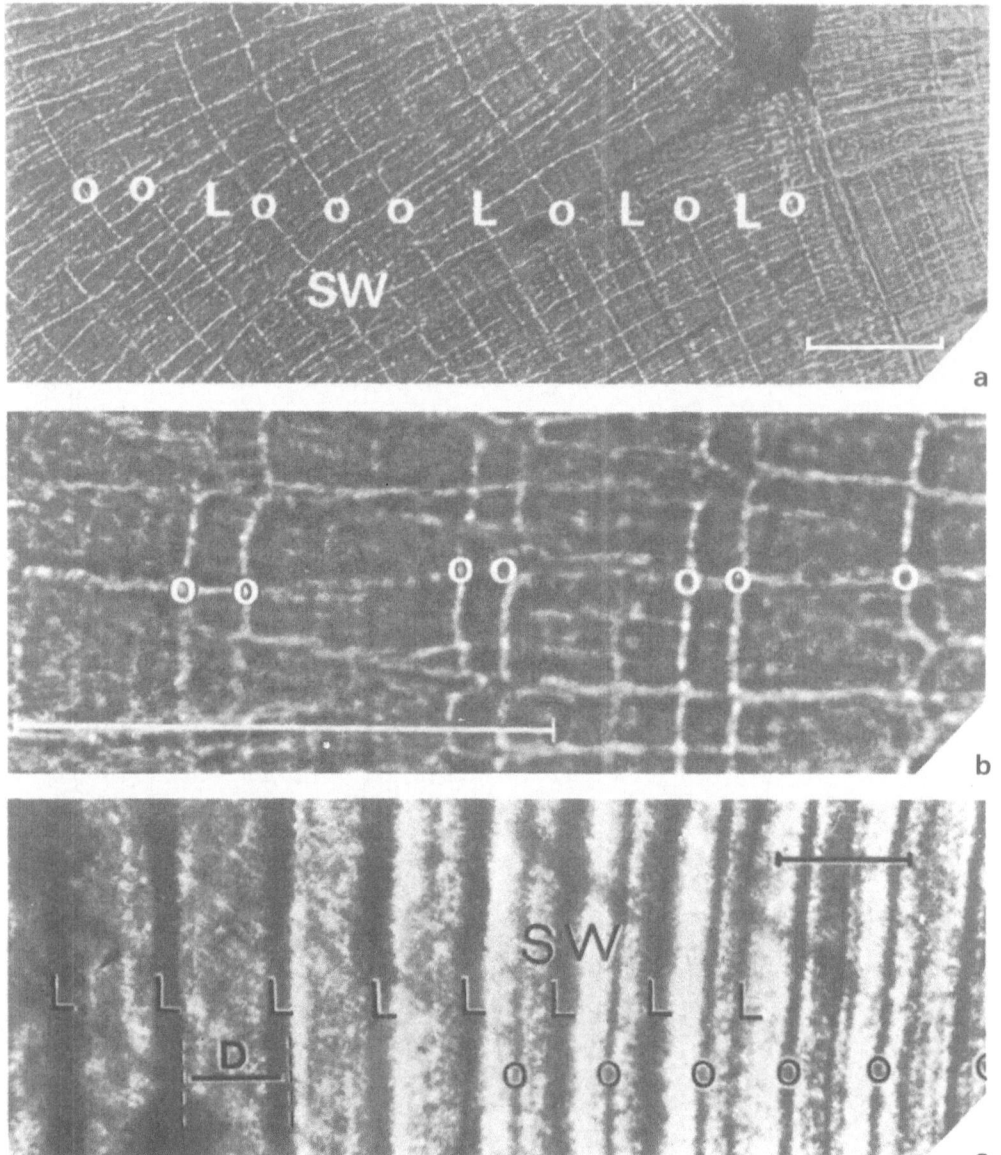

Fig. 6a-c. Types of growth increments in living bivalves. Direction of growth right to left. All scales 100 μm. (*a*) and (*b*) *Mercenaria mercenaria*, same intertidal transplanted individual. (*a*) Symmetrical increments and switch zone (*SW*). Organic-rich layers (*L*) sandwiched between two organic layers (*O*) are characteristic. In switch zone one or two half-lunar days are lost from the growth record. (*b*) Asymmetrical increments. Three organic layers (*O*) define a lunar increment. (*c*) *Tridacna squamosa*, solar daily increments (*D*) and switch zone (*SW*). Organic-rich layers (*L*) represent periods of slow calcification. Organic layers (*O*) are due to low spring tides interrupting calcification. Reprinted by kind permission of Dr. Giorgio Pannella and the publishers from Rosenberg and Runcorn *"Growth rhythms and the history of the Earth's rotation"*, p. 258. Copyright © 1975 by John Wiley and Sons Limited

spring tides. In *Mercenaria mercenaria* Pannella and MacClintock (1968)
termed the groups of closer spaced neap tide increments "clusters".
A pair of clusters normally represents a synodical month of 29.53 solar days (or 28.54 lunar days) in areas of semi-diurnal tides. A similar pattern is seen in Evan's (1972, 1975) illustrations of *Clinocardium nuttalli* in which the effect is enhanced by the bidaily tidal
exposure at neaps (see Fig. 5). In this case, however, tides on the
Oregon coast are anomalistic with a monthly cycle of 27.55 solar days
or 53.24 tide intervals. Clearly, therefore, the precise form of the
neap-spring tidal growth pattern will vary depending on the local tidal regime, as well as the species and its environment of growth. Not
only must the daily increment record be clearly understood but the tidal type must also be unambiguously decoded for the extraction of useful data from fossil material. With the analysis of sufficient contemporary samples from widely spaced localities this may be possible, although the process will be complex and a potential source of error.

Pannella and MacClintock (1968) record breeding-induced features in
shells of *Mercenaria mercenaria* with spawning related to a sharp break
in growth followed by a sequence of thin daily increments. Two groups
are recorded; in one, breaks occur only once or twice a year while in
the other, breaks are more frequent and often show a lunar monthly
periodicity. Where sex has been determined, the former are females and
the latter males.

Finally both semi-annual and annual growth rhythms are recorded. Little
has been added to the inferential equinoctial periodicity of Barker
(1964) but Pannella and MacClintock (1968) have described the winter
slow-down in growth in more detail and have related actual breaks in
the record to periods of particularly low temperature (see Fig. 4).
Farrow (1971, 1972) also produced evidence supporting a link between
frosts and interruptions to growth in the winter in *Cerastoderma edule*
although mean monthly growth did not correlate well with sea temperature. Whyte (1975) has suggested that with nocturnal calcification,
the relative lengths of day and night between summer and winter, affecting the time available for calcification, may act in opposition
to the temperature-controlled growth rate to produce the patterns of
mean monthly growth obtained by Farrow (1971). Whereas Pannella and
MacClintock showed continuous if very slow growth in inner parts of
the shell corresponding to sharp breaks closer to the outer surface,
Farrow considered some of the winter breaks in his material to represent a complete cessation of growth. Hall (1974, 1975) called the annual period of slower growth the biocheck. On limited evidence he considered the biocheck to begin abruptly, following the period of spawning and to preceed the onset of the coldest winter sea-surface temperatures. He also showed that the number of increments per annual cycle
varied with the age of the individual, latitude, and water depth. Progressively fewer increments are apparently formed during successive
years in older individuals, matching the well-known drop in linear
growth rate in bivalves with age. Although not all his figures were
presented in this form, all Hall's data suggest that the species he
studied, *Tivela stultorum* and *Callista chione*, only deposit close to as many
increments as elapsed days when young and growing under optimum conditions. Pannella (1975, p. 278) also commented on the effects of latitude on shell growth and noted that growth records in Arctic bivalves
extend for only 3 - 5 months of the year. He considered the length of
the growth stoppage during the winter to be a function of latitude,
at least in latitudes higher than 40° - 45° N at the present day. The
evidence suggests, therefore, that although temperature seems to be
an important overall controlling parameter in growth over the year,
other factors may also be involved in regulating the annual cyclicity

in growth rate. Again the effects vary between species and in different environments.

Internal shell structure in bivalves, therefore, may record environmental stimuli with great detail and precision. Diurnal, fortnightly, lunar monthly, equinoctial, and annual periodicities in growth occur as well as records of random disturbances. The basic increment record is now seen to be potentially complex, reflecting either the solar day or the lunar daily tidal cycle or a combination of the two, depending upon the species and the characteristics of its environment of growth. The possible role of biological clocks is considered in a later section. The cause and expression of most of the other periodicities appear fairly well established, but it seems necessary to watch for the effects of latitude and age class on the completeness of the annual cycle.

2.3 Growth Increments in Stromatolites

The strong internal laminations so characteristic of well-preserved stromatolites (Fig. 7) suggest considerable potential in growth increment studies. Stromatolites have also been the subject of considerable general interest in recent years (e.g., Walter, 1976), which has helped a great deal to assess their reliability as environmental recorders and thus their likely contribution to data on the rotational history of the Earth. At the moment, they are the only likely source of data for some 2400 m.y. of middle and late Precambrian time.

Stromatolites are formed by the activity of sediment-trapping algae and bacteria, although some may be non-biogenic geyserites (Walter, 1972). The control of the basic lamination in stromatolites is the effect of the solar daily cycle on algal and/or bacterial photosynthesis and growth, and the regulation of the sediment supply which in intertidal and shallow water environments is tidal (Monty, 1965, 1967; Gebelein, 1969; Sharp, 1969). In modern mats formed of filamentous algae the filaments extend vertically during the early part of the day, forming an organic-rich zone with only scattered sedimentary particles, but late in the day and during the night the filaments grow prostrate, and if sediment supply is continued at the same rate, a distinct sedimentary lamina is formed, which can be up to 600 µm thick (Monty, 1967). This simple model is disrupted, however, by variations in sediment supply and wave action in the upper sublittoral and intertidal zones with mat growth being discontinuous. In supratidal mats, laminae may not be daily at all but may reflect less frequent periods of flooding. It is also possible that with semi-daily tides two laminae per day could form at certain times in the tidal cycle.

Fig. 7a and b. Growth increments in stromatolites. (*a*) Upper Cambrian digitate ▶ stromatolite, Conococheaque Fm., Maryland, U.S.A. Diurnal laminations are alternations of thin dolomitic layers and thick calcium carbonate layers. Groupings into tidal patterns clearly visible. Reproduced by kind permission of Dr. Giorgio Pannella and the publishers from Pannella, 1972, *Astrophys. Space Sci.*, 16, plate 4. Copyright © 1972 by D. Reidel Publishing Co. (*b*) Precambrian digitate stromatolite, Gunflint Fm., Mink Mountain, Ontario, Canada. The number of laminae per tidal band (*t*) below the level *A* ranges from 14 to 17, above the level from 3 - 9. Boundaries between annual bands are marked with an *X*. Scale 1 cm. Reproduced by kind permission of Dr. Giorgio Pannella and the publishers from Pannella, 1976, in M.R. Walter (Ed.) *Stromatolites* (Developments in Sedimentology, 20), Fig. 4. Copyright © 1976 by Elsevier Scientific Publishing Co.

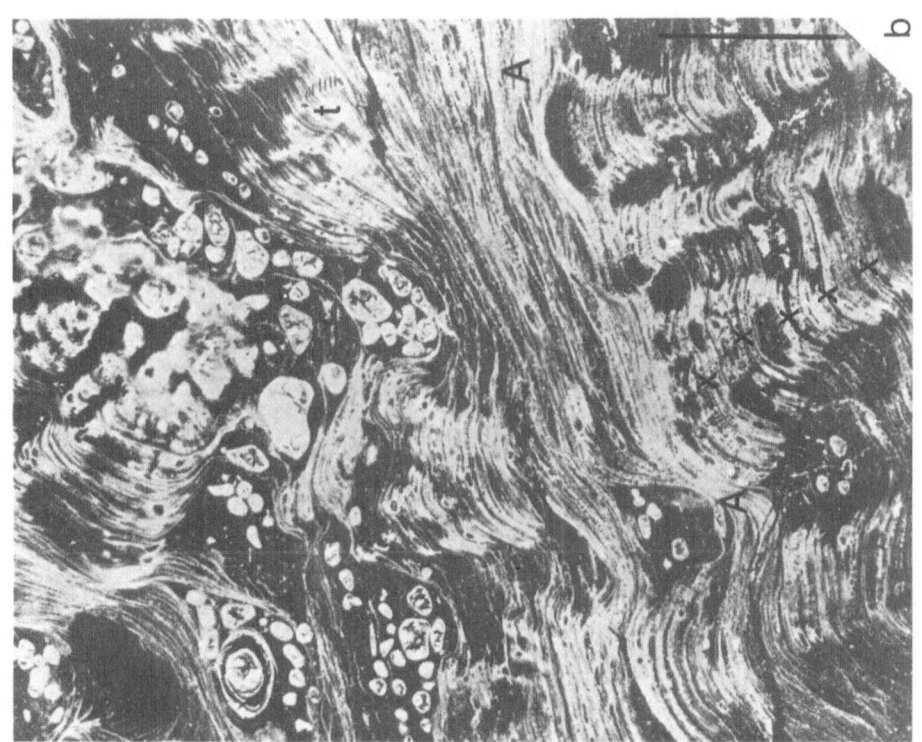

In many modern stromatolites calcium carbonate is actively precipitated on and in the mat and solution and recrystallisation of carbonate may modify the microstructure. Laminae may be ill defined or modified and may be completely destroyed. Although most modern stromatolites are composed of calcium carbonate, Walter et al. (1972) report the lamination of non-biogenically produced silica in bacterial and algal mats associated with hot springs at the present day and suggests a similar origin for some Precambrian stromatolites. The laminae in hot spring and geyser mats are also daily where deposition of material is continuous but this seems highly unlikely in any one mat for any length of time.

Very little precise information seems to be available on groupings of the daily laminae in modern algal stromatolites. Pannella (1972b) published some preliminary data from supratidal algal mats and biscuits from the Bahamas to show that tidal patterns are distinguishable. The highest counts of daily increments per band approach the actual value of days per fortnight and the numeralogical relationship is the basis for the identification of the pattern. A complete heirarchy of periodicities has been assembled by Pannella (1972a) largely on the basis of Precambrian stromatolites. Besides daily laminations and tidal bands, he distinguishes monthly bands consisting of paired tidal bands, one better developed than the other, seasonal bands consisting of groups of monthly bands of similar texture and annual bands consisting of a pair of seasonal bands, one organic-rich and dark, the other organic-poor and light. These are respectively I to V order periodicities in his terminology (Pannella, 1972a,b). Although this scheme was originally applied to all groups (and indeed adapted from Barker, 1964), Pannella (1975, p. 260) later set out this numerical heirarchy of periodicities descriptively, without temporal labels, specifically for use with stromatolites. There is presumably some advantage in this device in a group where ambiguities of interpretation may arise with incomplete sequences, as in Pannella's Bulawayan stromatolites (1972b, p. 234, and see Table 4 herein).

Observations on living algal mats suggest that they are unlikely to record elapsed days reliable for any significant lenght of time. Although most studied Recent algal mats are intertidal or supratidal, evidence is accumulating to suggest that subtidal stromatolites can be recognised in the geological record (Playford and Cockbain, 1969; Walter, 1970; Monty, 1973). In addition, the absence of competition and particularly of disturbance (e.g., by feeding) from more highly evolved organisms is reflected in the abundance of Proterozoic stromatolites and their acme in the late Precambrian. Subtidal forms were probably much more important then (Awramik, 1971; Monty, 1973). Exceptionally, therefore, useful data may be obtained from subtidal stromatolites in the Precambrian although there is no direct way in which the accuracy of their growth records can be experimentally assessed. Such forms, which are morphologically small, digitate, columnar stromatolites (Pannella, 1976a) (Fig. 7), may be expected to escape the more severe disruptions to continuity of growth experienced by intertidal and supratidal mats. Pannella (1976a) found longer sequences and higher numbers of laminae per periodic band in this growth form when compared to other growth forms in the Precambrian. Walter (1972), however, showed the small digitate growth form to also occur among modern geyserites, distinguished by very fine internal laminae 0.2 - 6.0 µm thick. He regards morphologically similar Precambrian material lacking microfossils that occurs among Gunflint Formation stromatolites and including Pannella's (1976a) Mink specimens (Fig. 7b), to be possibly of similar origin. In this case, rather than growing in a subtidal environment, they would indicate subaerial conditions (Walter, 1972, p. 969), and there would be much less confidence in the accuracy of their growth increment record. The oldest recorded stromatolites with tidal patterns are from the Bulawayan Group, considered to be at least 3000 m.y. old.

The current situation with stromatolites, therefore, is rather confused. There is no doubt that laminae are arranged into a heirarchy of periodicities, but the temporal significance of each grouping is established numeralogically and is somewhat ambiguous. In most stromatolites laminae do not record elapsed days very closely. Although Precambrian subtidal stromatolites may be much more reliable in this respect, the growth form considered characteristic of this environment is also found in subaerial geyserites which are less likely to experience continuous growth.

2.4 Biological Clocks

In recent years, considerable interest has been shown in the significance and control of endogenous rhythms, called biological clocks, in organisms. Clearly this is of considerable importance in the field of growth increment analysis. Current ideas are therefore briefly reviewed here.

It appears that in many cases, organisms reflect the regularly recurrent circadian environmental stimuli through entrained endogenous metabolic rhythms. Some claim that these internal clocks, independently of the environment that originally set them, are the principal control of all biological rhythms. Certainly many experiments have shown that rhythmic activity persist in organisms kept under constant conditions, or at least conditions under which all obvious environmental variables have been eliminated (Aschoff, 1965; Brown et al., 1970; Thompson, 1975; Clark, 1975; Hastings and Schweiger, 1976).

If growth increment formation was under the control of such biological clocks over considerable periods of geological time, unmoderated by changes in astronomical cycles, then data from fossil organisms would be of litte use in measuring these cycles. In fact it would seem likely from the complexities of circadian growth-increment patterns in bivalves and from the way in which they are now seen to match solar and tidal daily cycles that under natural conditions direct environmental stimulation strongly overshadows any effects of endogenous rhythms on shell growth in this group. Furthermore under constant conditions many rhythms do damp, so that perhaps in the longer term, environmental stimuli are necessary to keep the biological clock "wound up" and presumably therefore accurately entrained.

Other experiments also tend to support constant updating of biological clock rhythms in bivalves. In Clark's (1975) experiments with pectinids, specimens kept in aquaria were found to form growth ridges with a solar day periodicity, although they recorded fewer than the elapsed number of days. Other specimens kept in constant conditions (either total darkness or continuous light) maintained circadian growth line formation with as much accuracy as under conditions of normal illumination, but shells subjected to an artifical 16-h day of 8 h darkness and 8 h illumination apparently switched growth line deposition to this new periodicity immediately and with comparable accuracy. Of equal interest, Brown (1954) and Thompson (1975) both found that in bivalves transferred to aquaria where the lunar cycle was not in phase with that of the collecting site, the specimens rephased their activities to the local cycle even in the absence of tides or moonlight.

The latter observations tend to suggest that organisms are sensitive to environmental stimuli more subtle than those which so-called constant conditions experiments eliminate. Thompson (1975) remarks that these experiments invite comparison with conditions in abyssal deeps, where the column of sea water must considerably modify or eliminate the obvious environmental stimuli of near-surface waters. Cyclic variations in deep ocean environments are known, however, such as those of the gravitational and electromagnetic fields. Bivalves and corals from abyssal depths both show fine growth increments, although those of bivalves are reported to differ in their uniformity and lack of sharpness from increments in shallow subtidal and intertidal environments (Rhoads and Pannella, 1970). Perhaps these are the result of growth rhythms entrained by these more subtle geophysical cycles (Wells, 1970; Thompson, 1975).

The evidence suggests that, at least in bivalves, the operation of biological clocks does not invalidate the use of data from growth increment patterns to measure past astronomical cycles. It is difficult, however, to be certain in all cases. Consistent results from different groups of organisms would be some support, because it is unlikely that they would all show the same settings for their endogenous rhythms at any particular time in the past if they were independent of contemporary environmental stimuli.

3. The Data

3.1 Recording the Data

The published data on Recent and fossil growth increment records have either been assembled by counting on specimens or photographs (e.g., Wells, 1963) or by the measurement of increment thickness (Clark, 1968; House and Farrow, 1968; Farrow, 1971, 1972) or by combinations of the two (Mohr, 1975; Pannella, 1976a). Counting methods have involved either the direct recording of increments between successive higher order features (Scrutton, 1965; Pannella and MacClintock, 1968) or the continuous counting of increments over a number of cycles with the mean number of increments per cycle calculated by division (Berry and Barker, 1975). The latter method may be marginally more objective but the former allows an assessment of variation in the length of the cycle in the specimen. These methods, however, all require the observer to directly analyse and identify the units that are being counted and/or measured and are to this extent subjective. It has been suggested by various workers that some indirect or mechanical means of recording the data would be advantageous (Runcorn, 1966).

Attempts to record growth increments mechanically have so far proved unsuccessful (Hipkin, 1972; Scrutton and Hipkin, 1973). The noise in the record, such as the effects of random events like storms and the change of growth rate with age, tends to obscure any pattern. It is clear too that the complexity of the record requires a much more subtle approach than any originally envisaged. In the first place any counting technique requires the recognition of random or non-random imperfections such as pauses in growth (as far as these can be identified). Secondly the significance of the growth pattern must be established, and this must at least entail the identification of a basic unit or units and their temporal significance.

The measurement of each successive increment once identified does have major advantages. The data can then be subjected to Fourier analysis as described by Dolman (1975) although incomplete records present problems. There is also the problem that input needs to be homogeneous, so that the recorder is still faced with subjective manipulations to convert mixed semi-diurnal and diurnal tidal growth lines to a diurnal time series (Dolman, 1975). Pannella (1976a) also used a graphical representation of data of this sort, using the running mean to smooth the noise, to obtain a visual representation of the growth pattern in stromatolites for comparison with curves of tidal amplitude. Computer analysis of such curves may prove a powerful tool for reconstructing tidal patterns (Pannella, 1976b).

Up to the present, however, most data have been presented as means of a number of "feature-ratio" counts (Dolman, 1975), for example, the number of presumed daily increments in a presumed lunar monthly increment. Because of evidence suggesting a tendency for organisms to record less than the number of elapsed days (e.g., Clark, 1968), some have favoured the use of maximum counts (Mazzullo, 1971). Because organisms, at least in some cases, are only recording an approximation to the relevant astronomical cycle (as in annual cycles, or possibly the lunar cycle in corals), this method could lead to serious overestimates of the length of the cycle (Scrutton and Hipkin, 1973). Another consequence of this tendency for the biological cycle only to approximate to the length of any particular astronomical cycle is that the averaging of consecutive feature-ratio counts will be more accurate than the averaging of a series of non-consecutive counts (Scrutton, 1970; Scrutton and Hipkin, 1973). I have suggested that one method of attempting to compensate for the sort of recording deficiencies reported by Clark (1968) in these cases would be to use the mode rather than the mean of the data (Scrutton, 1970). This solution, however, cannot compensate for a strong and consistent tendency to underrecord.

Pannella (1975), on the other hand, has argued that the incompleteness of the record in bivalves has been overemphasised and that because pectinids do not take well to laboratory conditions, Clark's results should be interpreted with caution. He reports that although in high-latitude bivalves today, growth may cease over periods of several months, none of the fossil bivalves he has studied appear to have winter growth stoppages. Most of their increment records suggested life in more equable environments of relatively little seasonal variation in growth rates. It therefore seems most reasonable to take the mean of a series of consecutive counts. It has been strongly suggested by Pannella (1972a, b, 1975, 1976a, b), however, that stromatolites are an exceptional case where maximum counts should be used. Evidence from living stromatolites suggests that underrecording is likely to be a serious problem in this group at all levels. Whether or not this can be safely extrapolated to all Precambrian and early Palaeozoic stromatolites is another matter and is discussed further below.

The recent work on bivalves suggests that such an accurate growth increment record of solar and lunar daily stimuli is possible that generalised statistical manipulations may be inappropriate. Provided that stoppages in growth can be recognised and that the increment pattern can be unambiguously decoded, then theoretically, short complete growth records should yield precise information on the lunar cycle. Unfortunately it is not yet clear to what extent these prerequisites can be met. A high level of compatability between results from different species in different environments at particular times in the past would be required to give confidence in the data.

Work on the detection of growth increments by chemical means has been persued by Rosenberg (1973) and Rosenberg and Jones (1975). Using the electron microprobe Rosenberg and Jones reported regular fluctuations

in Ca and S in some bivalves correlating with the fortnightly tidal
rhythm. In addition a Precambrian stromatolite showed cyclicity in
Ca, Mg, Fe, and Si concentrations, but the results were not linked to
visually recorded growth patterns. In both cases results were repro-
ducible although adjacent traverses on the stromatolite correlated
only weakly. These techniques require considerable refinement at the
moment. Even with an accurate chemical reflection of the growth increment record, however, the results will require the same analysis as
the increment record itself.

3.2   The Published Data

The data that have been published so far are collected together in Tables
1 - 4 and illustrated graphically in Figures 8 and 9. Some of the
doubts and uncertainties surrounding these figures should be clear
from the comments in preceding sections of this review. A few additional general and specific comments will be made here to underline
the caution with which any conclusions should be drawn from these figures.

a) Data from Corals (Table 1)

Much of the coral data was accumulated before Barnes (1972) had shown
how the presumed daily increments formed in living corals. In addition,
the monthly periodicity still lacks any real experimental or observational basis among living corals. None of the authors who have published figures comment on the sort of substructures described by Hipkin
(1972) although if the daily increment is influenced by the tidal cycle
this needs to be taken into account in the interpretation of the record.

Wells (1963, 1970) does not describe how his counts were made and it
is clear in his early work that he meant no great mathematical precision by his data. Mazzullo (1971) and Johnson and Nudds (1975) also
give insufficient details of their methods and basic data. In addition
Mazzullo listed only means of maximum counts which may give overestimates for the astronomical cycles. The fact that his figure for months
per year was exactly 13 in all cases is surprising.

The manner in which Johnson and Nudds derive the number of months in
the year and hence days in the year for the Carboniferous should be
treated with caution. They take Scrutton's (1970) figures of 13.01
lunar months in the Devonian year and calculate the Carboniferous figure by assuming a constant rate of decrease in months per year to
the present day. Scrutton's figure was obtained, however, via an estimate for the rate of deceleration of the Earth's spin on its axis from
the calculations of Munk and MacDonald (1960) as reported by Wells
(1963). It is this deceleration that Johnson and Nudds (1975, p. 35)
go on to test.

b) Data from Bivalves (Table 2)

Pannella et al. (1968) were aware from the beginning of a problem with
the bivalve data when the mean of counts of days per lunar month from
freshly killed *Mercenaria mercenaria* was about 1 % too low. At the time
they were unable to explain this discrepancy although it is now clear
that confusion in how increments with substructure should be counted
had been caused by the switch zone. All the data published by Pannella
(1972b) predates the present understanding of the complexities of the

Table 1. Summary of published data from corals

Period	Material			d/f (I/II)	d/m (I/III)	m/y (III/V)	d/y (I/V)	Comment
	No. specimens	No. genera or species	No. patterns counted					
Wells (1963, 1970)								
Carboniferous (Penn.)	2	1 gen.					385	mean
Carboniferous (Miss.)	1						398	
Devonian (M.)	12	4 gen.					398	mean
Silurian (M.)	1						ca. 400	
Ordovician (Ur.)	1						ca. 412	
Scrutton (1965, 1970)								
Devonian (M.)	10	6 gen.	113		30.66	(13.01)	399 calc. from geophys. data	mean
Mazzullo (1971)								
Devonian (M.)	?	2 sp.			31.5	13	410	max.
Silurian (Lr.)	?	1 sp.			32.4	13	421	max.
Johnson and Nudds (1975)								
Carboniferous (Visean)	6	1 sp.			30.2	(12.95)	391.09 not directly counted	mean
Pannella (1972)								
Ordovician (Lr.)	1		46		30.70 (±1.60)			

Abbreviations: d – day, f – fortnight, m – month, y – year. I, II, III, V – equivalent growth pattern orders in stromatolites. ± values in parentheses are standard deviations.

Table 2. Summary of published data from bivalves. Abbreviations as for Table 1

Period	Material No. speci- mens	No.genera or species	No. patterns counted	d/f (I/II)	d/m (I/III)	m/y (II/v)	d/y (I/v)	Comment
Pannella and MacClintock (1968) Pannella et al. (1968) Pannella (1972)								
Recent	11	2 sp.	186		29.22 (±1.15)	12.34	359.3	means
Miocene (Ur.)	3	3 gen.	140		29.52 (±1.00)			mean
Miocene (M.)	1		26		29.43 (±1.08)			mean
Miocene (Lr)	1		31		29.22 (±0.99)			mean
Oligocene (Lr.)	1		38		29.63 (±0.97)			mean
Eocene (M.)	1		12		29.91 (±0.90)			mean
Eocene (Lr.)	1		66		29.41 (±1.03)			mean
			25		29.60 (±1.55)			mean
Palaeocene (Ur.)	1		50		29.96 (±0.88)			mean
Cretaceous (Ur.)	8	6 gen.	159		29.85 (±1.24)	12.6 (on 4 spec, 16 patterns only)	375	means
Triassic (M.)	3	2 gen.	77		29.66 (±1.25)	12.55 (on 7 patterns)	371.6	means
Carboniferous (Ur. Penn.)	3	1 gen.	50		30.16 (±0.68)		383 (11 patterns)	means
Carboniferous (Lr. Miss.)	2	1 sp.	138		30.37 (±1.28)		398 (9 patterns)	means
Devonian	1		30		30.53 (±1.25)			mean
	1		?		30.22 (±1.56)			mean
	1		6		-		405.5	mean

Berry and Barker (1968)

Cretaceous (Ur.)	?	1 sp.	?			mean
	8	1 sp.	37	29.65 (±0.18)	12.49 370.3 (calc.)	mean

Berry and Barker (1975)

Pleistocene	19	3 sp.	115	14.75	mean
Pliocene	9	2 gen.	40	14.83	mean
Oligocene	10	2 gen.	49	14.79	mean
Eocene	3	1 sp.	14	14.86	mean
Palaeocene	6	1 fam.	20	14.85	mean
Cretaceous	17	4 gen.	80	14.86	mean
Jurassic	4	1 sp.	12	14.92	mean
Triassic	4	1 sp.	17	14.88	mean
Carboniferous	8	3 gen.	29	15.06	mean
Devonian	1	1 sp.	8	15.25	mean

growth increment record, and these data are in the process of re-examination (Pannella, 1975, p. 280) although the results have not yet been published.

Berry and Barker (1968, 1975) have been the other major source of data. Although it is clear from their later work that they are aware of the importance of tides in the development of the growth increment record, they do not analyse these patterns in detail but rely on "different responses in terms of shell growth of each individual animal to environmental factors" to be averaged in the calculations (Berry and Barker, 1975, p. 20). They go on to note the relatively imprecise nature of their data. Although they stress the tide-related growth increment, they do not discuss whether they might be recording lunar days or solar days in their growth increment counts. In fact the close agreement between their data and Pannella's (1972b) data suggests that in both cases it is an interaction of the two.

Most of the bivalve data consists of days per fortnight or month rather than days per year. This reflects the difficulty of finding long sequences of well-preserved growth increments in fossil material. Another advantage of working with shorter periodicities is that in material showing evidence of a significant slow-down in growth with suspected missing growth increments over the winter period, yearly counts may be suspect whereas fortnightly or monthly counts over the summer growth period may be more reliable. Berry and Barker (1968) used a method of obtaining an estimate for the number of days per year in the Upper Cretaceous by counting days per month on specimens of one species of bivalve and multiplying the mean of the result by half the mean number of fortnights (tidal clusters) per year derived from specimens of a different but contemporaneous species (see Table 2). This method has important possibilities for extending the data on days per year if it can be used more widely (Pannella, 1975, p. 276).

c) Data from Other Invertebrates (Table 3)

Relatively little data have been published from other invertebrates. In both of the two groups involved, brachiopods and cephalopods, the counts have apparently been made on external surfaces. No details are available of increment morphology in brachiopods although they might be expected to reflect similar complexities in their growth increment record to those in bivalves from comparable environmental situations. Cephalopods, however, with a necktonic or nektobenthonic life style may not be directly comparable to benthonic organisms in the character of their growth increment records. The Pennsylvanian Kendrick Shale specimen illustrated by Pannella and MacClintock (1968, pl. 9) has an external appearance remarkably like coral epitheca and has yielded data which are closely comparable to those from contempory corals. These authors report similar patterns on Recent *Nautilus*, which supports their interpretation of the fine increments on the fossil material as daily. At the moment, however, more detailed observations are needed. In both brachiopods and cephalopods it is difficult to evaluate the worth of the data without more evidence on the activities of living representatives in response to natural cycles.

The published cephalopod data are in the form of means but the data for the brachiopods are given as the means of maxima. Numeralogically, the same objection seems to be applicable to the use of maximum counts in brachiopods as for corals and bivalves.

Table 3. Summary of published data from invertebrates other than corals and bivales. Abbreviations as for Table 1

Period	Material		No. patterns counted	d/f (I/II)	d/m (I/III)	m/y (II/V)	d/y (I/V)	Comment
	No. speci-mens	No. genera or species						
Pannella and MacClintock (1968)								
Pannella et al. (1968)								
Pannella (1972)								
Carboniferous (Penn.)	1	1 sp.	9		30.22 (±1.20)			cephalopod mean
Silurian (Ur.)	1	1 sp.	38		29.84 (±1.40)			nautiloid mean
Mazzullo (1971)								
Devonian (M.)	?	2 sp.			31.38	13	4O7.75	brachiopods means of maxima
Silurian (M.)	?	2 sp.			32.25	13	419	brachiopods means of maxima

d) Data from Stromatolites (Table 4)

It is clearly much more difficult to obtain meaningful data from stromatolites than from the invertebrate groups. Sequences are commonly incomplete and show clear signs of interruptions and erosion, so much so that Pannella (1976a) now regards all but one set of data among those published as unusable. Because of physical and biological disruptions, only Precambrian subtidal stromatolites are likely to approach a full complement of elapsed days in their internal laminations. In this group therefore maximum counts have been favoured so far as more likely to be meaningful.

Pannella (1976a) picks out the Mink Mountain stromatolites of the Gunflint Formation, about 2000 m.y. old, as the most promising of those he has studied. As well as providing maximum counts for days per year and days per month, harmonic analysis of the growth increment spacing resulted in a curve strikingly similar in shape to a semi-diurnal tidal amplitude curve, tending to confirm the II and III order periodicities as fortnightly and monthly tidal-controlled growth rhythms.

As well as Pannella's Gunflint Formation data, his other stromatolitic data are listed in Table 4 for interest, and those points falling within the respective fields are plotted on the graphs in Figures 8 and 9. It is interesting to note that the Jurassic maximum figure for days per month agrees well with other data. In the case of the Cambrian data, however, the mean (Pannella et al., 1968) fits a projection of other, younger means better than the maximum (Pannella, 1972b), which agrees more closely with a projection of Siluro-Devonian coral-brachiopod maxima. In view of Pannella's (1976a, p. 677) comments on reliability, these data need reassessing, but it is difficult to avoid speculating on the wisdom of always regarding stromatolite maximum counts as underestimates. Of course, a rapid fall in the number of days per month in the early Phanerozoic is a possibility. It is also possible, however, that Precambrian and earliest Palaeozoic subtidal stromatolites were capable of much more accurate records than their recent analogues, as suggested earlier. The Upper Cambrian stromatolites in question are small digitate forms (Pannella, 1972b, pl. 4) and thus may well be subtidal (but see reservations noted in Sect. 2.3).

The only other Precambrian data provided by stromatolites was published by Mohr (1975) and underlines the problem of how to interpret the ancient stromatolite growth records. The figures, obtained from stromatolites from the Biwabik Formation, which is regarded as a correlative of the Gunflint Formation, contrast strongly with Pannella's data and indeed suggest a lower value for days per synodic month 2000 m.y. ago than at the present day. Mohr, however, calculated a mean value from a frequency histogram of laminae per band. The basic data were assembled by a method involving separate counting of laminae and bands with a view to increasing objectivity. The results were compared with direct counts of laminae per band and the two methods agreed fairly closely.

Much of the discrepancy between Mohr's and Pannella's figures for this same time interval is the difference between mean and maximum counts, a view supported by the shape of Mohr's histograms. Even so, Pannella's (1972a,b, Fig. 2) histogram shows a principal peak somewhat higher in value than those of Mohr (1975, Figs. 1 and 2). It was argued that the Biwabik stromatolites, which have associated ooliths might have been more prone to disturbance and have a less complete record than the Mink material (see discussion of Mohr, 1975, p. 56). Mohr himself (1975, p. 50), recorded the presence of "many minor unconformities"

Table 4. Summary of published data from stromatolites. Abbreviations as for Table 1

Period	Material		d/f (I/II)	d/m (I/III)	f/y (II/v)	d/y (I/v)	Comment
	No. speci-mens	No. patterns counted					
McGugan (1967)							
Cambrian (M.)	1	1				424	
Pannella et al. (1968)							
Cambrian (Ur.)	1[a]	18 III order	-	31.56 (±3.15)	-	-	mean
Pannella (1972a, b)							
Recent	1	67 III order	6	10	10	43	max.
	3	30 III order	14	27	9	-	max.
Eocene	3	33 III order	14	24	21		max.
Jurassic (Ur.)	1	17 III order	15	30			max.
Jurassic (M.)	1	5 III order	15	26			max.
Cambrian (Ur.)	1[a]	18 III order	17	33	26	156	max.
Prec. (Belt Supergroup)	1	16 III order	16	31	26	180	max.
Prec. (Gunflint Fm.)	2	97 III order	19	39	28	448	max.
Prec. (Biwabik Fm.)	1	-	17	-	26	310 (442 calc.)	max.
Prec. (Gt. Slave Supergroup)	2	-	17	-	26	238 (442 calc.)	max.
Prec. (Steeprock Group)	1	-	-	-	-	245	max.
Prec. (Bulawayan Group)	1	118 or 59 III order	11/22	22/41	-	-	max.
Mohr (1975)							
Prec. (Biwabik Fm.)	?	341	12.8	25.6			mean

[a]Alternative data for same specimen

in his material although in discussion he regarded the digitate forms as seldom exposed at low tide, and Pannella (1976a, p. 677) certainly regards the small digitate growth form of the Biwabik Formation as promising material. It seems, however, that both digitate and mat-like forms were included with no separate treatment or consideration in Mohr's material. The latter probably indicate growth in shallower water and probably possess a less complete record than the former (see Pannella, 1976a, Table 1 and pp. 681-682). This could also have contributed to the low mean figure obtained by Mohr.

The fact that Mohr's particular mean figure may be an underestimate because of the environmental and material selection factors mentioned above, does not necessarily invalidate the use of the mean for data from carefully selected subtidal Precambrian stromatolites. Although Pannella has argued persuasively for the use of maximum counts, further work on those forms interpreted as subtidal and detailed analysis of their growth increment records may suggest a different approach.

e) General Comments

Although the increment records are now known to be more complex than previously thought, all authors explicitly or implicitly have considered or intentioned their data to reflect the solar day and synodical lunar month. None of the published data are, however, wholly satisfactory. At the present time, all the available figures should be treated as approximations rather than as precise quantities for mathematical analysis.

The most promising source of data is from bivalves in which the controlling factors of the growth increment record are currently best understood. The present data are likely to be inaccurate because of earlier misunderstanding of the interplay between lunar and solar day stimuli and the effects of tide type. It is dangerous to suggest an overall correcting factor (Pannella et al., 1968) and basically the data need re-recording with scrupulous regard to the details of the increment series.

The data from corals and more particularly other invertebrate groups are more difficult to assess for accuracy. A good deal of further work on their living representatives is necessary as a guide. The general agreement, however, between data from different sources in the Phanerozoic is encouraging. Data from stromatolites on the other hand must be viewed with considerable reserve at the moment.

Much of the published data have not been accompanied by sufficient supporting information. In the future all new contributions need to be accompanied by a full and well-illustrated account of the identification, interpretation, and recording techniques of basic increments and their rhythmic modulations.

4. Implications of the Growth Increment Data

In addition to their contribution in the astronomical field, growth increment data have considerable value in other areas such as palaeoecology (e.g., Craig and Hallam, 1963; Rhoads and Pannella, 1970) and archaeology (Coutts, 1975). Also Ma's use of the annual growth increment in corals to predict the position of past equators is well known (Ma, 1936 and many other papers; Fischer, 1964). This review, however, is primarily concerned with the use of the palaeontological data in the astronomical field, and these other applications are not considered further here.

The major stimulus to growth increment studies over the last thirteen years has been the potential value of the data to an understanding of the history of the Earth's rotation and the Earth-Moon system (Runcorn, 1964, 1975; Scrutton and Hipkin, 1973; Rosenberg and Runcorn, 1975). As others have pointed out (e.g., Dicke, 1966), in order to make a major contribution as an arbiter of the various theories in this field, the data must be of very great accuracy. It is doubtful if this is true of currently available data although the potential is there. Some broad conclusions can, however, be drawn from these figures.

At its most basic the data support a number of general observations. Firstly, on the basis of the stromatolite records, there is evidence for the existence of the Moon in orbit round the Earth extending back about 3000 m.y. The identification of tidal sediments in the Swaziland Supergroup extends this to 3300 m.y. (Brunn and Hobday, 1976). Secondly, during the Phanerozoic at least, there has been a gradual decrease in the number of days per year (Fig. 8). If the period of the Earth's rotation around the Sun has remained constant as is generally assumed, this can be translated into a gradual lengthening of the day. Thirdly, over the same period, the number of days per lunar month has also gradually decreased (Fig. 9) although the number of lunar months per year has apparently not changed significantly.

The slowing of the Earth's axial spin over geological time suggested by the palaeontological data lends support to the hypothesis of lunar tidal friction (e.g. Wells, 1963; Scrutton and Hipkin, 1973,). Furthermore, since the early Phanerozoic, a graphical representation of the palaeontological data for days per year and a projection back of days per year based on Muller and Stephenson's (1975) figure for the slow-down of the Earth's axial spin as an independent parameter based on the ancient eclipse records are more or less parallel in slope (Fig. 8).

It has been suggested by Pannella et al. (1968) that their data support a variable lunar tidal torque with relatively rapid slowing of the Earth's rotation rate during the early and mid Palaeozoic and from the Upper Cretaceous to the Present Day (see Runcorn, 1969). Similarly Runcorn (1968) on the basis of data from Wells (1963), Scrutton (1965), and Berry and Barker (1968) calculated the mean tidal torque between the Cretaceous and the present and between the Devonian and the present and found the former to be approximately twice the latter. Pannella et al. (1968) related changes in the tidal torque to the past distribution of shallow seas where the bulk of tidal energy dissipation has been considered to take place (Miller, 1966; Tarling, 1975; but see Brosche in discussion of Tarling, 1975).

More data than were used by Pannella et al. (1968), though not necessarily better, are now available. Several of their points were based on single specimens, as noted by Hazel and Waller (1969), who also pointed out small errors and discrepancies in their stratigraphic and radiometric data. Runcorn's (1968) analysis was also based on very limited data. Inspection of Figure 9 shows that if the maximum counts and the Triassic points are ignored, the justification for fitting a curved rather than a straight line is not strong, especially in view of the likely inaccuracies in the remaining data. Both Pannella et al. (1968) and Berry and Barker (1975) expressly considered their Triassic data to be suspect, so it is reasonable not to give them much weight at the present time. It must be concluded that although theoretical considerations suggest that the lunar torque may have varied in the Phanerozoic,

184

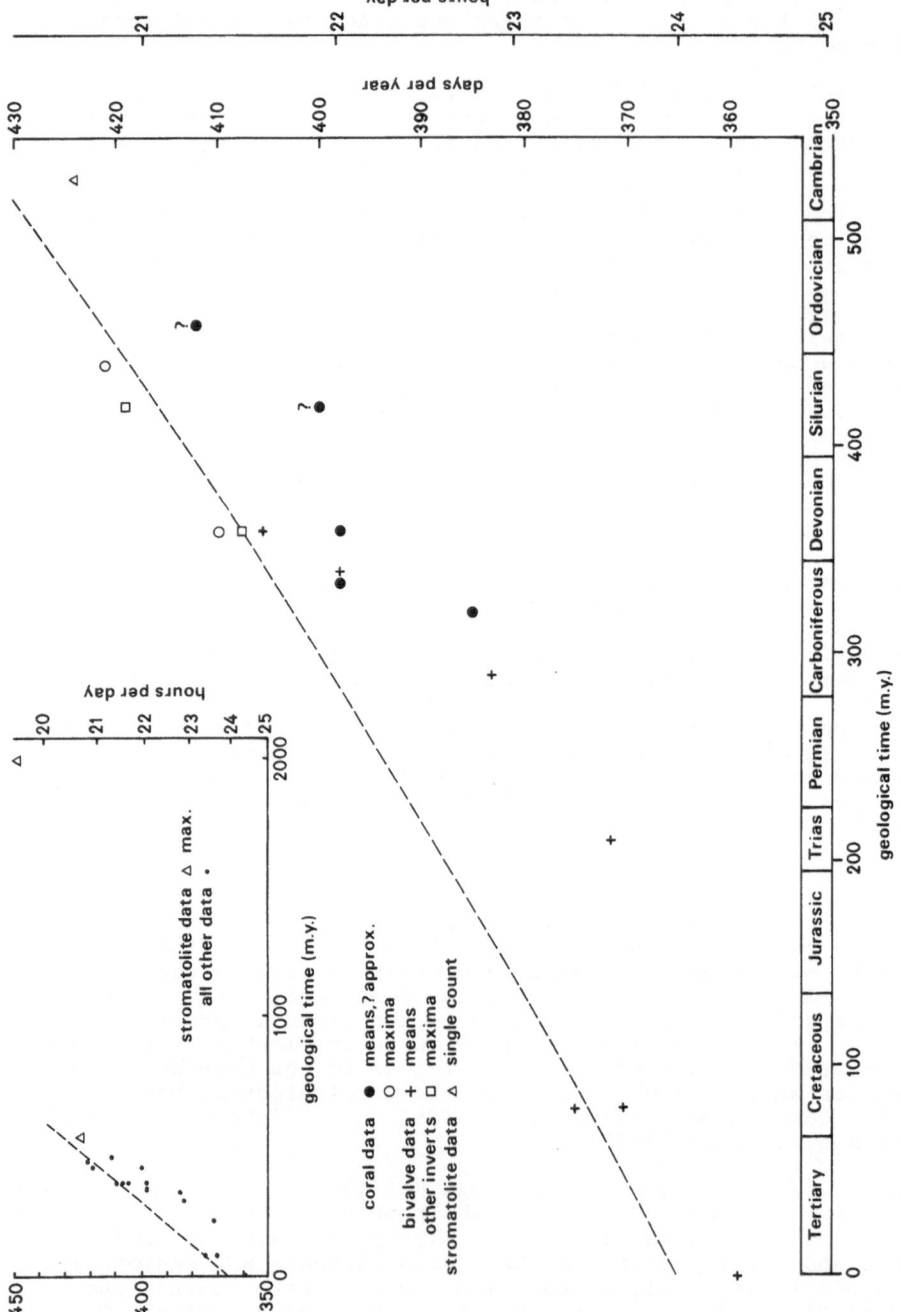

Fig. 8. Graphical representation of data for days per year (from Tables 1 – 4).
The *dashed line* is based on a lengthening of the day by 2 ms cent[-1] (Muller and Stephenson, 1975)

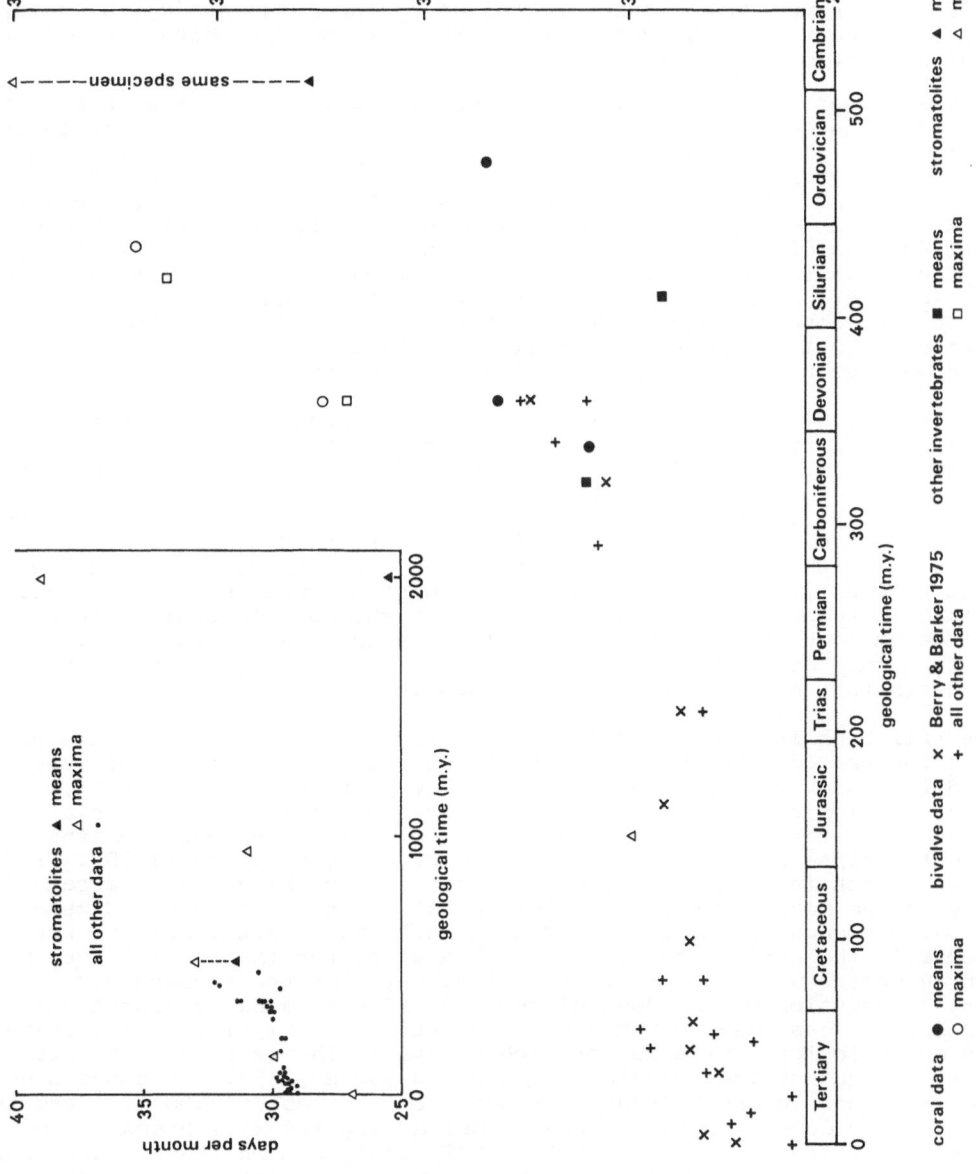

Fig. 9. Graphical representation of data for days per presumed synodic month (from Tables 1-4)

it is premature to consider this to be strongly supported by the palaeontological data. Calculations show, however, that the mean tidal
torque over the Phanerozoic, suggested by the palaeontological data,
is only about half the value calculated for the historical period from
the ancient eclipse data (Runcorn, 1964).

The interesting possibility suggested by Creer (1975) of a correlation
between changes in the value of the lunar tidal torque and disconti-
nuities in the geomagentic polarity bias must therefore be treated
with reservation. This correlation has been incorporated by Whyte (1977)
into a broad synthesis of Phanerozoic ocean ridge activity, sea level
and climatic changes, and their effects on extinction rates. The present data hardly justify the correlation, and certainly it is far too
speculative to use the relatively weak Triassic and Silurian data to
suggest periods of acceleration of the Earth's rate of axial spin as
Creer has done and Whyte accepts. Future data, however, should be capable of testing these possibilities, at least to the extent of detect-
ing more securely any past variation in the lunar tidal torque.

Runcorn (1964) also used the early data to separate the effects of the
lunar tidal torque on the Earth's rotation rate from those due to
changes in its internal state. He concluded that there had been little
if any change in the Earth's moment of inertia since the Devonian.
Creer (1975, Table 3) extended the calculations on the basis of later
data for the whole Phanerozoic but no clear trend emerged. He agreed,
however, with Scrutton and Hipkin (1973), that the computations could
not be considered very reliable in view of the weakness of the basic
data. It seems, however, that changes in the Earth's state requiring
large changes in the moment of inertia, such as a substantial expansion
during the Mesozoic (Egyed, 1956; Creer, 1965; Carey, 1975; Owen, 1976
and others), are unlikely unless G has varied.

Further back in geological time, the Precambrian and early Phanerozoic
are the periods over which palaeontological data could be most useful
and for which they have at the moment least to say. The reduction in
the rotation rate of the Earth by lunar tidal friction implies a recession of the Moon calculated at 5.8 cm/year on the basis of the ancient eclipse records analysed by Muller and Stephenson (1975). Pro-
jected back in time, assuming no change in the obliquity or eccentricity of the Moon's orbit, a recession rate of this order would mean a
close approach of the Earth and Moon such that crustal melting would
occur in the early Phanerozoic. With a value for the tidal torque more
in agreement with the palaeontological data for the Phanerozoic imply-
ing a recession of the Moon of about 3 cm/year, a close approach is
predicted less than 2000 m.y. ago (Scrutton and Hipkin, 1973). There
is no geological evidence for such an event. The fact that the stromatolite and sedimentological (Brunn and Hobday, 1976) evidence suggests a Moon in orbit around the Earth back to about 3300 m.y. and the
lack of evidence for exceptional tides during the late Precambrian and
earliest Palaeozoic (Klein, 1971, 1972) imply a much lower lunar tidal
torque in the remote past. Tarling (1975) has recently discussed this
problem. At the moment, the palaeontological data cannot contribute
much to its solution, because over the period 510 - 3000 m.y. ago, only
a handful of figures are available from stromatolites, most of them
doubtful (Pannella, 1976a) and their interpretation controversial
(Mohr, 1975). If we accept Pannella's (1976a) Mink data as a minimum
value this in turn suggests a minimum value for the lunar tidal torque,
but it also requires changes of state independent of lunar tidal friction to accommodate the data as discussed below.

Other possible manipulations of the data relating the length of the
month to the number of days per year, or hours per day, have been sug-

gested by Lamar and Merifield (1966, 1967), Weinstein and Keeny (1975), and Mohr (1975). Lamar and Merifield (1966) originally proposed that the length of the solar day could be calculated directly from the number of solar days per synodic month, assuming the hypothesis of lunar tidal friction and conservation of angular momentum. It is necessary, however, to take into account the solar tidal torque (Lamar and Merifield, 1967), and the method is only valid if it is assumed that the Earth's moment of inertia among other factors, is constant. The various theories of the Earth's evolution, however, imply a variety of different changes in the moment of inertia (Runcorn, 1964), and Lamar and Merifield (1967) modified their approach to show that Well's (1963) and Scrutton's (1965) data are consistent with a change of up to about 1 % since the Middle Devonian, depending on the value chosen for the ratio of the solar to lunar torque. The same reservation applies to this result as to the earlier result of Runcorn (1964).

Weinstein and Keeney (1975) used essentially the same approach to plot the variation in days per year as a function of days per month as required by the conservation of angular momentum in the Earth-Moon system, assuming the Earth's moment of inertia and G to be constant and neglecting solar tides, orbital eccentricities, and Hubble's constant. They showed that suitable data published by Pannella et al. (1968) and by Berry and Barker (1968) did not deviate substantially from the curve but that Mazzullo's (1971) data and Pannella's (1972b) Gunflint data did. The latter data, however, are regarded as particularly unreliable and their deductions of possible values of Hubble's constant based on these figures cannot be given much confidence.

Mohr (1975) plotted a family of curves similar to that of Weinstein and Keeney to estimate the number of solar days per year from the number of solar days per synodic month yielded by his Biwabik stromatolites. His curves, assuming a constant solar torque and moment of inertia through time, suggest a year of 880 solar days for Biwabik times. As discussed earlier, it can be argued that Mohr's mean value for days per month may be too low for various reasons. He points out, however, that Pannella's much higher figures would not fit on his curves. Mohr was here concerned about Pannella's Bulawayan data, which were considered capable of alternative interpretations (Pannella, 1972b, p. 234). Nevertheless the situation is similar with Pannella's latest Mink data, which he (Pannella) now regards as the best of the stromatolite data. If one accepts the Mink maximum figures in preference to Mohr's figures then from Mohr's curves either a lower moment of inertia and/or a higher solar torque and/or the operation of other factors is necessary to accommodate the data. However it is worth noting, as Mohr pointed out, that the mean value of days per month and the maximum value of days per year calculated from Pannella's (1972a,b) original Gunflint and Biwabik data would plot on the appropriate curve. From previous discussion, this need not be an unacceptable treatment of the data.

The situation at the moment is thus very unsatisfactory, because the stromatolite figures are so difficult to interpret for reliability. As Pannella (1976a) remarks, the odds are very low that stromatolites will ever provide data with any degree of accuracy. There is also the problem of recognising when the data are accurate, which may be more difficult to resolve in this group than in some of the invertebrates.

A slightly different use of the palaeontological data is in the reconstruction of tidal patterns. There appears to be considerable scope in this field particularly with bivalve data. To some extent it should be a natural consequence of extracting accurate data from intertidal and

immediately subtidal bivalve growth patterns in the future, because
for this the interference of the solar and tidal cycles will have to
be unravelled. Pannella (1976b) constructed a qualitative curve of
tidal amplitude for the late Cretaceous interior sea of the U.S.A.
from the analysis of the growth increment record of a group of bivalves
from the Fox Hills Formation of South Dakota (Fig. 10). He has also
suggested that stromatolites, despite their problems, may allow the
reconstruction of palaeotide patterns for the Precambrian, and indeed
the running mean curve of increment thickness for the Mink stromatolites is an encouraging approximation to a curve of tidal amplitude
(Pannella, 1976a, Fig. 2).

The problem of calculating an actual figure for tidal amplitude is
more difficult. Cloud (1968), following Logan's (1961) interpretation
of modern intertidal stromatolites in Shark Bay, Australia, took the
maximum growth relief of fossil stromatolites as equal to tidal amplitude in the environment of growth. He interpreted figures available
to him as indicating that Precambrian stromatolites, particularly those
older than 1000 m.y., grew in environments of greater tidal amplitude
than at present. This indication that the Moon was in orbit and possibly closer to the Earth in the mid Precambrian he concluded was consistent with the hypothesis of lunar capture and closest approach in
the early Precambrian. Cloud quoted records of stromatolites up to 6 m
high but as Walter (1970) pointed out, not only are these probably not
mid but latest Precambrian in age, but stromatolites up to 15 m high
had been recorded in early to middle Cambrian rocks near Lake Baikal
in Siberia. If growth height in stromatolites was a reliable indicator
of tidal amplitude this would suggest the closest approach of the Moon
in the early Cambrian. Walter, however, has shown that this supposition
is unwarranted. Not only is there rarely firm evidence that Precambrian
stromatolites were intertidal, but Playford and Cockbain (1969) demonstrated that some Devonian fore-reef "mound-shaped" stromatolites
grew in water depths of up to 45 m. In addition, isolated data on tidal amplitude could be unreliable as an indication of the tidal torque,
because amplitudes may vary considerably with different environmental
situations. It is interesting to note, however, that although Cloud's
stromatolitic tidal gauge must be discarded, recent sedimentological
work suggests higher (12 - 25 m) tides about 3000 m.y. ago (Brunn and
Hobday, 1976).

Pannella (1976a) suggests that stromatolites may still yield information on tidal amplitude. He produced a table of theoretical growth
pattern frequencies in sublittoral, littoral, and supralittoral environments and proposed that such changes in pattern frequency should
be sought for by tracing bedding planes of stromatolites perpendicu-
larly to the shore line of the time. Pannella (1976b, p. 540) also
suggests that other organisms, such as bivalves, could be used in the
same way. This method, although laborious, might provide the width of
the intertidal zone, but it is difficult to see how this could be translated into a tidal amplitude unless a very accurate means of determining
the palaeoslope was available. Pannella (1976a, p. 681) goes on to note
that vertical variations are often easier to study than horizontal ones
and quotes examples from two vertical sequences. One from the early
Proterozoic, containing repeated rhythms of columnar stromatolites —
stratiform stromatolites — finely laminated dolomites, is interpreted
as reflecting intertidal (or slightly subtidal) to supratidal conditions. The cycles are consistently 1.5 m thick, which Pannella considers to represent the tidal amplitude of the locality. His second example is from the Mink stromatolites (Fig. 7b). Some of these show an
upwards sequence of form changes from erect columns to coalescent and
linked mats with increasing abundance of ooliths and decreasing numbers

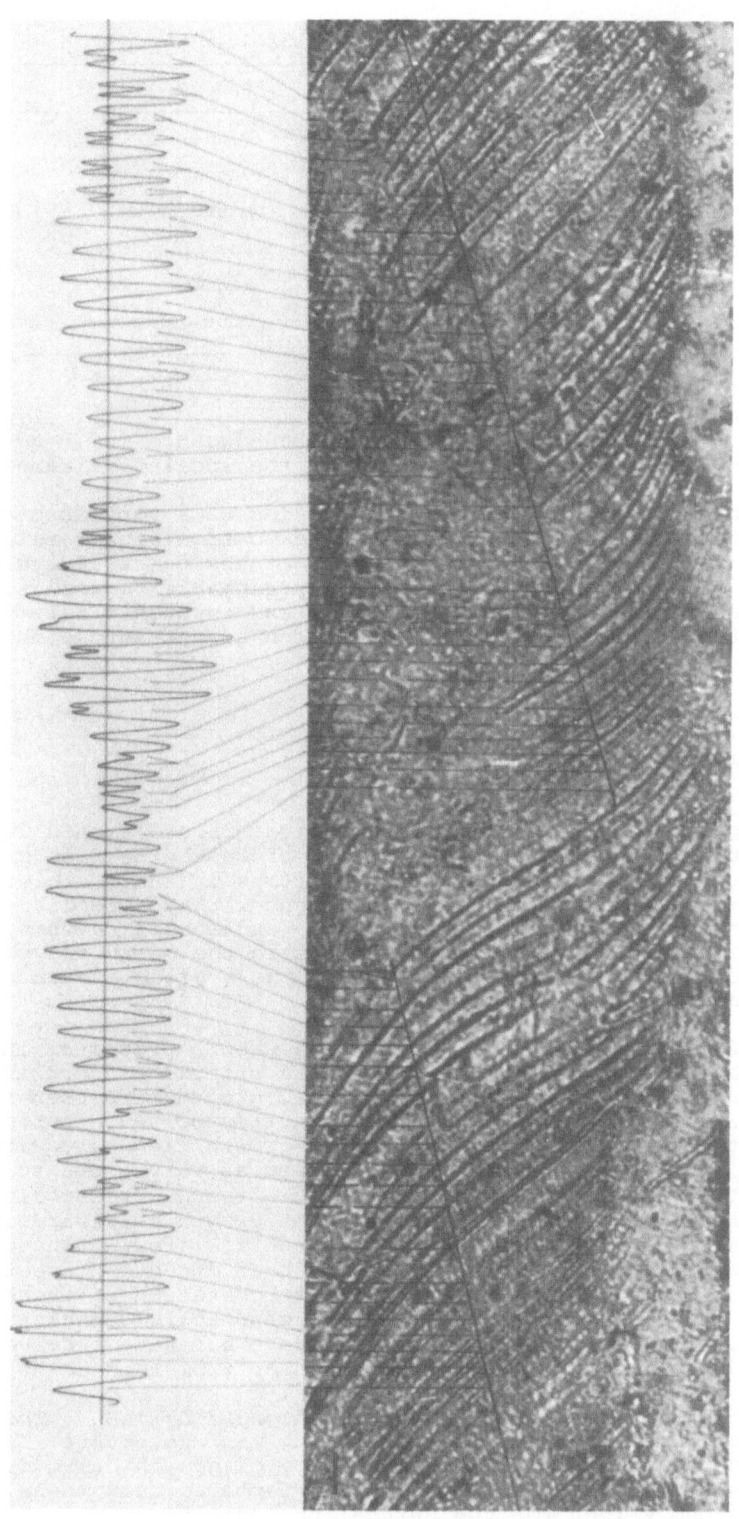

Fig. 10. Tide-curve reconstruction from growth patterns in the bivalve *Limopsis striatus-punctatus*. Upper Cretaceous, Fox Hills Fm; South Dakota, U.S.A. Scale 20 μm. Reproduced by kind permission of Dr. Giorgio Pannella and the publishers from Pannella, 1976, *Naturwissenschaften*, 63, fig. 4. Copyright © 1976 by Springer Verlag

Table 5. Approximate ages of Precambrian stratigraphic units

Unit	Approximate age	Source
Belt Supergroup	950 m.y.	Pannella, 1972b
Gunflint Formation	2000 m.y.	Pannella, 1976a
Biwabik Formation	2000 m.y.	Mohr, 1975
Gt. Slave Supergroup	1845 – 2370 m.y.	Pannella, 1972a
Steeprock Group	2500 m.y.	Pannella, 1972a
Bulawayan Group	3000 m.y.	Truswell and Eriksson, 1975

of laminae per fortnightly tidal band. The sequence is 15 cm thick, which again Pannella regards as possibly the local tidal amplitude.

Although this method has potential it relies on a very accurate ecological interpretation of growth form and no significant relative rise or fall in sea level during the build-up of the vertical cycle. The initiation of new cycles, however, does presumably require some eustatic or tectonic adjustment of base level, and in such a setting, the assumption of uniform conditions between each adjustment seems rather doubtful.

5. Conclusions

Various organisms with accretionary skeletons record the influence of environmental rhythms on their metabolic processes as fine growth increments grouped into a hierarchy of patterns in their skeletons. Based on an identification of the basic increment with a circadian rhythm and the patterns with tidal, equinoctial, and annual rhythms, increment counts give values for the number of days in the month and year in the past. Contributors have assumed that all the available data reflect the solar day and the synodical month.

Of the organisms that have provided data, growth processes in bivalves and their ability to reflect environmental stimuli are probably those now best understood. Some aspects of coral growth have been studied, but the origin of the monthly rhythm is not supported experimentally and the detailed morphology of the daily growth ridges requires further investigation. Stromatolites are recognised as very prone to environmental disturbance at the present day and although Precambrian examples are not directly comparable, discontinuous growth is regarded as likely in most cases although least serious in subtidal forms. The origin of tidal and annual growth patterns has not been directly investigated in living stromatolites, although the interpretation of these patterns seems theoretically and numeralogically reasonable. Cephalopod molluscs and brachiopods have also provided some data but have received little direct study of their shell growth processes from this point of view.

Most data from invertebrates have been recorded as means. This is regarded as more satisfactory than recording maximum counts. In stromatolites, on the other hand, the likelihood of growth being discontinuous has resulted in maximum counts being favoured over means, and then only as minimum values for the natural periodicities. This approach, however, may not always be justified. Mechanical means of recording data have not proved successful. All the available data have been obtained by the direct counting of increments by the observer. The measurement of individual increment thickness, which has been done in some cases, allows better manipulation of the data.

The published data are regarded as approximate rather than precise reflections of days per month and year in the past for various reasons. Some data were obtained with insufficiently rigorous methods and others with misinterpretations of the detailed growth increment record. Some data are recognised as less reliable than others, particularly those from stromatolites. There is a possibility that data published in the future, particularly from bivalves, could be very accurate if based on painstaking and complete analysis of increment patterns from carefully selected growth sequences. All future published data should be accompanied by the fullest details of material selection, methodology, and basis for growth record interpretation.

The available data allow some general conclusions on the past history of the rotation rate of the Earth and the Earth-Moon system. (1) Tidal patterns in Precambrian stromatolites suggest that the Moon has been in orbit around the Earth for the last ca. 3000 m.y. at least. (2) There has been a gradual decrease in the number of days per year during the Phanerozoic and probably before. This implies a gradual and concomitant lengthening of the day. (3) There has been a gradual decrease in the number of days per lunar month during the Phanerozoic and possibly before. The number of lunar months in the year has apparently remained fairly constant. (4) The decrease in the rotation rate of the Earth over geological time generally supports the hypothesis of lunar tidal friction. (3) The rate of deceleration of the Earth's axial spin since the Ordovician suggested by palaeontological data agrees broadly with a back projection of the Earth's acceleration as an independent parameter computed from eclipse data over the historical period. (6) It is likely that the lunar tidal torque has varied with time, but the present data are not sufficiently refined to confirm this. (7) Palaeontological data from the Precambrian and earliest Phanerozoic are considered largely unreliable. (8) Computations suggest that large-scale changes in the Earth's moment of inertia during the Phanerozoic are unlikely.

In addition, growth increment records can be used to plot qualitative tidal curves for the environment of growth, and thus the tidal characteristics of ancient seas may be deduced. Calculation of tidal amplitude is more difficult but horizontal or vertical sequences of environmentally linked growth forms and increment patterns, with certain assumptions, may provide estimates.

References

Aschoff, J. (Ed.): Circadian Clocks. 462pp. Amsterdam: North Holland Publ. Co., 1965

Awramik, S.A.: Precambrian columnar stromatolite diversity: reflection of metazoan appearance. Science 174, 825-827 (1971)

Barker, R.M.: Microtextural variations in pelecypod shells. Malacologia 2, 69-86 (1964)

Barnes, D.J.: The structure and formation of growth-ridges in scleractinian coral skeletons. Proc. R. Soc. London B, 182, 331-350 (1972)

Berry, W.B.N., Barker, R.M.: Fossil bivalve shells indicate longer month, year in Cretaceous than present. Nature (London) 217, 938-939 (1968)

Berry, W.B.N., Barker, R.M.: Growth increments in fossil and modern bivalves. In: Growth Rhythms and the History of the Earth's Roation. Rosenberg, G.D, Runcorn, S.K. (eds.). London: Wiley, 1975, p. 9-25

Brown, F.A., Jr.: Persistent activity rhythms in the oyster. Am. J. Physiol. 178, 510-514 (1954)

Brown, F.A., Jr., Hastings, J.W., Palmer, J.D.: The Biological Clock; Two Views. New York: Academic Press, 1970, pp. viii + 1-94

Brunn, V. von, Hobday, D.K.: Early Precambrian tidal sedimentation in the Pongola Supergroup of South Africa. J. Sedim. Petrol. 46, 670-679 (1976)

Buddemeier, R.W.: Environmental controls over annual and lunar monthly cycles in hermatypic coral calcification. Proc. 2nd Int. Symp. Coral Reefs 2, 259-267 (1974)

Buddemeier, R.W., Kinzie, R.A., III: The chronometric reliability of contemporary corals. In: Growth Rhythms and the History of the Earth's Rotation. Rosenberg, G.D., Runcorn, S.K. (eds.). London: Wiley, 1975, p. 135-147

Buddemeier, R.W., Maragos, J.E., Knutson, D.W.: Radiographic studies of reef coral exoskeletons: rates and patterns of coral growth. J. Exp. Mar. Biol. Ecol. 14, 179-200 (1974)

Carey, W.S.: The expanding Earth — an essay review. Earth-Sci. Rev. 11, 105-143 (1975)

Clark, G.R., II: Mollusk shell: daily growth lines. Science 161, 800-802 (1968)

Clark, G.R., II: Calcification on an unstable substrate: marginal growth in the mollusk *Pecten diegensis*. Science 183, 968-970 (1974a)

Clark, G.R., II: Growth lines in invertebrate skeletons. A. Rev. Earth Planet Sci. 2, 77-99 (1974b)

Clark, G.R., II: Periodic growth and biological rhythms in experimentally grown bivalves. In: Growth Rhythms and the History of the Earth's Rotation. Rosenberg, G.D., Runcorn, S.K. (eds.). London: Wiley, 1975, p. 103-117

Cloud, P.E.: Atmospheric and hydrospheric evolution on the primitive Earth. Science 160, 729-736 (1968)

Coutts, P.J.F.: The seasonal perspective of marine-orientated Prehistoric hunter-gatherers. In: Growth Rhythms and the History of the Earth's Rotation. Rosenberg, G.D., Runcorn, S.K. (eds.). London: Wiley, 1975, p. 243-252

Craig, G.Y., Hallam, A.: Size-frequency and growth-ring analyses of *Mytilus edulis* and *Cardium edule*, and their palaeoecological significance. Palaeontology 6, 731-750 (1963)

Creer, K.M.: An expanding Earth. Nature (London) 205, 539-544 (1965)

Creer, K.M.: On a tentative correlation between changes in the geomagnetic polarity bias and reversal frequency and the Earth's rotation through Phanerozoic time. In: Growth Rhythms and the History of the Earth's Rotation. Rosenberg, G.D., Runcorn, S.K. (eds.). London: Wiley, 1975, p. 293-318

Davenport, D.B.: Growth lines in fossil pectens as indicators of past climates. J. Plaeont. 12, 514-515 (1938)

Dicke, R.H.: The secular acceleration of the Earth's rotation and cosmology. In: The Earth-Moon System. Marsden, B.G., Cameron, A.G.W. (eds.). New York: Plenum, 1966, p. 98-164

Dodge, R.E., Thomson, J.: The natural radiochemical and growth records in contemporary hermatypic corals from the Atlantic and Caribbean. Earth Planet Sci. Lett. 23, 313-322 (1974)

Dodge, R.E., Vaisnys, J.R.: Hermatypic coral growth banding as environmental recorder. Nature (London) 258, 706-708 (1975)

Dolman, J.: A technique for the extraction of environmental and geophysical information from growth records in invertebrates and stromatolites. In: Growth Rhythms and the History of the Earth's Rotation. Rosenberg, G.D., Runcorn, S.K. (eds.). London: Wiley, 1975, p. 191-222

Egyed, L.: Determination of changes in the dimensions of the Earth from palaeo-graphical data. Nature (London) 178, 534 (1956)

Evans, J.W.: Tidal growth increments in the cockle *Clinocardium nuttalli*. Science 176, 416-417 (1972)

Evans, J.W.: Growth and micromorphology of two bivalves exhibiting nondaily growth lines. In: Growth Rhythms and the History of the Earth's Rotation. Rosenberg, G.D., Runcorn, S.K. (eds.). London: Wiley, 1975, p. 119-134

Farrow, G.E.: Periodicity structures in the bivalve shell: experiments to establish growth controls in *Cerastoderma edule* from the Thames estuary. Palaeontology 14, 571-588 (1971)

Farrow, G.E.: Periodicity structures in the bivalve shell: analysis of stunting in *Cerastoderma edule* from the Burry Inlet (South Wales). Palaeontology 15, 61-72 (1972)

Fischer, A.G.: Growth patterns of Silurian Tabulata as palaeoclimatologic and palaeographic tools. In: Problems in Palaeoclimatology. Nairn, A.E.M. (ed.) London: Wiley, 1964, p. 608-617

Gebelein, C.D.: Distribution, morphology and accretion rate of recent subtidal stromatolites (Bermuda). J. Sediment. Petrol. 39, 49-69 (1969)

Goreau, T.F.: The physiology of skeleton formation in corals. I. A method for measuring the rate of calcium deposition by corals under different conditions. Biol. Bull. Mar. Biol. Lab. Woods Hole 116, 59-75 (1959)

Goreau, T.F., Goreau, N.I.: The physiology of skeleton formation in corals. II. Calcium deposition by hermatypic corals under various conditions in the reefs. Biol. Bull. Mar. Biol. Lab. Woods Hole 117, 239-250 (1959)

Hall, C.A. Jr.: Shell growth in *Tivela stultorum* (Mawe, 1823) and *Callista chione* (Linnaeus) (Bivalvia): Annual periodicity, latitudinal differences, and diminution with age. Palaeogeogr. Palaeoclimatol. Palaeoecol. 15, 33-61 (1974)

Hall, C.A. Jr.: Latitudinal variation in shell growth patterns of bivalve molluscs: implications and problems. In: Growth Rhythms and the History of the Earth's Rotation. Rosenberg, G.D., Runcorn, S.K. (eds.). London: Wiley, 1975, p. 163-175

Hastings, J.W., Schweiger, H.-G. (eds.): The Molecular Basis for Circadian Rhythms. Berlin: Dahlem Konferenzen, 1976, 464pp.

Hazel, J.E., Waller, T.R.: Stratigraphic data and length of the synodic month. Science 164, 201-202 (1969)

Hipkin, R.G.: Some Aspects of the Dynamical History of the Earth-Moon System. Ph.D. Thesis, University of Newcastle upon Tyne, 1972, 220pp.

House, M.R., Farrow, G.E.: Daily growth banding in the shell of the cockle, *Cardium edule*. Nature (London) 219, 1384-1386 (1968)

Jell, J.S., Hill, D.: The microstructure of corals. In: Drevnie Cnidaria. Sokolov, B.S. (ed.). Novosibirsk: Nauka, 1974, 1, 8-14

Johnson, G.A.L., Nudds, J.R.: Carboniferous coral geochronometers. In: Growth Rhythms and the History of the Earth's Rotation. Rosenberg, G.D., Runcorn, S.K. (eds.). London: Wiley, 1975, p. 27-42

Klein, G. de V.: A sedimentary model for determining paleotidal range. Bull. Geol. Soc. Am. 82, 2585-2592 (1971)

Klein, G. de V.: Determination of paleotidal range in clastic sedimentary rocks. Int. Geol. Congr. 24, Sect. 6, 397-405 (1972)

Knutson, D.W., Buddemeier, R.W., Smith, S.V.: Coral chronometers: seasonal growth bands in reef corals. Science 177, 270-272 (1972)

Lamar, D.L., Merifield, P.M.: Length of Devonian day from Scrutton's coral data. J. Geophys. Res. 71, 4429-4430 (1966)

Lamar, D.L., Merifield, P.M.: Influence of solar tidal torque on length of day and synodic month. J. Geophys. Res. 72, 3734-3735 (1967)

Logan, B.W.: Cryptozoan and associated stromatolites from the Recent, Shark Bay, Western Australia. J. Geol. 69, 517-533 (1961)

Ma, T.Y.H.: On the seasonal change of growth in some Palaeozoic corals. Proc. Imp. Akad. Tokyo 9, 407-409 (1933)

Ma, T.Y.H.: On the seasonal change of growth in a reef coral, *Favia speciosa* (Dana), and the water-temperature of the Japanese seas during the latest geological times. Proc. Imp. Acad. Tokyo 10, 353-356 (1934a)

Ma, T.Y.H.: On the growth rate of reef corals and the sea water temperature in the Japanese Islands during the latest geological times. Sci. Rep. Tohoku Imp. Univ., 2nd Ser. (Geol.) 16, 166-189 (1934b)

Ma, T.Y.H.: On the Devonian equator located by the growth rate of tetracorals. J. Geol. Soc. Japan. 43, 353-356 (1936)

Ma, T.Y.H.: Growth rate of reef corals and its relation to sea water temperature. Palaeontol. Sin., Ser. B, 21, 326pp. (1937)

Mazzullo, S.J.: Length of the year during the Silurian and Devonian period: new values. Bull. Geol. Soc. Am. 82, 1085-1086 (1971)

McGugan, A.: Possible use of algal stromatolite rhythms in geochronology. Abstr. Mtg. Geol. Soc. Am. 1967, 145 (1967)

Miller, R.G.: The flux of tidal energy out of the deep ocean. J. Geophys. Res. 71, 2485-2489 (1966)

Mohr, R.E.: Measured periodicities of the Biwabik (Precambrian) stromatolites and their geophysical significance. In: Growth Rhythms and the History of the Earth's Roation. Rosenberg, G.D., Runcorn, S.K. (eds.). London: Wiley, 1975, p. 43-56

Monty, C.L.V.: Recent algal stromatolites in the windward lagoon, Andros, Island, Bahamas. Ann. Soc. Geol. Belg. 88, 271-276 (1965)

Monty, C.L.V.: Distribution and structure of Recent stromatolitic algal mats, Eastern Andros Island, Bahamas. Ann. Soc. Geol. Belg. 90, 55-100 (1967)

Monty, C.L.V.: Precambrian background and Phanerozoic history of stromatolitic communities, an overview. Ann. Soc. Geol. Belg. 96, 585-624 (1973)

Muller, P.M., Stephenson, F.R.: The acceleration of the Earth and Moon from early astronomical observations. In: Growth Rhythms and the History of the Earth's Rotation. Rosenberg, G.D., Runcorn, S.K. (eds.). London: Wiley, 1975, p. 459-534

Munk, W.H., MacDonald, G.J.F.: The Rotation of the Earth. London: Cambridge University Press, 1960, 323pp.

Owen, H.G.: Continental displacement and expansion of the Earth during the Mesozoic and Cenozoic. Philos. Trans. R. Soc., A, 281, 223-291 (1976)

Pannella, G.: Fish otoliths: daily growth layers and periodical patterns. Science 173, 1124-1127 (1971)

Pannella, G.: Precambrian stromatolites as palaeontological clocks. Int. Geol. Congr. 24, Sect. 1, 50-57 (1972a)

Pannella, G.: Paleontological evidence on the Earth's rotational history since early Precambrian. Astrophys. Space Sci. 16, 212-237 (1972b)

Pannella, G.: Paleontological clocks and the history of the Earth's rotation. In: Growth Rhythms and the History of the Earth's Rotation. Rosenberg, G.D., Runcorn, S.K. (eds.). London: Wiley, 1975, p. 253-284

Pannella, G.: Geophysical inferences from stromatolite lamination. In: Stromatolites. Walter, M.R. (ed.). Amsterdam: Elsevier, 1976a, p. 673-685

Pannella, G.: Tidal growth patterns in Recent and fossil mollusc bivalve shells: a tool for the reconstruction of paleotides. Naturwissenschaften 63, 539-543 (1976b)

Pannella, G., MacClintock, C.: Biological and environmental rhythms reflected in molluscan shell growth. J. Paleontol. Mem. 42, 64-80 (1968)

Pannella, G., MacClintock, C., Thompson, M.N.: Paleontological evidence of variations in length of synodic month since late Cambrian. Science 162, 792-796 (1968)

Playford, P.E., Cockbain, A.E.: Algal stromatolites: deepwater forms in the Devonian of Western Australia. Science 165, 1008-1010 (1969)

Rhoads, D.C., Pannella, G.: The use of molluscan shell growth patterns in ecology and palaeoecology. Lethaia 3, 143-161 (1970)

Rosenberg, G.D.: Calcium concentration in the bivalve *Chione undatella* Sowerby. Nature (London) 244, 155-156 (1973)

Rosenberg, G.D., Jones, C.B.: Approaches to chemical periodicities in molluscs and stromatolites. In: Growth Rhythms and the History of the Earth's Rotation. Rosenberg, G.D., Runcorns, S.K. (eds.). London: Wiley, 1975, p. 223-242

Rosenberg, G.D., Runcorn, S.K. (eds.): Growth Rhythms and the History of the Earth's Rotation. London: Wiley, 1975, pp.xvi + 1-559

Runcorn, S.K.: Changes in the Earth's moment of inertia. Nature (London) 204, 823-825 (1964)

Runcorn, S.K.: Corals as paleontological clocks. Sci. Am. 215, (4), 26-33 (1966)

Runcorn, S.K.: Fossil bivalve shells and the length of the month and year in the Cretaceous. Nature (London) 218, 459 (1968)

Runcorn, S.K.: Tidal friction and time. Science 163, 1227 (1969)

Runcorn, S.K.: Palaeontological and astronomical observations on the rotational history of the Earth and Moon. In: Growth Rhythms and the History of the Earth's Rotation. Rosenberg, G.D., Runcorn, S.K. (eds.). London: Wiley, 1975, p. 285-291

Scrutton, C.T.: Periodicity in Devonian coral growth. Palaeontology 7, 552-558 (1965)

Scrutton, C.T.: Evidence for a monthly periodicity in the growth of some corals. In: Palaeogeophysics. Runcorn, S.K. (ed.). London: Academic Press, 1970, p. 11-16

Scrutton, C.T., Hipkin, R.G.: Long-term changes in the rotation rate of the Earth. Earth-Sci. Rev. 9, 259-274 (1973)

Sharp, J.H.: Blue-green algae and carbonates. *Schizothrix calcicola* and algal stromatolites from Bermuda. Limnol. Oceanogr. 14, 568-578 (1969)

Sorauf, J.E., Jell, J.S.: Structure and incremental growth in the ahermatypic coral *Desmophyllum cristagalli* from the North Atlantic. Palaeontology 20, 1-19 (1977)

Stearn, C.W., Scoffin, T.P., Martindale, W.: Calcium carbonate budget of a fringing reef on the west coast of Barbados, Part I. Zonation and productivity. Bull. Mar. Sci. 27, 479-510 (1977)

Stubbs, P.: Coral timekeepers of the slowing Earth. New Sci. 29 (489), 828-829 (1966)

Tarling, D.H.: Geological processes and the Earth's rotation in the past. In: Growth Rhythms and the History of the Earth's Rotation. Rosenberg, G.D., Runcorn, S.K. (eds.). London: Wiley, 1975, p. 397-412

Thompson, I.: Biological clocks and shell growth in bivalves. In: Growth Rhythms and the History of the Earth's Rotation. Rosenberg, G.D., Runcorn, S.K. (eds.). London: Wiley, 1975, p. 149-162

Truswell, J.F., Eriksson, K.A.: Facies and laminations in the Lower Proterozoic Transvaal Dolomite, South Africa. In: Growth Rhythms and the History of the Earth's Rotation. Rosenberg, G.D., Runcorn, S.K. (eds.). London: Wiley, 1975, p. 57-73

Walter, M.R.: Stromatolites used to determine the time of nearest approach of Earth and Moon. Science 170, 1331-1332 (1970)

Walter, M.R.: A hot spring analog for the depositional environment of Precambrian Iron Formation of the Lake Superior region. Econ. Geol. 67, 965-980 (1972)

Walter, M.R.: Stromatolites. Amsterdam: Elsevier, 1976

Walter, M.R., Bauld, J., Brock, T.D.: Siliceous algal and bacterial stromatolites in hot springs and geyser effluents of Yellowstone National Park. Science 178, 402-405 (1972)

Weber, J.N., Deines, P., White, E.W., Weber, P.H.: Seasonal high and low density bands in reef coral skeletons. Nature (London) 255, 697-698 (1975a)

Weber, J.N., White, E.W., Weber, P.H.: Correlation of density banding in reef coral skeletons with environmental parameters: the basis for interpretation of chronological records preserved in the coralla of corals. Paleobiology 1, 137-149 (1975b)

Weinstein, D.H., Keeney, J.: Palaeontology and the dynamic history of the Sun-Earth-Moon system. In: Growth Rhythms and the History of the Earth's Rotation. Rosenberg, G.D., Runcorn, S.K. (eds.). London: Wiley, 1975, p. 377-384

Wells, J.W.: Individual variation in the rugose coral species *Heliophyllum halli*. Edwards and Haime. Palaeontogr. Am. 2, 1-22 (1937)

Wells, J.W.: Coral growth and geochronometry. Nature (London) 197, 948-950 (1963)

Wells, J.W.: Problems of annual and daily growth-rings in corals. In: Palaeogeophysics. Runcorn, S.K. (ed.). London: Academic Press, 1970, p. 3-9

Whitfield, R.P.: Notice of a remarkable specimen of the West Indian coral *Madrepora palmata*, Lk. Bull. Am. Mus. Nat. Hist. 10, 463-464 (1898)

Whyte, M.A.: Time, tide and the cockle. In: Growth Rhythms and the History of the Earth's Rotation. Rosenberg, G.D., Runcorn, S.K. (eds.). London: Wiley, 1975, p. 177-189

Whyte, M.A.: Turning points in Phanerozoic history. Nature (London) 267, 679-682 (1977)

Addendum

Since this paper was written, the Proceedings of the 12th International Conference of the International Society for Chronobiology held in Washington, August 1975 has become available. The book contains a number of articles which are of relevance to this review:

Pannella, G.: Periodical growth patterns in some calcified tissues   p.649

Berry, W.B.N., Barker, M.: Shell growth in bivalves: its relationship
  to environmental factors   p.657

196

Kennish, M.J.: Growth increment analysis of *Mercenaria mercenaria* from
   artificially heated coastal marine waters: a practical monitoring method   p.663
Weber, J.N., White, E.W., Deines, P., Weber, P.H., Baker, P.A.: Skeleto-
   genic biorhythms in reef corals   p.671
Clark II, G.R.: Growth lines as biological rhythms: evidence from the
   bivalve family Pectinidae   p.677
Rosenberg, G.D.: A classification of growth rhythm terminology and its
   significance   p.685

These papers report new experimental work, mainly on living bivalves, and review
previous contributions on the environmental and temporal significance of growth
rhythms and their terminology. None of the papers is particularly directed towards
the astronomical significance of growth increment data, and the conclusions of the
present review are not materially affected.

Halberg, F. (ed.): Proc. 12th Int. Conf. Int. Soc. Chronobiol. pp. **xxxii** + 782.
   Milan: Il Ponte, 1977

Geological and Geophysical Evidence Relating to Continental Growth and Dynamics and the Hydrosphere in Precambrian Times: a Review and Analysis

J. D. A. Piper

1. Introduction

A very large number of geological and geophysical observations have been forthcoming in recent years from the Precambrian areas of the globe. Although these studies are unevenly distributed and come largely from the cratonic regions of Laurentia (the Canadian Shield, Greenland, and NW Scotland), southern Africa, and western Australia, they have shed considerable light on the early history of the Earth. They provide a sound basis for the tectonic subdivision of Precambrian times into the Archaean (> ca. 2700 m.y.) and Proterozoic (2700 - 600 m.y.) eras and show for example that a hydrosphere and shallow seas have existed on the Earth since 3800 m.y. while the history of the Earth-Moon system extends back to at least 3000 m.y. We can, however, only surmise as to the nature of the oceanic (simatic) components of the lithosphere in pre-Palaeozoic times because, with a few possible exceptions, this has been entirely destroyed. According to the concepts of the New Global Tectonics (Isacks et al., 1968) we understand that this oceanic component is generated at the mid-ocean ridges and subducted below the continental lithosphere by the action of a mobile circulating asthenosphere (Hess, 1962; Dietz, 1961). The nature of the ocean crust, which must have covered at least 70 % of the globe in post-Archaean times (< 2500 m.y.), is therefore a matter of speculation only.

This contribution reviews the evidence relating to the growth and distribution of the continents and to the hydrosphere in Precambrian times, and shows how the collective data may in a qualitative way explain observations relating to the Precambrian history of the Earth-Moon system, the terrestrial evidence for which is also outlined.

2. The Hydrosphere

Since recognition that the noble gases are depleted by factors of 10^{-7}-10^{-11} compared with their cosmic abundances, it has been appreciated that the atmosphere and hydrosphere have a secondary origin from degassing of the mantle. For the hydrosphere this is confirmed by the paucity of terrestrial neon: Although neon has a higher molecular weight than water, it was apparently entirely lost in the early history of the Earth when the constituents of the present hydrosphere were presumably still locked in silicate mineral phases. The early melting of the Moon led to the formation of anorthositic crust, but in the Earth differentiation of a molten mantle would have produced much more water together with alkali metals giving a calcalkaline frac-

Sub-Department of Geophysics, University of Liverpool, P.O. Box 147, Oxford Street, Liverpool L69 3BX, England

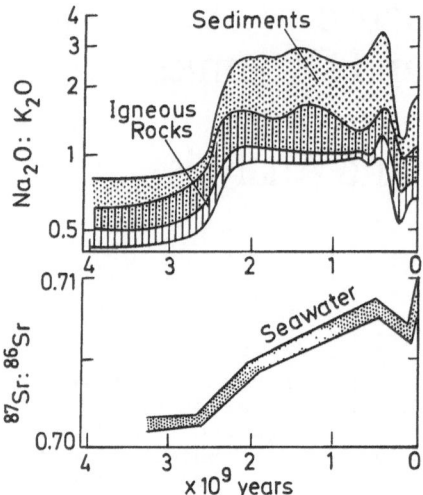

(a) Changes in the Na_2/K_2O ratios of igneous and sedimentary rocks during geological time after Engel et al. (1974), and (b) changes in the $^{87}Sr/^{86}Sr$ ratio of sea water during geological time after Veizer (1976)

tionation trend to igneous derivatives. Present discussion focuses on the rate at which the hydrosphere has accumulated, with models relating to catastrophic (Walker, 1976) or continuous growth (Fisher, 1976). The oldest sedimentary rocks comprise water-laid clastic and chemical deposits requiring the existence of both an atmosphere (for weathering but in an anoxygenic environment) and a hydrosphere (for deposition). The occurrence of deposits of probable shallow water origin in the Isua supracrustal sequence of West Greenland ($3,760 \pm 70$ m.y., Moorbath et al., 1973, 1975) and quartzite/marble/pelite associations in other Archaean terrains, together with widespread development of shallow marine deposits in Proterozoic (2700 - 570 m.y.) terrains, implies that the continents have presented a low freeboard to the hydrosphere through most of geological time in all but tectonically disturbed regions. The growth of the hydrosphere must therefore be related to growth of the continents although the mechanism of this equilibrium is unknown.

The ^{18}O isotopic composition of sedimentary carbonates is presumed to reflect the isotopic composition of sea water and has been examined from 3,200 m.y. to the present by Schidlowski et al. (1975) who demonstrate a progressive increase in ^{18}O through geological time. Entirely consistent interpretations have not, however, been placed on these data: Perry and Tan (1972) suggest an Archaean ocean of about half the volume of the present ocean while Chase and Perry (1972) argue for negligible change in ocean volume since 3200 m.y. The change in $^{87}Sr/^{86}Sr$ ratio of sea water as indicated by measurements on carbonate rocks is more readily interpreted. Measurements show a marked increase in this ratio at about 2500 m.y. followed by little change after 2100 m.y. (Veizer, 1976), a pattern that is reflected in the trend of K_2O/Na_2 ratios in igneous and sedimentary rocks (Fig. 1). The collective data suggest that the major episodes of crustal and hydrosphere evolution took place prior to 2500 m.y., and are consistent with the view that ocean volume has increased little since the last major episode of crustal growth (Armstrong, 1968) ending at about 2500 m.y. The restriction of major mantle fractionation to the Archaean is also independently supported by measurements of crustal thickness in Archaean regions; Condie (1973) for example found that all cratonic nuclei > 2500 m.y. in age have crustal thicknesses falling in the range 38 - 40 km (Sect. 3).

It is not possible to estimate the present rate of volatile addition to the atmosphere/hydrosphere system with any degree of precision because most of the volatiles (predominantly water vapour and CO_2) released by the ca. 490 exposed active volcanoes on the Earth include much recirculated ground water. However, even if only 0.5 % of this water is juvenile, volcanism at the present rate would be sufficient to produce the present hydrosphere since 3800 m.y. The recognition of sea floor-spreading (Hess, 1962) introduced a new dimension to this problem because there is no doubt that the magnitude of submarine magmatic activity along the 54,000 km mid-ocean ridge system greatly exceeds continental volcanism. However, it is unlikely that this magmatism contributes appreciably to the atmosphere/hydrosphere volatile content for two reasons: Firstly, only perhaps 20 % of this magmatism is extrusive volcanism on the sea floor, the remaining additions to the crust being represented by intrusive material in which the volatiles will be largely or completely contained; secondly most of this extrusive volcanism takes place in water depths greater than 200 metres and under such hydrostatic pressures that volatile release (depending on the initial water content of the magma) is unlikely to take place (McBirney, 1963). Thus the hydrothermal systems at the present mid-ocean ridges are probably predominantly mobilised sea water, and only partial fusion at the subducting margin of the plate will be able to add significant juvenile content to the atmosphere/hydrosphere system. This implies a low accretion rate for the hydrosphere at the present time, which is also suggested by small changes in dissolved salt content of the oceans over the last 200 m.y. (McIntyre, 1970).

In theory an upper limit can be placed on the time required for formation and recycling the hydrosphere because water is gradually lost from the system. Dissociation of water vapour in the upper atmosphere results from 2100 - 1500 Å ultraviolet radiation of the solar spectrum. If all hydrogen produced in this way escaped from the atmosphere, the hydrosphere would disappear within about 1000 m.y. in the absence of replenishment, according to estimates by Buetner (1977).

3. The Continental Crust

The two extreme views of continental growth consider that the continental crust has grown by addition of fractionated material from the mantle early in Precambrian times, or progressively by incremental addition through geological time either to the underside of the continents (underplating) or in the form of peripheral orogenic belts.

The heat produced by the decay of radioactive isotopes within the Earth is critical to these considerations because it provides the energy for mantle motion and differentiation. Dickinson and Luth (1971) calculated this using the K/U and Th/U radiogenic compositions of chondritic meteorites and the Wasserburg model for terrestrial materials; they showed that the ratio of the present to the initial heat balance in the Earth probably ranged from 1 : 4.3 to 1 : 7.6. More recent calculations (Lambert, 1976) for specific crustal compositions show that the radiogenic heat production lay between 3 and 6 times the present value at 3800 m.y. and between 2 and 4 times this value at the start of Proterozoic times (Fig. 2). However, the effects of this radiogenic heat on the continental crust have certainly not diminished progressively as predicted by these curves. We can be rigorous about this conclusion because the sum total of radiometric age determinations from the continental crust

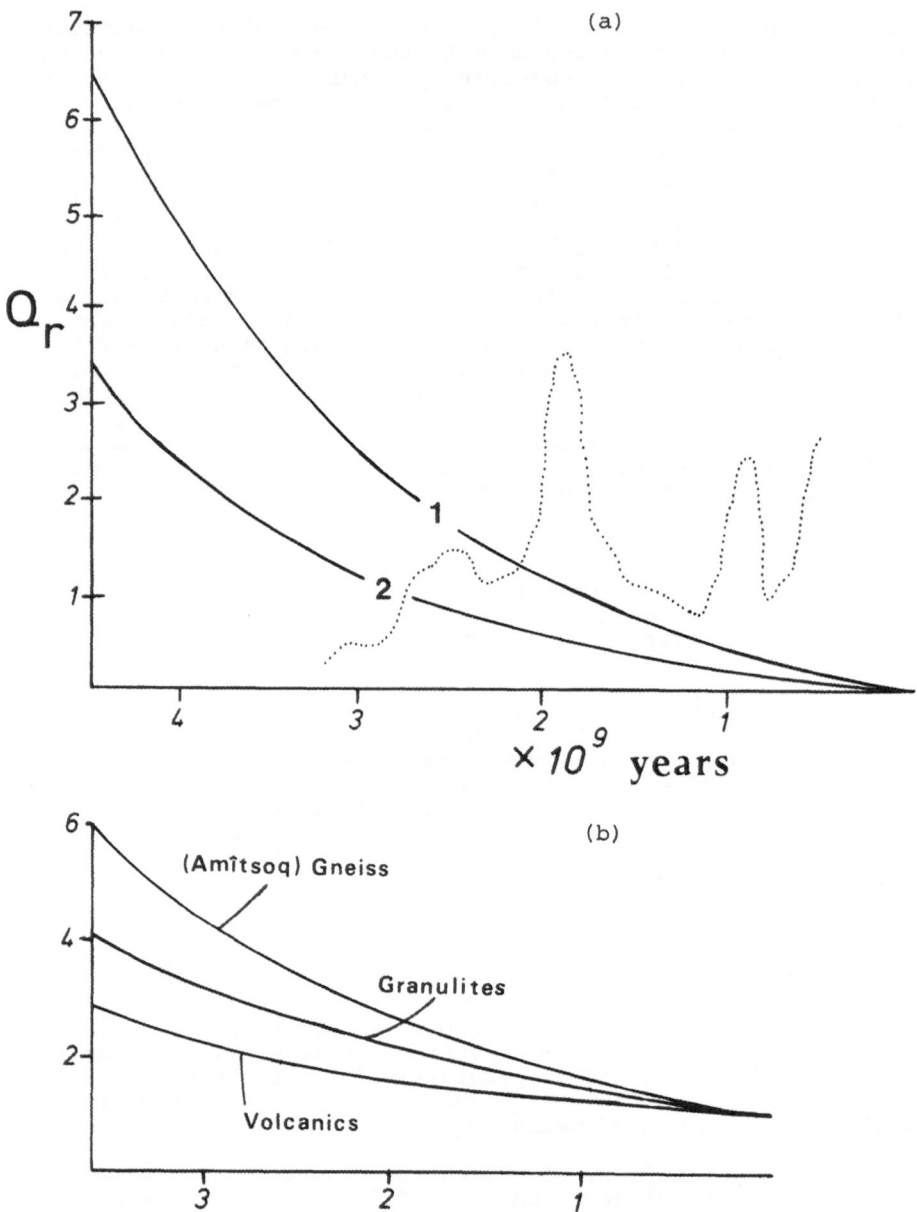

Fig. 2a and b. (a) The decrease in radiogenic heat production within the Earth during geological time as predicted from the K, U, and Th composition (1) for the Wasserberg model for terrestrial materials and (2) for chondritic meteorites. The heat production (Q_r) is expressed as a multiple of the present heat production and the curves are from Dickinson and Luth (1971). The frequency distribution of radiometric age determinations from Precambrian rocks is shown by the dotted line after Dearnley (1965); more recent data would emphasise the 3000 – 2600 m.y. interval but would not appreciably alter the shape of the remainder of this curve. (b) The decrease in heat production during geological time predicted from the radiogenic compositions of some specific crustal materials (after Lambert, 1976)

(Gastil, 1960; Dearnley, 1965 and Fig. 2) show that it experienced widespread heating and partial melting during intervals of intense thermal/magmatic activity separated by periods of quiescence. In a more recent assessment Moorbath (1976) confirms the general conclusions of these authors and argues for relatively short periods of accelerated crustal growth ("accretion superevents") occurring episodically in Earth history. These episodes of activity (notably at 2100 - 1800 m.y. and 1100 - 900 m.y.) also coincide with intervals of rapid apparent polar movement, and the periods of quiescence with low polar movement (Piper, 1974 and next Sections); but their explanation remains speculative.

Present subduction (ocean-continent collision) zones yield evidence relevant to the present growth and development of the crust. Both intrusive and extrusive igneous materials forming in these zones have uniformly low $^{87}Sr/^{86}Sr$ ratios in the range 0.703 - 0.708, which contrast with ratios of 0.71 - 0.78 found in regions of reworked ancient sialic crust. Crustal thicknesses in active subduction zones in the circum-Pacific region are much greater than average, although these regions (in contrast to continent-continent collisions) are often characterised by a lack of compressive tectonics (Brown, 1977). In western North America a wide range of geological and geochemical evidence suggests that the crust has been thickening, probably by in situ aggregation of sial, during Mesozoic and Tertiary times (Rogers and Novitsky-Evans, 1977). Contemporaneous addition of mantle material to the 40,000 km of active subduction zones is estimated to be about 0.5 km^3/year (Brown, 1977) with a ratio of extrusive to intrusive igneous rocks of the order of 1 : 5 or more. It would thus require an average rate of approximately twice this value to produce the $5 \cdot 10^9$ km^3 of continental crust since 3800 m.y. We are thus faced with the question of whether or not these present crustal additions derive from virgin mantle or represent crustal material recycled through the mantle; this question can only be answered directly by consideration of the Precambrian record.

Although Precambrian rocks are known to outcrop or underlie most of the present continental area, when the areal distribution of crust is plotted as a function of geological age (Hurley and Rand, 1969, Veizer, 1976), it shows an exponential increase up to the present. There is no doubt that this is an unrealistic result and simply reflects the extensive reworking of older crust in areas of later mobility. It is generally considered that plutonic magmatic activity has been largely responsible for the growth of the Archaean crust. Not only is the presently preserved sedimentary contribution small, but the low initial $^{87}Sr/^{86}Sr$ ratios of Archaean igneous rocks indicate a mantle origin shortly before addition to the crust (Moorbath, 1976). Large sections of the Archaean crust appear to represent igneous activity over a relatively short period of time. Kalsbeek (1976), for example, finds that 84 % of a 4000 km^2 area of the Greenland Archaean craton comprises gneisses of igneous origin intruded between 2900 and 2600 m.y. Gneisses of comparable origin make up a very similar percentage of the high grade terrains in Rhodesia, NW Scotland, and the Superior Province of North America (Goodwin, 1976; Sutton, 1977). Windley and Smith (1976) consider that the Archaean crust grew rapidly by igneous addition of granodiorite-tonalite suite rocks in environments with some analogies in present-day Cordilleran subduction zones. Sutton considers that 80 % of the preserved Archaean crust was injected and consolidated during a peak of igneous activity between 3000 and 2600 m.y. Only small islands of crust older than 3000 m.y. have survived; the remainder has presumably been consumed and recycled.

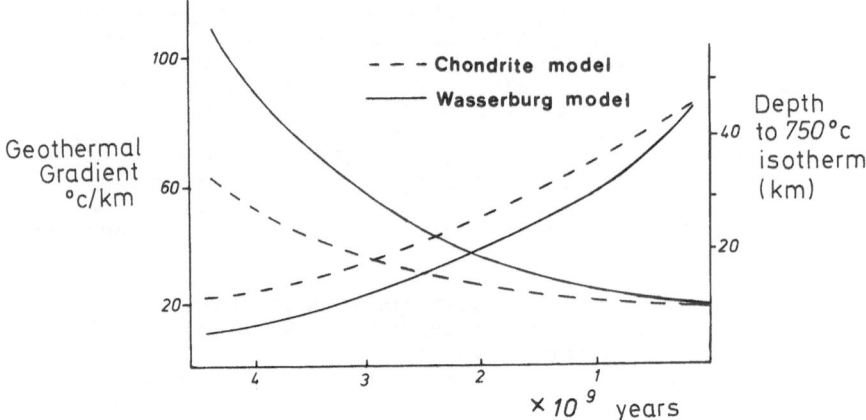

<u>Fig. 3.</u> Geothermal gradients and depth to the 750°C isotherm as a fraction of time calculated for a model of the sialic crust with a conductivity typical of granite ($8.0 \cdot 10^{-3}$ cal cm^{-1} s^{-1} °C^{-1}) and assuming radiogenic compositions of the chondrite and Wasserburg models (after Hargraves, 1977)

This large scale reworking of the Archaean crust reflects the high temperature gradients prevailing at that time. The minumum granite melting point isotherm (ca. 750°C) would probably have been within 10 to 20 km of the surface at 3000 m.y. (Fig. 3) or well within probable crustal thicknesses (Condie, 1973). Thus the crust/mantle interface would have been a zone of permanent instability, and at this early stage of crustal evolution repeated melting of the low melting point fraction would leave behind a depleted granulite facies rock and produce the decrease in such elements as K, Th, U, and Rb with depth that is identified in the granite-gneiss terrains of several cratons (Eade and Fahrig, 1971; Glikson and Lambert, 1973). In addition, Armstrong (1968) argues that low ^{87}Sr/^{86}Sr ratios in Archaean rocks indicate a much more effective exchange of material between mantle and crust prior to 2500 m.y.

With a few exceptions, all workers are agreed that the interval 3000 - 2200 m.y. was the most important in the evolution of the present crust, and temperature gradients had diminished sufficiently by this time for large units of sial to achieve a measure of stability. Assuming a simple fractionation model for the mantle following the models of Figure 2 between 60 % and 90 % of the present crustal volume could have formed by the end of Archaean times. Although Anhaeusser et al. (1969) regarded the Archaean crust as relatively thin (15 - 20 km) other workers have used Rb-Sr ratios and seismic data to infer that thicknesses were comparable to, or greater than, the present crust (Condie and Potts, 1969; Condie, 1973). Naqvi et al. (1974) use a chemical index to suggest a rapid increase in thickness of the Indian Shield from 10 to 35 km between 2500 and 2100 m.y. Armstrong (1968) concludes from Pb isotope data that the bulk of the crust was fractionated by 2500 m.y. Jahn and Nyquist (1976) emphasise the widespread thermal activity at ca. 2700 m.y., and argue for extensive recycling of the Archaean crust/ mantle system to produce a constant Rb/Sr ratio in much of the Upper Mantle. The combined work of Veizer (1976), Jahn and Nyquist (1976), and others suggests that two-thirds of the present crust was formed by the interval 2500 - 2100 m.y., while Armstrong suggests a figure near 90 %. The ^{87}Sr/^{86}Sr isotope ratio in carbonates and the K_2O/Na_2O (Fig. 1) ratio also show a change at about 2500 m.y., indicating a major fractionation episode and followed by little subsequent change. The K_2O/N_2O ratio is paralleled by several other chemical trends, but the explanation is uncertain: It is possible that they reflect the progressive increase in depth of melting that would result from falling lithosphere temperature gradients.

Given then that these varied lines of evidence imply that the bulk of the crust (and hydrosphere) had formed by 2500 m.y., they have to be reconciled with two other areas of observational evidence which might initially have led to the contrary conclusion. At the present time we can only begin to suggest answers to these contradictions.

Firstly, extensive recycling of crustal material through the mantle by an unspecified process, presumably related to subduction, must have continued to take place to account for the material of mantle derivation accumulating at constructive plate boundaries (Armstrong, 1968); also volatile addition to the hydrosphere by either juvenile or recycled water has succeeded in maintaining its volume.

Secondly, we have to account for changing thermal regimes which are specifically indicated by the exclusive presence of blueschist facies (high pressure — low temperature) rocks in Phanerozoic orogenic belts, the importance of greenschist facies rocks in Archaean times, and the exclusive presence of granulite facies rocks (at present erosion levels) in Precambrian terrains. The absence of Precambrian blueschists may result either from the absence of ophiolite obduction in Precambrian times, or from steeper geothermal gradients, or from the involvment of thinner crustal plates in subduction so that the inclination of the descending slab was relatively shallow. Despite the higher heat flux it is by no means certain that temperature gradients were higher throughout the continental crust in Precambrian times, and the heat may have been largely dispersed by more rapid ocean floor spreading (next Section). Thus many thick greenstone successions are only slightly metamorphosed throughout (Glikson, 1972) and suggest low temperature gradients. It is possible that the prevalence of Precambrian granulite terrains reflects higher temperature gradients in the crust (Heier, 1973), but even here we do not have definitive information. The formation of most exposed granulites was confined to the intervals 3000 - 2700 m.y. and 1300 - 900 m.y. If these granulite terrains were indeed formed at great depths comparable to, or greater than, the base of the present crust (O'Hara, 1977) they imply that very large vertical movements of these terrains have since taken place. In NW Scotland Dickinson and Watson (1976) link an estimated 15 - 20 km of slow uplift of granite gneiss terrain to the Laxfordian plutonism at ca. 1800 m.y. The recent identification of nappe structures in the Archaean, the juxtaposition of high and low grade terrains, and contrasting metamorphic grades across major transcurrent faults in granulite belts all provide indirect evidence for large vertical movements of these terrains (Sutton, 1977). The cause of these movements is unknown: We may speculate that they resulted from the last major accretion of material by underplating to the lower levels of the crust, a process which effectively ceased in Proterozoic times (O'Hara, 1977), but detailed lithospheric studies will be necessary to resolve their significance.

4. Precambrian Dynamics of the Continental Crust

A coherent crust could not have developed on the Earth until the outer layers cooled sufficiently for the granite minimum melting isotherm to reach the base of early sialic differentiation products (Fyfe, 1976).

Theoretical considerations suggest that this stage was reached between 300 and 1500 m.y. of the formation of the solid Earth (Hargraves, 1977), at which stage the thin protocrust would have become welded to the mantle, and the highly mobile tectonics which must have characterised the early history of the Earth, and for which all evidence has since been obliterated, would have ceased.

The Archaean crust (ca. 3800 - 2700 m.y.) is of two fundamentally contrasting types (Windley and Bridgwater, 1971): High-grade metamorphic terrains, predominantly of granulite or higher amphibolite facies gneisses, and greenstone belts of volcanic-sedimentary assemblages. The initial nature of the former is difficult to decipher, but it appears to contain a minimum of volcanic and sedimentary suites and to comprise predominantly plutonic tonalites, anorthosites, and ultrabasic associations. By contrast the greenstone belts, for which the record starts with the 3500 m.y. Barberton belt of Rhodesia, occur as thick volcanic-sedimentary basins vertically graded from primarily basic volcanics at the lowest levels to primarily sediments at the top, and subject to low grade greenschist facies metamorphism (Anhaeusser et al., 1969). They have long been considered to be structural downwarps, but Coward et al. (1976) demonstrated major lateral displacements with nappe development in examples from southern Africa, and it is possible that the very large stratigraphic thicknesses of ca. 20,000 metres, which have sometimes been quoted, may result from tectonic repetition.

The origin of these belts is a matter of most active debate at the present time, and is beyond the scope of this paper to summarise. However all workers are agreed that the Archaean represents a more mobile phase of crustal development than the ensuing Proterozoic (2700 - 600 m.y.) interval, with relative movements of crustal blocks covering most of the cratonic nuclei. Whether these relative movements were minor or major is more difficult to resolve, and arguments centre around the chemical and lithological affinities of the greenstones and their structural relationship to the high-grade terrains. Windley and Smith (1976) show that the greenstones have lithologic and chemical affinities with present-day accreting (continent-ocean) plate margins, and draw comparisons between the 3100 - 2800 m.y. tonalite gneisses and those forming at present continental leading edges. Although such parameters as heat flow, plate dimensions, and the bulk of pre-existing crust were obviously quite different in the Archaean, there do seem to be grounds for drawing some general analogies with this modern environment. Tarney et al. (1976) were able to relate rock types forming in the back arc trench of South Chile with greenstone belts. However, Shackleton (1976) and Coward et al. (1976) find structural continuity between high-grade gneiss terrains and greenstone belts separated from them by unconformity, and they may best be explained by a model such as that of Windley (1973) involving crustal attenuation or limited separation with symmetrical development of a basin including much mantle-derived material [as evidenced by Sr isotope ratios (Moorbath, 1976)] and subject to subsequent deformation between bordering regions of high-grade gneiss.

McKenzie and Weiss (1975) developed a model for the early thermal history of the Earth assuming a temperature distribution to ensure mantle-wide convection and a two-tier mantle convection system. Essentially these authors accept a range of evidence indicating that the mantle convection system has two regimes separated by the transition zone (Bullen and Haddon, 1967; Stacey, 1969), and their model predicts a small-scale flow system in the upper mantle in Archaean times separating large-scale flow throughout the deep mantle. Because radiogenic heat production in the Archaean mantle was two to three times its pres-

ent value, and because the bulk of the Earth's present heat is lost
from the mid-ocean ridge system, it is probable that oceanic lithosphere was produced and consumed at two to three times its present
volume in Archaean times. This may have taken place along a longer
ridge system separating many small plates (Sutton, 1977; McKenzie and
Weiss, 1975) with the greenstone belts developing in areas of upwelling
accompanied by tensional cracking, while compression and thickening
of the sial took place under cooler downwelling areas coincident with
the high grade terrains.

After about 3000 m.y. the release of heat as thermal/magmatic activity
became more localised so that large areas of the crust became sufficiently stabilised to allow the development of thick shelf sedimentary
sequences. This fundamental change from Archaean to Proterozoic tectonics was essentially diachronous (Sutton, 1973). In southern Africa
for example, thousands of metres of sediments and volcanics were laid
down on the Kaapvaal craton between 3200 and 1800 m.y. comprising the
Swaziland, Pongola, Dominion Reef, Witwatersrand, and Ventersdorp Systems (Anhaeusser, 1973) while greenstone belts were still forming within the Rhodesia craton at 2200 m.y., and folding and metamorphism of
the Birrimian, a major greenstone belt in West Africa, was not complete
until 2150 m.y. In the Canadian shield the first sedimentary-volcanic
platform sequence, the Huronian, was deposited on eroded greenstone
terrain at ca. 2300 m.y.

In both the well-studied regions of Greenland and South Africa a marked
localisation of tectonic activity also becomes evident at about the
time of this transition: Within the Archaean craton of Greenland several discrete belts of simple shear developed at 3000 - 2600 m.y. (Bak
et al., 1975) at a comparable time to the Limpopo belt between the
Rhodesia and Kaapvaal cratons. This restriction of tectonic activity
to linear mobile belts between stabilised cratonic nuclei, which were
themselves subject to brittle fracture and widespread intrusion by
basic dyke swarms, is a characteristic of Proterozoic times.

According to Strong and Stevens (1974) the contrasting geological regimes associated with Archaean and Proterozoic times are due to the
changing intersection of phase boundaries with geothermal gradients
(Fig. 4). In Archaean times extensive melting of the Upper Mantle would
be able to take place because the geothermal gradient intersects the
wet peridotite solidus; numerous mantle diapirs would be produced to
give the extensive small-scale convection systems envisaged by McKenzie
and Weiss (1975) and others. However, with progressive cooling of the
mantle the geotherms would migrate downwards while the peridotite solidus would move upwards with progressive degassing of the mantle.
Eventually a point would be reached at which the geotherm and solidus
no longer intersect, there would be no more mantle-wide melting, and
the Proterozoic regime with activity restricted to zones of large-
scale convective heat release would take over.

The subject of crustal tectonics in Proterozoic times has generated a
comparable amount of debate to that of Archaean tectonics. Inevitably,
by employing the Principle of Uniformitarianism, the concepts of the
New Global Tectonics have in recent years been widely applied to re-
interpret earlier studies of the Proterozoic crust. The most ardent
exponents of a plate tectonic analysis for the Proterozoic are Burke
and Dewey (e.g., 1973) who have, with little discrimination, drawn
Proterozoic sutures (ancient continental collisions) over most of the
Precambrian areas of the globe. The present-day situation in which we
have many plates, some of them entirely oceanic and some of them containing only minor pieces of continental lithosphere, suggests that
if the conclusions of these workers are essentially correct, then we

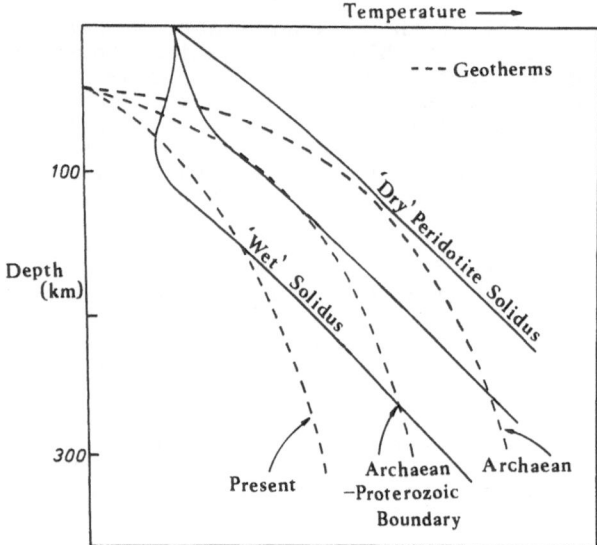

Temperature ⟶

--- Geotherms

100

Depth (km)

'Wet' Solidus

'Dry' Peridotite Solidus

300

Present

Archaean
-Proterozoic
Boundary

Archaean

Fig. 4. Schematic relationships between the geotherms at different times in Earth history and the peridotite solidus at different water pressures. Simplified after Strong and Stevens (1974)

have little or no hope of reconstructing the form of continents in the past. However, even these authors concede that there are major differences between Proterozoic mobile belts and belts formed at continent-continent or continent-ocean lithosphere collisions under the scheme of global tectonics in Mesozoic and Tertiary times.

Perhaps the most fundamental difference between Proterozoic mobile belts and present-day orogenic belts is the great geometrical complexity and strike variation of the latter, reflecting the irregular shapes and varying times of impingement of the colliding margins. Even neglecting the palaeomagnetic evidence, which is potentially decisive and is documented in the next section, there have long been grounds for believing that many Proterozoic mobile belts are ensialic and developed essentially in place, and thus that continental plates were very large in Proterozoic times.

We may document under several headings the evidence conflicting with the operation of a tectonics analogous to plate tectonics in Proterozoic times. This compilation is based largely on analyses by Shackleton (1973), Tanner (1973), Kröner (1977), and Sutton (1977):

1. Since the reworking of older crust in Proterozoic mobile belts is generally incomplete and has not succeeded in obliterating older Archaean or Proterozoic structures and lithologies, it is often possible to demonstrate a continuity of older structures through and beyond younger mobile belts. Examples from the well-exposed Proterozoic terrain of central and southern Africa have been documented by Shackleton (1973) and Kröner (1977).

2. There is a general absence of fundamental structural or stratigraphic breaks between cratons and mobile belts. Basic rocks that might be interpreted as ophiolite slices are absent or rare, and no case for a cryptic suture in a Proterozoic mobile belt has yet been substantiated by field examination. The case for Africa has been given by Kröner (1977); some disagreement surrounds examples from the Laurentian Shield, but it is established that there is no break across the Grenville Front

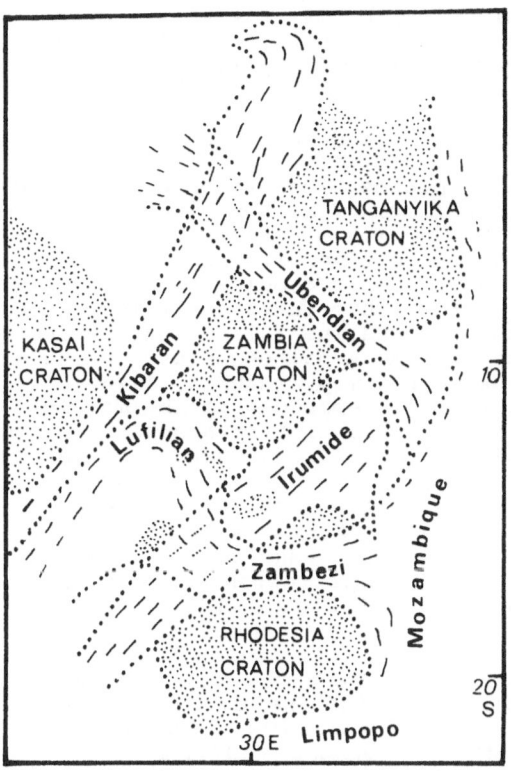

Fig. 5. The pattern of intersecting Proterozoic mobile belts in central and southern Africa modified after Shackleton (1973) and Kröner (1977). Although sporadically covered by younger rocks,this represents the largest single exposed area of Archaean-Proterozoic crust at the present time

separating the ca. 1100 m.y. Grenville mobile belt from the remainder of the Shield. In Greenland the Nagssugtoqidian belt ($\sim$ 2600 m.y.) involves the localised reworking of Archaean craton, and supracrustal rocks lying unconformably on the Archaean can be followed continuously into the Ketilidian mobile belt ($\sim$ 1800 m.y.).

3. Structural interpretations vary, but all are consistent with limited crustal compression; vertical and transcurrent motions dominate the structural evolution of these belts and structural changes are generally transitional rather than abrupt. Barr (1976) estimates a maximum shortening of 25 % across the Zambezi belt (ca. 650 m.y.). The Limpopo belt of southern Africa (> 2600 m.y.) has traditionally been regarded as a zone of shear separated by stable cratons; although this interpretation is now regarded as over simplified and the boundaries with the craton are highly diffuse (Coward et al., 1976), there is no doubt that the belt is ensialic. Similarly, sialic floor is exposed across the entire width of the Irumide and Zambezi belts in central Africa (Fig.5).

4. Mobile belts are frequently symmetrical and they do not as a rule include thick geosynclinal sequences. Indeed there are documented examples from the margins of the Kasai, Rhodesia, and Tanganyika Shields of Africa where older, shallow-water platform sedimentary sequences can be followed into younger mobile belts without a major break. Their preservation within the belts precludes large post-tectonic uplift. Similarly molasse deposits are absent and again preclude rapid post-tectonic uplift (Watson, 1976) that would be expected if the belts had resulted from continent-continent collisions and underthrusting of sialic plates according to the Tibetan situation (Burke and Dewey, 1973).

5. It has been established that several mobile belts terminate at one or both ends within older cratonic regions. Examples from Africa include the Limpopo (> 2600 m.y.), Ubendian (> 2000 m.y.), and Kibaran (ca. 1300 m.y.) belts (Tanner, 1973; Kröner, 1977).

This evidence leaves no doubt that Proterozoic mobile belts developed as ensialic belts, and hence that Proterozoic tectonics relate to restricted heat release within very large plates. The evidence from Late Precambrian times is, however, more conflicting. Although the extensive Pan African belts of Gondwanaland (750 - 450 m.y.) developed largely or exclusively as mobile belts, they overlap in time with the development of the Appalachian-Caledonian belt. The tectonic complexity and widespread obduction of ophiolite slices within this latter belt leave no doubt that it resulted as a consequence of continental separation and collision. Early phases in development of the Appalachian orogen indicating continental rupture extend back to 820 m.y. (Rankin, 1976). Similarly post-850 m.y. magmatic and tectonic episodes at the margin of the Arabian Shield can be related to a subducting lithospheric plate (Greenwood et al., 1976); fully developed ophiolite sequences of Late Precambrian age occur in Morocco (Leblanc, 1976); blueschist facies are present in early Caledonian (ca. 600 m.y.) ophiolites of Anglesey, Wales, and the Baikalian suture of Siberia dates from about 850 m.y. Indeed there are even a number of features in the Pan African mobile belts that suggest a transition towards orogenic tectonics processes. Thus the Buem Series incorporated in the Pan African mobile belt in West Africa (ca. 500 m.y.) includes greywackes and (spilitic) basic volcanics analogous to miogeosynclinal and ophiolitic rocks respectively (Burke and Dewey, 1973). The Damaran belt (650 - 580 m.y.) of SW Africa includes a great thickness of geosynclinal sediments and some submarine lavas, although an origin by plate collision is apparently not tenable in either of these two examples (Kröner, 1977; Piper et al., 1973). It seems probable that, in the same way as the transition from Archaean to Proterozoic tectonics was a gradual one, so too was the transition from Proterozoic tectonics to Phanerozoic plate tectonics (Fig. 6 (B)).

This analysis leaves the nature of the continent-ocean crustal boundaries in Proterozoic times unanswered. Unfortunately there are very few clues about the nature of this boundary. In the NW Canadian Shield the Coronation geosyncline developed between 2100 and 1700 m.y.; it comprises a maximum of over 10,000 metres of sediments deposited on a subsiding continental shelf of Archaean rocks (Hoffman, 1973) and subsequently overthrust onto the Archaean platform with development of molasse deposits; the comparisons with Phanerozoic Cordilleran orogenic belts are very compelling. A second possible example comes from the Ketilidian belt of south Greenland where 4000 metres of submarine volcanics were erupted onto the Archaean craton at ca. 2000 m.y.; this belt was subsequently intruded by the Julianehaab granite with strong structural and chemical similarities with many plate margin granites (Bridgwater et al. 1973).

5. Palaeomagnetic Analysis of Precambrian Crustal Movements

Studies of rocks as old as 2700 m.y. suggest that the Precambrian geomagnetic field possessed properties comparable to the present field (e.g., McElhinny, 1973), and although we have little specific palaeoclimatic evidence to show that this field was consistently axial, on general grounds any protracted departures are considered to be unlikely. Hence palaeomagnetic studies offer the only completely quantitative

method for reconstructing movements of the continents in Precambrian
times when we have neither sea floor nor faunal data to provide cor-
roborative evidence.

However, there is a large measure of disagreement about the correct
interpretation of Precambrian data which is reflected in the large number of different apparent polar wander (a.p.w.) curves[1] in the literature and hence in the number of widely different geological deductions
made from them. These differences reflect individual assessments of a
wide range of empirical data of varying reliability and arise from three
fundamental problems:

1. The record inevitably has many gaps because rocks were not continuously forming through Precambrian times, and the limited exposure that
we now have available for study is never going to provide a complete
sample of the geomagnetic field through Precambrian time. Hence we can
seldom be sure that we are really justified in connecting two separate
palaeomagnetic pole determinations together to define the path of the
pole in the time interval between them. An exception to this generali-
sation appears to be the case of certain slowly cooled metamorphic
rocks which have averaged magnetic field behaviour over a long period
of time (Morgan, 1976). Even in the case of widespread igneous events
it is likely that magmatism was concentrated into relatively short periods of time. The widespread ca. 1250 - 1200 m.y. MacKenzie igneous
episode in the Laurentian Shield and the contemporaneous Jotnian episode in the Baltic Shield are examples. In the Gardar Igneous Province
of South Greenland (Piper, 1977) detailed Rb-Sr studies have shown
that activity attributed to the interval 1400 - 1160 m.y. was concentrated into three short episodes each probably lasting not more than
a few million years; this major event can therefore only provide us
with a few spot records of the geomagnetic field over a long time interval.

2. The Precambrian shields were subject to very widespread thermal/
tectonic events at intervals of Precambrian time (Dearnley, 1965;
Sutton, 1973, 1977) and post-tectonic uplift and cooling took place
very slowly (Watson, 1976) at rates an order or more less than those
associated with Phanerozoic orogenic belts. Under conditions of protracted cooling the critical range of blocking temperatures of magnetic
minerals is greatly lowered (Pullaiah et al., 1975). Hence it is difficult to assign ages to the magnetisations recovered from these rocks:
They certainly lie below the Rb-Sr determinations and may be comparable
to the temperatures at which certain minerals closed to argon, but the
question of which mineral often remains. An example of this problem
comes from palaeomagnetic studies of the Archaean craton in central-
southern Greenland. Fahrig and Bridgwater (1976) derive magnetisations
from this region which they relate to Rb-Sr determinations and regard
as Archaean in age. However, Morgan (1976) found that magnetisations
of rocks from the same region showed characteristics of acquisition
during slow regional uplift and cooling, and relates them to Proterozoic K-Ar uplift ages. It is therefore suspected that the whole Archaean craton and adjoining mobile belts underwent magnetisation during
final uplift and cooling at times suggested by K-Ar ages of 1850 - 1650
m.y. and the chances of "seeing through" this event to derive older
magnetisations are probably remote. Similar problems relate to palaeo-

---

[1]Because palaeomagnetic poles are statistically defined with error ovals typically
of the order of 5 - 15° at the 95 % confidence level, and also because they may include components representing real departures from a true axial geocentric dipole,
it is now usual to represent the time sequence of poles as a band or swathe about
20° wide.

magnetic data from regions affected by the Hudsonian activity at ca.
1800 m.y. in the Canadian Shield (Irving and McGlynn, 1976; Roy and
Lapointe, 1976) and the Grenville mobile belt at ca. 1100 m.y. (Mc-
Williams and Dunlop, 1975) along the southeast margin of the same
shield.

3. The age data associated with Precambrian palaeomagnetic poles have
their own associated literature. Even assuming that the technical problem of isolating stable primary and secondary magnetisations from a
rock body can be overcome, geochronological studies are often not sufficient to assign an age to these magnetisations except within broad
limits. Age estimates have associated errors varying from 10^7 to 10^8
years so that the last digit quoted is never significant, and the assigned errors become greater the farther we go back into Precambrian
times.

For reasons given under headings (2) and (3) it is unlikely that we
will ever achieve complete agreement on a.p.w. movements prior to
2000 m.y. All we can do is to make the best of an incomplete palaeomagnetic and radiometric record and derive models from the data which
must stand or fall by testing against the geological and geophysical
constraints from direct observation of the Precambrian crust. The problem is best approached by first considering the later parts of Precambrian time when we know that large areas of the crust had stabilised and cooled. In these times it is relatively easier to isolate
magnetisations and the age data have the lowest associated errors. We
can then extrapolate deductions made from the latest Precambrian to
determine whether they help to resolve uncertainties shrouding the
palaeomagnetic and radiometric studies of the older Precambrian. However, even in Late Precambrian times we encounter a major difficulty:
The a.p.w. paths connecting separate dated poles involve movements of
the order of 90° and are ambiguous because they might connect to either
normal or reversed poles; for this reason there is still some uncertainty surrounding the correct polarity of all Precambrian a.p.w. curves
(see p. 28).

It is customary to use the a.p.w. swathes connecting dated poles for
tectonic interpretations. Thus if the cratonic nuclei within a Proterozoic shield stabilised only after plate convergence along later mobile
belts, then the swathes for each craton will in general be quite different prior to the time of welding, after which they will be statistically the same within the limits of the completeness of the record.
On the other hand, if these mobile belts developed in situ as a consequence of some mechanism quite different from orogenic belts, the a.p.w.
swathes defined from bordering cratons will be similar before, during
and after the development of the mobile belts (Piper et al., 1973,
Fig. 1). This test for plate tectonic mechanisms in Precambrian times
is in practise limited by the factors discussed above: While in principle we should be able to identify a transition between the two models by an increase in the width of the swathe and inability to con-
strain all the poles within a single swathe, in practise the incompleteness of the age control generally leaves some ambiguity.

The practical problems in palaeomagnetism involve the isolation of
primary directions of magnetisation acquired during initial igneous
cooling or sedimentary deposition and lithification. Secondary components of magnetisation are frequently superimposed on the primary components by subsequent thermal and/or tectonic events, and whether or
not the primary magnetisation will survive is a function both of the
time and the duration of the later events. Because of the long periods
of geological time involved, Precambrian magnetisations are frequently
very complex and recognition of this fact has gone hand in hand with

increasing sophistication in the methods of field and laboratory analysis, and hence in the range of primary and secondary magnetisations
recovered. For further details the reader is referred to McElhinny
(1973). We now examine the palaeomagnetic evidence for each of the
Proterozoic shields, analysing first the degree of correlation between
constituent cratons and then the correlation between separate shields.

5.1   Gondwanaland — Late Precambrian and Lower Palaeozoic

A very large movement of the pole relative to the supercontinent of
Gondwanaland, is identified from Late Precambrian and Cambrian data
(Fig. 6). The outline of this path was first identified by the time
sequence of Australian poles (McElhinny and Embleton, 1976) commencing
with a magnetic palaeoinclination estimate of 85° from the 700 - 600 m.y.
Marinoan glacial deposits of the Adelaide Geosyncline. Other Gondwanaland data, notably from Africa, but with ancilliary results from South
America, India, and Antarctica (Fig. 6), show very close overall agreement with the Australian data and leave no doubt that a single a.p.w.
curve applies to the whole supercontinent during this interval. However, the distribution of poles is so great between ca. 750 - 650 m.y.
that the apparent polar wander path remains uncertain for this interval and we cannot even be sure of its polarity. The 650 m.y.-Ordovician
motion of Gondwanaland relative to the poles correlate closely with
the widespread occurrence of glacial rocks (McElhinny et al., 1974),
notably the Late Precambrian and Ordovician glacial sequences of north
and west Africa, which are separated by sediments of probable low latitude origin.

The interval defined by the a.p.w. movement of Figure 6 coincides with
the development of extensive mobile belts within Gondwanaland (Fig. 6,
inset B) identified by resetting the K-Ar record over wide areas and
tectonic activity over more restricted areas. These "Pan African" belts
are defined by ages in the range 750 - 450 m.y. (Clifford, 1970) and
clustering around 550 m.y.; they do, however, incorporate much older
crust, and the evidence for the ensialic nature of these belts within
Africa has been discussed by Shackleton (1973), Kröner (1977) and many
others. Kröner emphasises the transitional nature of the belts, which
appear to be the last major tectonic lineaments that can be correctly
described as mobile belts; within Gondwanaland, however, they were
contemporaneous with events related to a subducting plate margin within the Arabian Shield (Greenwood et al., 1976) and to relative plate
movements in the Tasman orogenic zone of eastern Australia. The palaeomagnetic results of Figure 6 restrict the possible motions of across
the Pan African mobile belts lying between the major continents to
minor rotations of the cratonic margins and preclude an origin by plate
convergence.

The polar motion in Figure 6 is the most rapid yet recognised in geological history (∿ 1°/m.y.), and we can be certain that it represents
continental translation over the globe rather than local rotation or
true polar wander, both because the main component of movement is in
palaeolatitude, and because the movement correlations with palaeoclimatic indicators. The age of the polar data is defined here by a combination of radiometric and palaeontologic data; in earlier geologic
times, we do not have the facility of the latter and the accuracy of
the former is proportionally less; hence if movement of this magnitude
took place earlier in Precambrian times it is likely that they would
be entirely unresolvable.

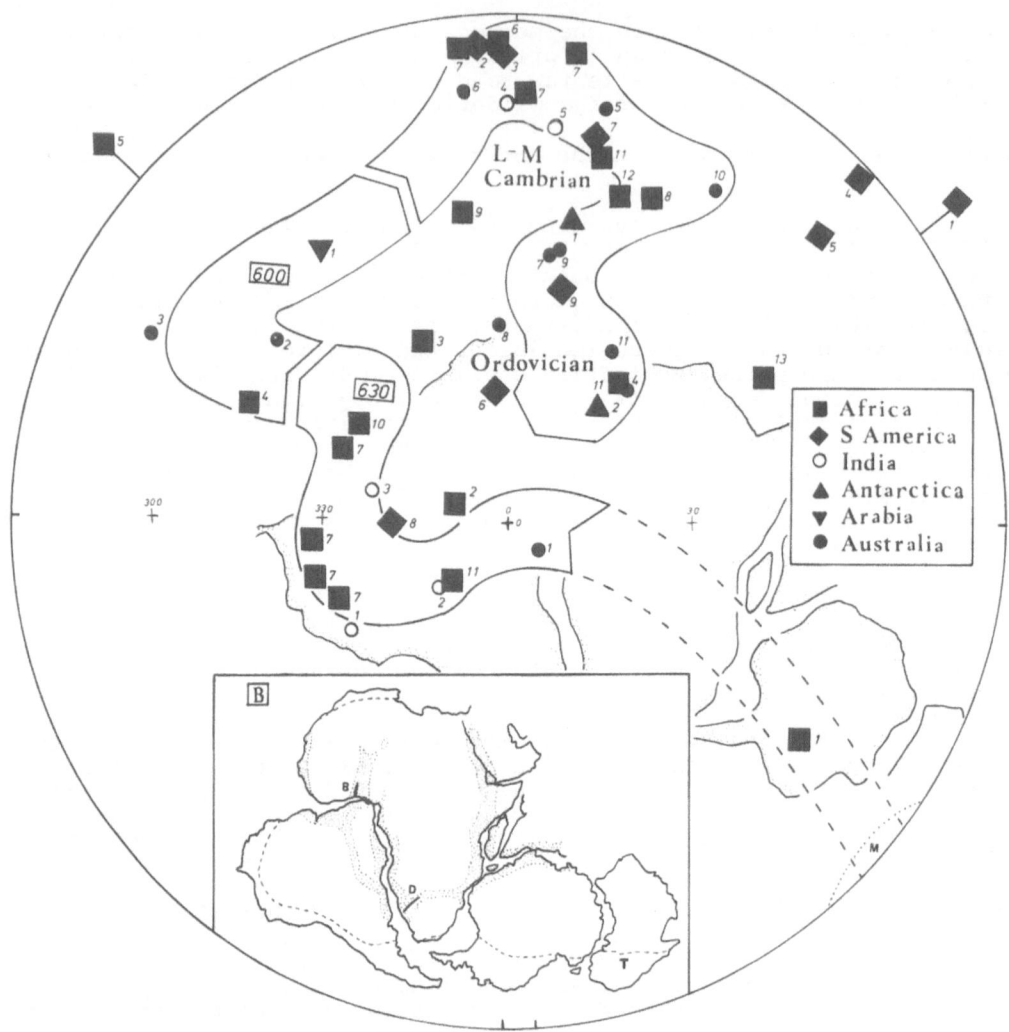

5.2 Laurentian Shield — Upper Proterozoic

The Laurentian Shield comprises the Precambrian basement of North
America together with Greenland and NW Scotland on the pre-Phanerozoic
drift reconstruction of Bullard et al. (1965). This shield is extensively exposed in Greenland and Northern Canada but is also known to
underlie younger rocks as far south as a line joining the California
coast and Alabama, and as far west as the eastern margin of the Cor-
dilleras. Five cratonic nuclei of Archaean age are identified: The
Slave, Superior and North Atlantic (Labrador coast, Greenland and NW
Scotland) structural provinces, the Bear Tooth zone of Wyoming and
Montana, and the Minnesota region. They are separated by a broad area
of Proterozoic mobility distinguished as the Hudsonian in North America
and the Nagssugtoquidian and Ketilidian in Greenland (inset, Fig. 9);
activity ceased in the latter region at about 1600 m.y. Along the southeast margin of the Shield is the Grenville mobile belt of medium-high
grade metamorphic rocks tectonised during the approximate interval
1250 - 1100 m.y.

Upper Proterozoic palaeomagnetic data for this region are summarised by Irving and McGlynn (1976); their assessment is updated with the addition of Greenlandic data by Piper (1977) and reinterpretation of Grenville Province data by McWilliams and Dunlop (1975). Since the whole of the Laurentian Shield with the exception of the Grenville Province had stabilised by this time the collective data provide an opportunity to test the agreement of palaeomagnetic poles across a large area of a Proterozoic shield. The collective data are plotted in Figure 7 and define a closed loop executed between 1200 and 1000 m.y. The form of this loop, named the "Great Logan Loop" by Robertson and Fahrig (1971), has been known in outline since the work of DuBois (1962). A notable feature of this interval is the group of poles ca. 1245 m.y. in age derived from rocks of the Mackenzie igneous province which were intruded, notably as basic dyke swarms, across the entire width of the Laurentian craton (Fig. 7, inset).

The Grenville Province has yielded the only systematically discordant sequence of Precambrian palaeomagnetic poles identified from the Laurentian Shield; these poles have been interpreted both in the context of plate collision and of complex a.p.w. movements. In the first model (Irving et al., 1972) these data are regarded as broadly contemporaneous with quite different poles from the remainder of the Shield and are accomodated in terms of convergent plate motion (Irving and McGlynn, 1976, Fig. 18), although it was universally recognised from the geological evidence that the Grenville Front separating the Province from the Shield could not itself be the location of this suture. The second model (McWilliams and Dunlop, 1975; Morris and Roy, 1977), which has now received wide acceptance, considers that the poles from the Grenville Province represent a different time interval from existing

---

◀ Fig. 6. A synthesis of Late Precambrian-Lower Palaeozoic palaeomagnetic data from Gondwanaland assigned to the time interval approximately 700 - 450 m.y. The poles are plotted relative to Africa according to the continental reconstruction of Smith and Hallam (1970), and ages assigned to poles are Late Precambrian (LPc), Cambrian (C), and Ordovician (O), $l,m,u,$ = lower, middle, or upper divisions of the Cambrian and Ordovician. African poles (■) are: 1, Quarzazate volcanics (LPc/C); 2, Sijarira Group (LPc/C); 3, Ntonya ring structure (630); 4, Klipheuvel Formation (LPc/C); 5, Amouslek Tuffs (C_1); 6, Sabaloka ring structure (>540); 7, Fish River Series (7 poles of LPc-C age); 8, Moroccan lavas (C_m); 9, Table Mountain Series (O); 10, Hook intrusives (500); 11, Plateau Series (3 poles of Lower Palaeozoic age); 12, Tassili Sediments (Lower Palaeozoic); 13, Doornpoort Formation (550 - 500). The poles from other regions are rotated with Africa remaining fixed; poles from Australia (●) are: 1, Pound quartzite (LPc); 2, Antrim Plateau volcanics (LPc/C_1); 3, Arumbera Sandstone (LPc/C_1); 4, Aroona Dam sediments (C_1); 5, Hugh River Shale (C_{1-m}); 6, Hudson Formation (C_1); 7, Lake Froome Group (C_{m-u}); 8, Dundas Group (C_u); 9, Jinduckin Formation (O_l); 10, Stairway Sandstone (O_m); 11, Tumblagooda Sandstone (O); M represents the 80° palaeolatitude estimate from borehole material from the Marinoan glacial sediments (700 - 600). South American poles (◆) are: 1, Purmamarca Village (C); 2, South Tilcara (C); 3, North Tilcara (C); 4, Purmamarca (C); 5, Abra de Cajas (C); 6, Salta and Jujuy (C-O), 7, Salta (O); 8, Sedimenta, Bolivia (O); 9, Urucum Formation (O). Indian poles (○) are: 1, Bhander Sandstone (LPc/C); 2, Upper Rewa Sandstone (LPc/C); 3, Upper Bhander Sandstone (LPc/C); 4, Purple Sandstone (C_1); 5, Salt Pseudomorph Beds (C_m); Indian poles 4 and 5 are plotted after an additional local 75° rotation of the Salt Range (see McElhinny and Embleton, 1976). Antarctic poles (▲) are: 1, Charnockites (C_u-O_1); 2, Sør Rondale intrusions (485, O_{1-m}). The Arabian pole (▼) 1 is from Jordanian redbeds (C-O). The data plotted here are summarised in Piper (1976a) and McElhinny and Embleton (1976). Inset B illustrates the Smith-Hallam reconstruction of Gondwanaland; zones affected by the Pan African thermal-tectonic episodes (700 - 450) are stippled. B = Buem Series of West Africa, D = Damara geosyncline, T = Tasman geosyncline of eastern Australia

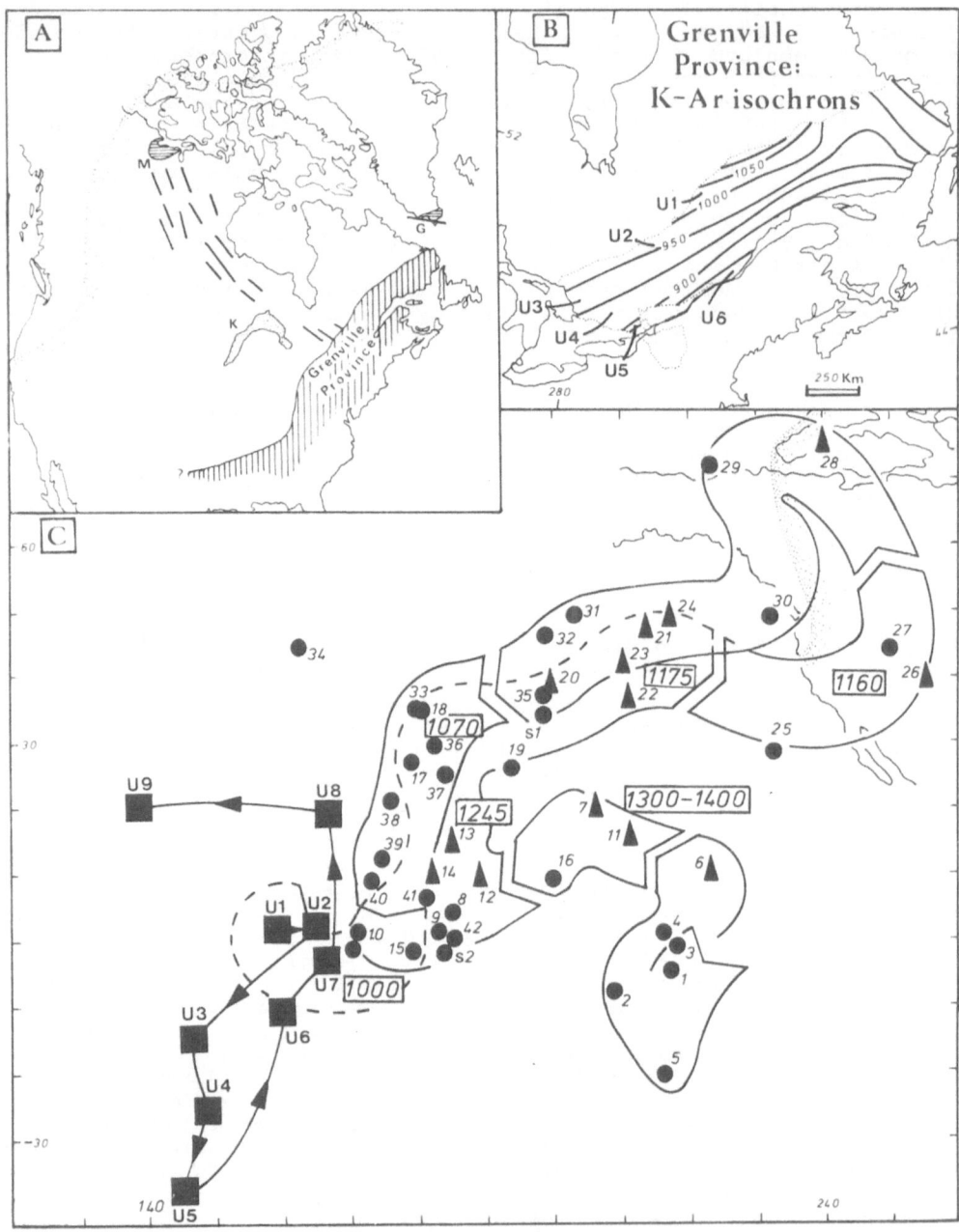

data for the Laurentian craton. The Grenville poles are derived mostly from medium to high-grade metamorphic rocks with a long cooling history during slow regional uplift. Under these conditions it is very unlikely that any pre-metamorphic magnetisations have survived, and the actual magnetisations will have been acquired after times indicated by the Rb-Sr isochrons and probably prior to certain K-Ar mineral ages (Pullaiah et al., 1975). McElhinny and McWilliams (1977) have calculated

mean magnetisations for all results lying between the K-Ar chrontours
of the Grenville Province as indicated in the inset diagram of Figure
7. The derived palaeomagnetic poles U1-U9 define a second closed loop
which connects with the end of the Great Logan loop at 1000 m.y. (Fig.
7); the K-Ar age estimates assigned to these poles are approximate only,
because the actual blocking temperatures of the rocks under conditions
of slow cooling are not precisely known. The continuity of this (Grenville) loop with the older curve and preliminary data from elsewhere
in North America (Irving and McGlynn, 1976) strongly support geological
and radiometric evidence (e.g., Wynne-Edwards, 1972), suggesting that

the Grenville Province formed an integral part of the Laurentian Shield
in Proterozoic times, but the data cannot of course exclude the possibility of plate collision prior to 1000 m.y.

5.3 African Shield — Upper Proterozoic

The present body of Upper Proterozoic palaeomagnetic evidence from
Africa suffers from poor age control and with the age uncertainties
assigned to poles 2 - 4 in Figure 8 it remains uncertain whether the
curve defines a loop as shown by the dashed section in this figure;
also because of uncertainties as to the ages of poles 9 and 10, for
which Rb-Sr estimates range from 1260 - 1020 m.y. (Piper, 1976a; Mc-
Elhinny and McWilliams, 1977), it is uncertain whether these poles
should be accommodated into the a.p.w. path.

The collective results of Figure 8 are derived from the Tanganyika
craton, the centre and western margin of the Rhodesia-Kaapvaal craton

---

◀ Fig. 7A-C. Upper Proterozoic palaeomagnetic poles from the Laurentian craton. Numbered poles with assigned ages given in brackets are: *1*, Croker Island complex
(1475); *2*, Sherman granite (1410); *3*, Michikamau anorthosite (1450 - 1400); *4*, St.
Francois rocks; *5*, Sibley group (1370); *6*, Lower Gardar lavas; *7*, Upper Gardar
lavas; *8*, Muskox intrusion; *9*, Coppermine group; *10*, Mackenzie diabase (1240); *11*,
BDO dykes (1245); *12*, Lamprophyre dykes (1278 - 1254); *13*, NNW dolerite dykes; *14*,
Kûngnât ring dyke (1245); *15*, Sudbury dykes (1225); *16*, South Trap Range lavas (i,
1220); *17*, Gila County diabase; *18*, Logan diabase, Canada (1160); *19*, Logan diabase,
U.S.A.; *21*, Hviddal Giant Dyke (1175); *21*, Tugtutôq giant dykes (1175 - 1168); *22*,
Narssaq gabbro (1175 - 1168); *23*, Tugtutôq NE dykes; *24*, Logan diabase, Canada (ii);
25, South Trap Range lavas (ii); *26*, Ilîmaussaq syenites (1168); *27*, Alona Bay lavas;
28, Ilîmaussaq fractionated rocks (1168 - 1150); *29*, Lower Gargantua volcanics; *30*,
Mamainse lavas, reversed (1076); *31*, Osler lavas (1115); *32*, North Shore volcanics,
reversed (1110); *33*, Upper Gargantua volcanics; *34*, Nemagosenda carbonatite (1036);
35, North Shore volcanics, normal (1105); *36*, Mamainse lavas, normal (1070); *37*,
Portage Lake lavas (1100); *38*, Copper Harbour lavas (1080); *39*, Nankoweap formation
(1000); *40*, Freda and Nonsuch sandstone (1040); *41*, Pikes Peak granite (1030); *42*,
Cardenas lavas (1090). U1 - U2 are mean uplift magnetisations from zones of the Grenville Province dated by reference to uplift and cooling K-Ar mineral ages at: *U1*
(1050 - 1000), *U2* (1000 - 950), *U3* (950 - 925), *U4* (925 - 900), *U5* (900 - 850), *U6* (850 -
750), *U7* (750 - 650); *U8* and *U9* are from secondary components of unknown age (see
McElhinny and McWilliams, 1977). The (*C*) inset map illustrates the K-Ar thermochrons
for the Grenville Province and the zones from which the magnetisations U1 - U6 are
calculated. Poles from the Precambrian of NW Scotland after correction for Phanerozoic drift are: *S1*, Stoer Group (991) and *S2*, Torridon Group (796). The *inset diagram* of the Laurentian craton (*A*) shows the distribution of major Upper Proterozoic
events (from which many of the palaeomagnetic results are derived); these are lettered: *M*, Muskox and Coppermine, *K*, Keweenawan basin, *G*, Gardar igneous province.
The palaeomagnetic data for this interval are summarised by Irving and McGlynn (1976)
and Piper (1977); circles are North American poles triangles are Greenlandic poles.

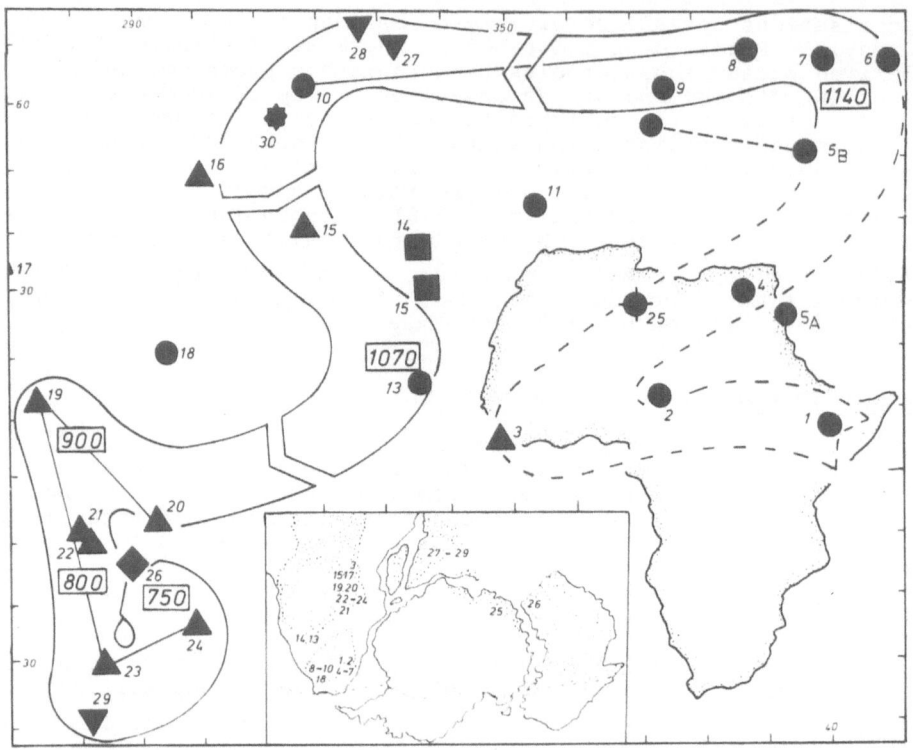

Fig. 8. Upper Proterozoic palaeomagnetic data for Africa and related regions of Gondwanaland. Numbered poles with the assigned radiometric ages in m.y. in brackets are for Africa: Rhodesia-Kaapvaal craton (●): *1*, Pilansberg dykes (1310); *2*, Van Dyk Mine dolerite; *3*, Kisii lavas; *4*, Vredefort ring structure; *5*, Premier Mine kimberlite (1115); *6*, Waterberg dolerites (1115); *7*, Umkondo dolerites (1140); *8*, Barby lavas (1265 – 1020); *9*, Umkondo lavas (1140); *10*, Guperas lavas (> 1250); *11*, Auborus formation; *12*, O'Okiep intrusions (1070); Kasai craton (■): *13*, Chela sediments; *14*, Nosib quartzite; poles from the Tanganyika craton (▲) are: *15*, Bukoba sandstone (1000); *16*, Abercorn sandstone; *17*, Ikorongo group; *18*, Klein Karas dykes; *19*, Kigonero flags (> 890); *20*, Malagarasi sandstone; *21*, Mbala dolerites; *22*, Bukoban dolerites (806); *23*, Gagwe lavas (813); *24*, Manyovu red beds (< 813); Other poles are from Antarctica (◆): *25*, Vestfold dykes (1030); Australia (◆): *26*, B group dykes (750); India (▼): *27*, Cuddapah sandstone; *28*, Cuddapah shale; *29*, Malani rhyolite (750); South America (✵): *30*, Bambui Formation (> 886); these poles are rotated relative to Africa according to the continental reconstruction of Smith and Hallam (1970) illustrated in the *inset diagram* on which the palaeomagnetic studies are located together with the area of crust (*stippled*) which had stabilised by 1000 m.y. The *straight lines* connect poles in stratigraphic sequence

and the Kasai craton (Fig. 8, inset), which are separated from one another by broad tracts of terrain affected by the later Pan-African tectonic episodes; all the data, however, define a single a.p.w. curve. There is also close agreement between ca. 750 m.y.-poles from Africa, India, and Australia on the Smith-Hallam reconstruction of Gondwanaland (poles 26 and 29 in Fig. 8), indicating that the greater crustal unit of Gondwanaland was already in existence by this time.

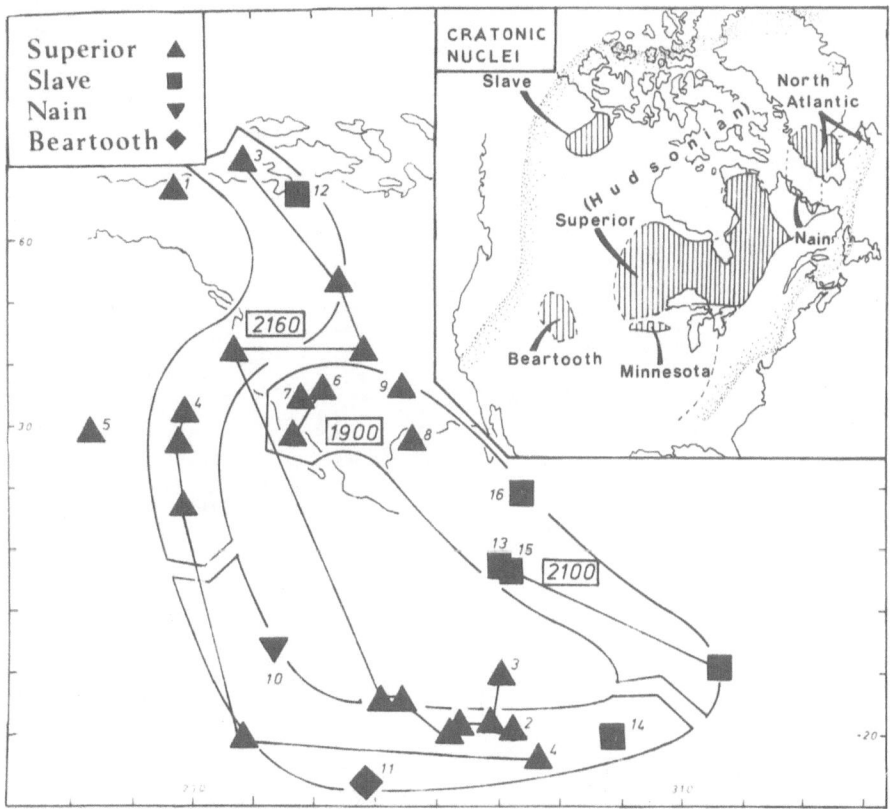

Fig. 9. Palaeomagnetic poles from the Laurentian craton assigned to the interval 2200 - 1900 m.y. The poles are from the Superior Province (▲) numbered: *1*, Otto Stock primary (2160); *2*, Otto Stock secondary; *3*, Nipissing diabase C-D magnetisations (2162, 2155); *4*, Abitibi dykes (2160 - 2147); *5*, Marathon dykes (2218); *6*, Molson dykes; *7*, Sudbury irruptive (1900, 1844); *8*, Gunflint formation (> 1635); *9*, Spanish River carbonatite (1892). Pole *10*, Indian Harbour dykes (2080*) is from the Labrador and Nain structural province (▼). Pole *11* from the Beartooth Uplift of Montana and Wyoming (◆) is from Owl Creek dykes (2110 - 1910*). Poles from the Slave Province (■) are: *12*, Big Spruce complex (2218 - 2170); *13*, Big Spruce complex, secondary; *14*, X dykes (2165); *15*, Slave Province SE magnetisations; *16*, Indin dykes (2093). The *inset map* shows the distribution of the cratonic nuclei (structural provinces) stabilised by 2200 m.y. with Greenland and NW Scotland rotated to correct for Phanerozoic drift. * = K-Ar age estimates, other datas are based on Rb-Sr isochrons or U-Pb determinations. Lines connect poles from the same rock suite

5.4 Laurentian Shield — Middle-Lower Proterozoic

A small number of studies (summarised by McElhinny, 1973, and Irving and Naldrett, 1977) relate to the earlier part of Proterozoic times (2700 - 2200 m.y.) but in view of the long time interval involved and uncertainties shrouding the nature and age of the magnetisations they cannot yet be used to derive realistic a.p.w. curves and hence are not considered further here.

Palaeomagnetic poles which can be dated with some degree of certainty commence at ca. 2200 m.y., with several independent investigations of

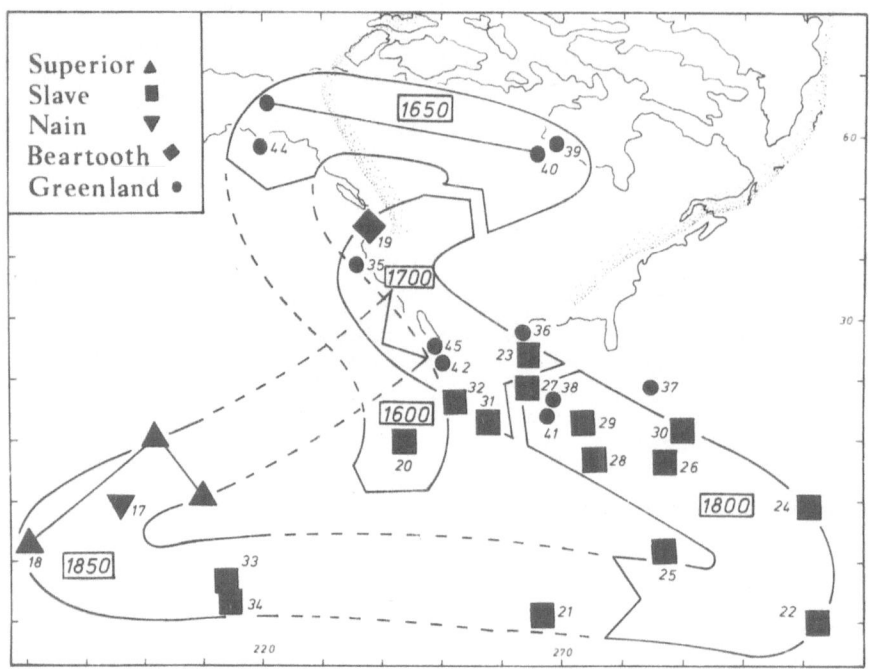

Fig. 10. Palaeomagnetic poles from the Laurentian Shield assigned to the Proterozoic interval 1900 - 1600 m.y. These data follow sequentially from Figure 9. Pole *17* is from the Labrador and Nain structural province (▼) for the Murdoch, Thompson Lake, and Merihek metavolcanics (1855). Pole *18* from the Superior Province (◆) is Nipissing diabase E magnetisations. Pole *19* from the Beartooth Uplift of Montana and Wyoming (◆) is for the Wind River dykes (1880 - 1680*). Pole *20* from the Slave Province is for the Western Channel diabase (1780 - 1400). Poles from the Churchill structural province (the terrain between the Slave and Superior Provinces affected by the Hudsonian thermal-tectonic episode,■) are: *21*, Metamorphosed Kaminak dykes (1892 - 1615); *22*, Flin-Flon secondary direction (1800); *23*, Flin-Flon tertiary directions (1700 - 1600*); *24*, Et-Then Group (1835 - 1630*); *25*, Martin Formation (1930 - 1635); *26*, Kahocella primary directions (1873); *27*, Kahocella secondary directions (< 1873); *28*, Dubawnt Group (1825); *29*, Nonacho sediments (2500 - 1700); *30*, Sparrow dykes (1700*); *31*, Daly Bay metamorphics (1622*); *32*, Cape Smith metavolcanics (1650 - 1450*); *33*, Stark Formation (1873 - 1845*); *34*, Tochatwi Formation (1873 - 1845*). Uplift and cooling magnetisations from the Archaean craton and Nagssugtoqidian, Ketilidian, and Laxfordian mobile belts of Scotland and Greenland (●) are: (i) Scotland: *35*, Scourie dykes (1650); (ii) Greenland: *36*, Nordre Stromfjord metamorphic rocks; *37*, Holsteinsborg metamorphic rocks; *38*, Itivdleq dykes (1830 - 1700*); *39*, Sagdlerssuaq metamorphic rocks (1650*); *40*, Ketilidian metamorphic rocks (1700 - 1600*); *41*, Itivdleq area dykes (1790 - 1650*); *42*, Kangamiut dykes; *44*, Amphibolite boudins; *45*, Godhab gneisses. The poles from Greenland and Scotland are plotted relative to North America after correction for Phanerozoic drift. The data for this interval are summarised by Cavanaugh and Seyfert (1977), Roy and Lapointe (1976), and Piper and Stearn (1976); poles *36* and *37* are given in Beckmann et al. (1977). * = age estimates from K-Ar mineral or whole rock ages, or K-Ar isochrons (*38*, *39*, *41*); other estimates are based on Rb-Sr or U-Pb determinations. The *straight lines* connect poles from the same rock suite

basic magmatism within the Superior craton (Nipissing and Abitibi dykes and sills, Otish gabbro). Contemporaneous poles from the Slave, Nain, and Bear Tooth Provinces define a single a.p.w. path with data from the Superior craton (Fig. 9). Laboratory studies employing a range of techniques (alternating fields, low and high temperature thermal demagnetisation, and chemical leaching experiments), accompanied by vector analysis of the various magnetisations have, in certain cases, revealed the presence of magnetisations acquired at different times and temperatures (e.g., Park, 1975; Roy and Lapointe, 1976); in some instances these can be dated relative to geologic events by field (baked contact or fold) tests. The Middle-Lower Proterozoic data of this kind have been analysed in relation to dated primary poles by Irving and McGlynn (1976), and Roy and Lapointe (1976). The studies of the latter authors confirm the sense of movement of the a.p.w. curve shown in Figure 9 at ca. 2150 m.y. from the C and D magnetisations of unit 3. The analyses of slightly younger data has, however, been a matter of some dispute. It has been customary (e.g., Irving and McGlynn, 1976) to fit all 2150 - 1600 m.y. poles to a single closed loop, but this is clearly an oversimplification because the same swathe is used to accommodate 2100 - 1900 m.y. data as 1800 - 1600 m.y. data. Also the discovery of poles 17, 33, and 34 (Fig. 10) from the Slave/Nain areas ca. 1850 m.y. in age and coincident with secondary magnetisations from the Nipissing diabase (E magnetisations) of the Superior craton has confirmed that this analysis is too simple (Roy and Lapointe, 1976; Piper, 1976b). The original assessment permitted Cavanaugh and Seyfert (1977) to argue for a joining of the Superior and Slave Provinces by relative plate motion during the Hudsonian episode (ca. 1800 m.y.). However, this hypothesis is now known to be untenable because the series of poles 10 - 16, mostly derived from 1850 - 1650 m.y. sediments of the Slave craton and used by these authors to derive a polar track for this craton, are coincident with metamorphic poles 30 - 32, 35, 36, to 38 from the opposite side of the Hudsonian mobile belt in west Greenland, and related to uplift and cooling ages in the range 1850 - 1650 m.y. (Fig. 10). The trend of this part of the a.p.w. curve is confirmed by the sequence of uplift magnetisations from the northern margin of the Archaean craton in west Greenland (Morgan, 1976).

The collective data from all parts of the Laurentian craton are summarised by Irving and McGlynn (1976), Roy and Lapointe (1976), Piper and Stearn (1976), and Cavenaugh and Seyfert (1977), and are completely accommodated by the three loops shown in Figures 9 and 10. Present data do not completely define the form of the loop at 1900 m.y., and the loop at ca. 1650 m.y. is currently only defined by Greenlandic data. However, since the results from all the cratonic nuclei define a single a.p.w. path, they confirm that these nuclei were close to their present configuration in Middle-Lower Proterozoic times; the Hudsonian mobile belt is clearly ensialic, as are the Aphebian and other geosynclinal suites (including the Labrador Trough) margining the Superior craton. A paucity of data in the interval 1600 - 1400 m.y. leaves the form of the connecting swathe between Figures 7 and 10 uncertain.

5.5 Africa — Middle-Lower Proterozoic

This data has been fully analysed by Piper (1976b) and is complemented by new results of Jones et al. (1977). As for the Laurentian Shield, a body of data sufficient to derive a meaningful a.p.w. curve is currently only available after 2200 m.y. The collective poles for the interval 2200 - 1850 m.y. (Fig. 11) come from the West Africa, Kasai, and

220

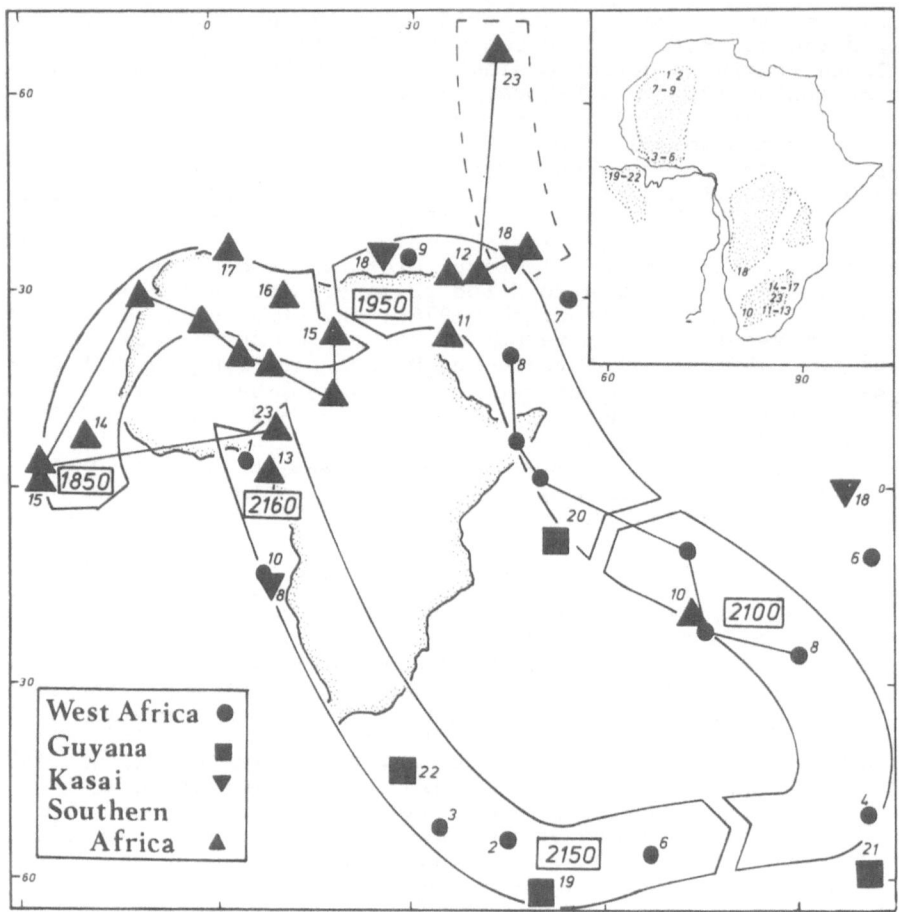

Fig. 11. Lower-Middle Proterozoic palaeomagnetic poles from Africa and South America assigned to the interval 2200-1750 m.y. Poles from the West Africa craton (●) are: 1, Tonalites, N.W. Sahara (2160); 3, Syntectonic igneous rocks, same area (2160); 3, Tarkwaian intrusions (2150*); 4, Obuasi greenstone (2150*); 5, Obuasi dolerite (2150-2070); 6, Ivory Coast dolerites (2150-2070); 7, Aftout gabbro (< 2090); 8, Sahara diorites, poles for six sites (2090-1950); 9, NW Sahara dolerite dykes (< 1920). Poles from the Rhodesia-Kaapvaal craton (▲) are: 10, Orange River Lavas (2100); 11, Bushveld gabbro (1954); 12, Losberg intrusion (1950); 13, Palabora complex (1900); 14, Mashonaland dolerites (1850); 15, Satellite dykes to Great Dyke of Rhodesia, Limpopo remagnetisation (1900*); 16, Bubi dyke swarm; 17, Crystal Springs dyke swarm; 17, Waterberg Sandstones (ca. 1760). Poles from the Kasai craton (▼) are: 18, Cunene anorthosite complex (2160-2050); and from the Guyana craton of South America (■) are: 19, 20, Roraima dolerites, Guyana (2090-2070); 21, 22, Kabaledo dolerites, Suriname. The *inset figures* shows the locations of the palaeomagnetic studies with respect to the regions (*stippled*) which had stabilised by 2100 m.y. * = K-Ar age estimates, others are Rb-Sr estimates. *Straight lines* connect different poles from the same rock suite

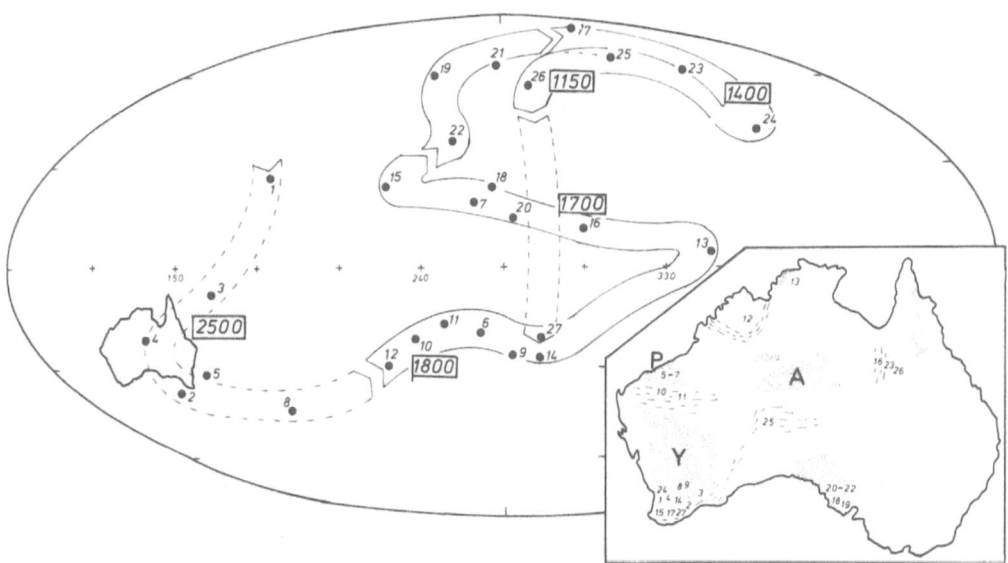

Fig. 12. Precambrian palaeomagnetic poles from Australia. The poles are numbered:
1, YE dykes (2500); *2*, Ravensthorpe dykes (2500); *3*, Widgiemooltha dykes (2420);
4, YA dykes (2500 - 1700); *5*, *6*, and *7*, Mt. Goldsworthy ores (3000 - 2000) numbers *3*,
1, and *2*, respectively; *8*, Koolyanobbing-Dowd's Hill ore (2750 - 2000); *9*, Koolyanob-
bing'A' ore; *10*, Mt. Tom Price ore (1800); *11*, Mt. Newman (1800); *12*, Hart dolerite
(1800); *13*, Edith River volcanics (1760); *14*, YF dykes (1700); *15*, YD dykes (1700);
16, YA dykes (1750); *17*, YC dykes (1500); *18*, GB dykes (1700); *19*, GA dykes (1500);
20, and *21*, Iron Monarch ore, positive and negative groups (1800 - 1500); *22*, Iron
Prince ore (1800 - 1500); *23*, Lunch Creek lopolith (1498); *24*, Morowa lavas (1390);
25, Giles complex (1250 - 1140); *26*, Lake View dolerite (1149); *27*, YB dykes (750).
The data are listed in McElhinny and Embleton (1976). The *inset map* of Australia
shows the location of the palaeomagnetic studies in relationship to the cratonic
regions which had stabilised by 1800 m.y. (*stippled*) and the mobile belts separating
them; the cratons are lettered: *P*, Pilbara, *Y*, Yilgarn, and *A*, Arunta

Rhodesia-Kaapvaal cratons and from the Guyana craton of South America
(after correction for Phanerozoic drift), and they define a single
closed loop as shown in Figure 11. From these data McElhinny and Mc-
Williams (1977) conclude that a single a.p.w. curve embraces "all the
data irrespective of the craton without violating any of the age constraints upon magnetisation times". These data confirm that the cratonic nuclei were approximately in their present relative positions
at 2200 - 1850 m.y. and indicate that the later mobile belts (illustrated in part in Fig. 5), which stabilised during the intervals 1350 -
1100 m.y. and 750 - 450 m.y., did not involve major relative movements.
Indeed, they restrict such later movements to the opening and closing
of small oceans (< 1000 km) under the very special conditions that the
cratonic nuclei always returned to their original relative positions
(Piper et al., 1973; McElhinny and McWilliams, 1977). For the case of
Africa the palaeomagnetic data thus leave no alternative to an ensialic
origin for the Proterozoic mobile belts as envisaged by Shackleton
(1973) and Kröner (1977).

5.6 Australia — 2600 - 1100 m.y.

Australian Proterozoic palaeomagnetic data are summarised by McElhinny
and Embleton (1976). Age control of most of the poles is poor but these

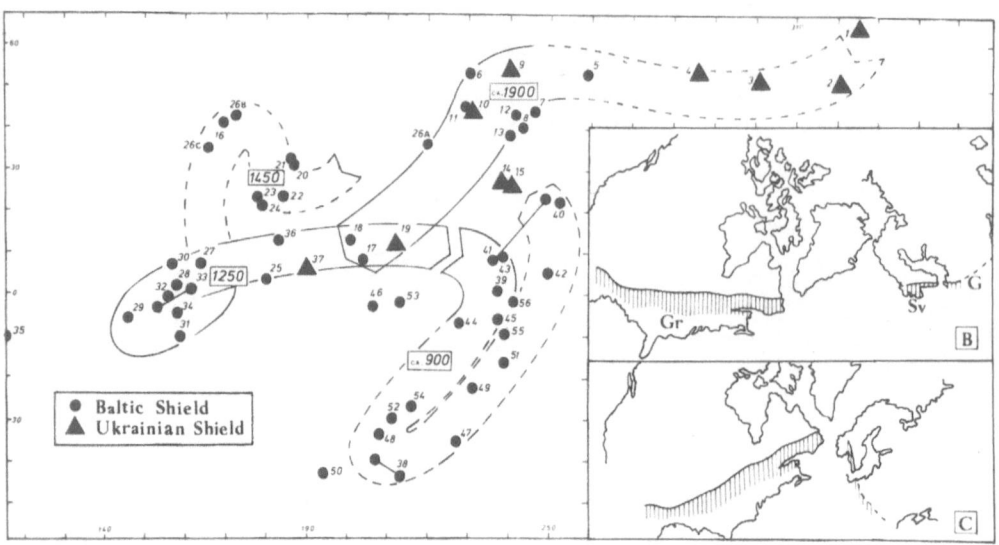

Fig. 13. Precambrian palaeomagnetic results from the Baltic (●) and Ukrainian (▲) Shields. Poles are from: *1*, Ingulets complex (2000 - 1700); *2*, Tokova complex; *3*, Bokovyansk-osnits complex; *4*, Bug Podolsky complex (2300 - 2000*); *5*, Tammeal gabbro (1900*); *6*, Mikkeli gabbro (1900*); *7*, Hyvinkäa gabbro (1900*); *8*, SW Finland gabbro diorites (1900*); *9* and *10*, Iotniy Sandstones (1950 - 1850*); *11*, Tärändo gabbro (1900*); *12*, Ylivieska gabbro (1900*); *13*, Pohjanma gabbro (1900*); *14*, Korosten complex (1750); *15*, Umam granite; *16*, Åva intrusives (1830); *17*, Kumlinge dolerites (1650*); *18*, Kumlinge and Åva dolerites (1830 - 1650); *19*, Turchinki gabbro (1400 - 1200); *20*, Föglö dolerite; *21*, Dalarna basalts; *22*, Upper Dala volcanics (1685 - 1490); *23*, Dalarna dolerites (> 1250); *24*, Lofthammer gabbro (> 1460); *25*, Satakunta (Jotnian) sandstones (1300); *26*, Rapakivi-type granites: *A*, Nordingrå (1415), *B*, Ragunda (1350), *C*, Gävle (1350); *27*, Satakunta dolerite; *28*, Vaasa dolerite; *29*, Märket dolerite; *30*, Gävle dolerite; *31*, Gnarp dolerite; *32*, Nordingrå anorthosite (Jotnian remagnetisation); *33*, Ulvo dolerite; *34*, Nordingrå dolerite; *35*, Dolerites of western Sweden (1200); *36*, Post-Ragunda dolerites; *37*, Azov granite (< 1000); *38*, Karlshamn dolerite; *39*, Tärnö dolerite; *40*, Bräkne-Hoby dolerite; *41*, Väby dolerite; *42*, Fäjö dolerite; *43*, Nilstorp dolerite; *44*, Årby dolerite; *45*, Falun dyke; *46*, S. Norway amphibolites; *47*, Basement rocks, Rogaland; *48*, Hunnedalen dolerites; *49*, Egersund dolerites; *50*, S. Norway farsundite; *51*, Tuve dolerite; *52 - 56*, S. Sweden hyperite intrusion. *Lines* connect poles for separate studies of the same rock unit. *, date assigned by reference to related rocks. It is suggested that ca. 900 m.y. magnetisations from units 38 - 55 which experienced slow uplift and cooling define closed loop-like contemporaneous African and North American data (Patchett and Bylund, 1977). *Inset B* shows the approximate continental reconstruction derived by superimposing 1900 - 1250 m.y. Baltic-Ukrainian and Laurentian palaeomagnetic poles; the Grenville (*Gr*), Sveconorwegian (*Sv*), and Galitsian (*G*) tectonic-magmatic lineaments (*vertical shading*) then lie on a single great circle lineament (Piper, 1976a; Poorter, 1976). *Inset C* shows the continental reconstruction derived by superimposing the < 1050 m.y. uplift and cooling magnetisations from the Grenville and Sveconorwegian Provinces (Patchett and Bylund, 1977): The collective data suggest relative movements between the two shield regions after the Jotnian and Grenville-Sveconorwegian tectonic-magmatic episodes. The polar data plotted here are given by Kruglyakova (1961), Poorter (1976), Patchett and Bylund (1977) and references given therein or are author's unpublished data

authors show that it is possible to construct a single a.p.w. curve
for the continent, and hence argue that the mobile belts separating
the cratonic nuclei are ensialic in origin. Poles which are known to
be certainly older than 1800 m.y. come only from the Yilgarn craton
of southwest Australia, and the critical evidence is the close agreement between ca. 1800 m.y. poles 10 - 12 (Fig. 12) separated by the ca.
1800 m.y. King Leopold mobile belt, and ca. 1700 m.y. poles 15, 16,
18, and 20 separated by a number of Middle and Upper Proterozoic belts.
Other poles, however, can either be connected by a simple path or lie
close to the path joining dated poles.

5.7 Baltic-Ukrainian Shield — 2000 - 1200 m.y.

Palaeomagnetic coverage of Precambrian rocks of the Baltic-Ukrainian
Shield is still very incomplete and age control of the available data
is mostly indirect (Fig. 13). Three intervals are however well covered:
Syntectonic gabbro-diorite intrusions of the Svecofennian mobile belt
in Finland and Sweden yield poles 5 - 13 with assigned ages of about
1900 m.y.; anorgenic magmatism in central Sweden dated 1650 - 1450 m.y.
yields poles 20 - 24, and finally, Jotnian dolerites widely distributed
over the Baltic Shield give poles near 160°E, 0°N; these latter rocks
are dated ca. 1250 - 1200 m.y. and are thus contemporaneous with the
Mackenzie igneous episode of the Laurentian craton. Because inter-
vening periods are so poorly covered any a.p.w. curve must be very
tentative, and data from the Ukrainian region (Kruglyakova, 1961) are
neither sufficiently numerous or reliable to preclude relative movements between the Baltic and Ukrainian regions in Proterozoic times.

A fourth group of poles in Figure 13 is derived from rocks of the
Sveconorwegian (Gothide) Province of southwest Sweden and Norway (Fig.
13, insets). This region has a structural and tectonic history comparable to that of the Grenville Province of North America, and the poles
are similarly related to slow uplift and cooling between 1100 and 800
m.y. (Patchett and Bylund, 1977); according to these workers they probably lie on a closed loop (Fig. 13) like contemporaneous uplift and
cooling poles from the Grenville Province. A comparison of Figures 7
and 13 shows, however, that the relationship between pre-1200 m.y.
poles and 1000-750 m.y. poles for the two shield is quite different.

6. Palaeomagnetic Correlations Between the Proterozoic Shields

The data reviewed in the preceding sections show conclusively that
Proterozoic mobile belts developed essentially in situ without large
relative movements of the adjoining cratons. On the other hand, Figures 6 - 13 also demonstrate that the Proterozoic shields moved continuously and often rapidly with respect to the poles during this era.
We are thus left to investigate two possibilities: Either a few large
Proterozoic shields moved independently over the surface of the globe
(presumably colliding occasionally) or these shield were coupled together, either loosely or rigidly, as a single supercontinent.

Comparison of Figures 7 and 8 shows that there is a very close correspondence between the Upper Proterozoic a.p.w. curves of Africa and
North America; this is particularly well defined by the succession of
dated poles 1150 - 1070 - 1000 - 900 m.y. terminating in the loops at
900 - 750 m.y. If we superimpose these curves we find that on the same
reconstruction the 2150 - 1850 m.y. a.p.w. curves of Africa (Fig. 11)
and North America (Fig. 9) coincide (Piper, 1976b). This close cor-

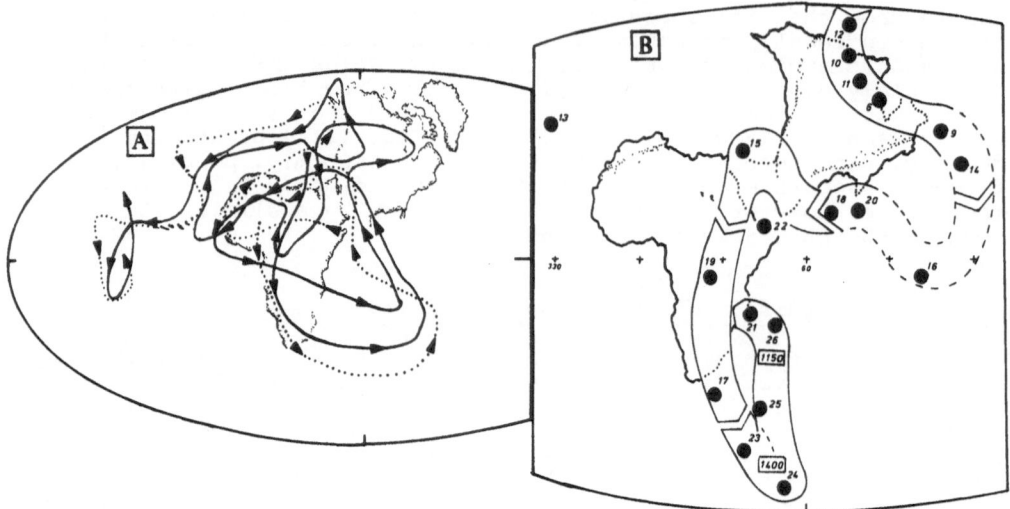

<u>Fig. 14A and B.</u> (*A*) The correlation between a.p.w. curves of Africa and North America using a single continental reconstruction for the two intervals of Proterozoic time between 2150 and 750 m.y. for which contemporaneous information is currently available. (*B*) Proterozoic palaeomagnetic data for Australia (see Fig. 12) plotted on this same model using the Smith-Hallam (1970) reconstruction of Gondwanaland; it is clear that no correlation exists between Upper Proterozoic data for the two regions; hence relative movements between the African and Australian Shields prior to 750 m.y. is inferred

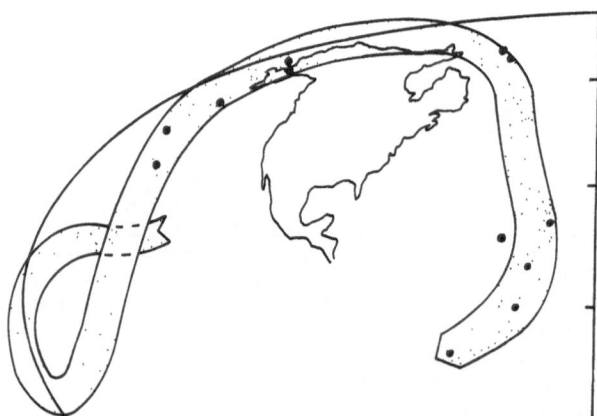

<u>Fig. 15.</u> The Hadrynian Loop as defined by Morris and Roy (1977). The central part of this loop is defined by results approximately 750 – 650 m.y. in age falling in present high latitudes, and the form of the a.p.w. path thus remains uncertain. A similar problem relates to contemporaneous Gondwanaland data (McElhinny and Embleton, 1976; Piper, 1976a)

respondence over the whole a.p.w. path for which we have contemporaneous data is illustrated in Figure 14(A): The two a.p.w. curves match very well and imply that the two shield were closely coupled as a single supercontinent during the interval ca. 2200 - 750 m.y. So close is this comparison that the two poles 8 and 10 in Figure 8 which cannot be matched with contemporaneous North American data (McElhinny and McWilliams, 1977) probably reflect inadequate age control rather than non-validity of the model; indeed, the limitations of the data discussed on pages 13 and 14 require that it is the large measure of overall agreement which is important in this comparison.

The reconstruction could be proved conclusively if it could be demonstrated that the two curves matched in Figure 14(A) have the same polarity. A small number of poles indicate that a very large polar shift took place relative to Gondwanaland (Piper, 1976a, Fig. 1; McElhinny and Embleton, 1976, Fig. 3) between ca. 750 m.y. (the end of the a.p.w. curve plotted in Fig. 11) and 600 - 700 m.y. (the beginning of the a.p.w. curve plotted in Fig. 6). These data suggest, but do not confirm, that the a.p.w. curve plotted in Figure 11 is a south a.p.w. curve. Recently Morris and Roy (1977) argued for a comparable large polar movement (the Hadrynian Track, Fig. 15) connecting the end of the Grenville Loop with Late Precambrian/Lower Palaeozoic poles. This implies that the curve plotted in Figure 7 is also a south a.p.w. curve. Morris and Roy accommodate a greater time range of data in the Hardynian Track than is permitted by the analysis of McElhinny and McWilliams (1977), but there is little doubt surrounding the outline of the path which is defined both by the sequence of magnetisations in slowly cooled metamorphic rocks and the magnetisations of sedimenatary sequences. Unfortunately there is currently insufficient data from North America to make a comparison with the Late Precambrian-Lower Palaeozoic results from Gondwanaland plotted in Figure 6.

A second comparison has been made between pre-1200 m.y. palaeomagnetic data from the Baltic and Laurentian Shields (Donaldson et al., 1973; Piper, 1976a; Poorter, 1976). Although it is not yet possible to identify a single a.p.w. curve from the former region (Fig. 13), the distribution of 2000 - 1200 m.y. poles for the two shields is similar such that when the two sets of data are superimposed the Grenville and Sveconorwegian mobile belts (ca. 1100 m.y.) are colinear (Fig. 13B; Piper, 1976a, Fig. 7; Poorter, 1976, Fig. 7). This reconstruction should be regarded as one for testing against the geological evidence only, and with our present understanding of Baltic-Ukrainian Shield palaeomagnetism it cannot be regarded as precluding relative movements between the Baltic and Laurentian Shields prior to 1200 m.y. Several workers have also compared the ca. 1100 - 800 m.y. results from the Grenville and Sveconorwegian Provinces of the two Shields. Patchett and Bylund (1977) and Morris and Roy (1977) derived closely similar positions comparable to that shown in Figure 13C with the two shields in juxtaposition but with the Grenville and Sveonorwegian lineaments now oblique to one another. As noted in Section 5.7, a simple comparison of the Upper Proterozoic palaeomagnetic data for the two shields demonstrates that relative movements occurred between 1200 and 1000 m.y.

It is equally clear that Australia moved relative to the African and Laurentian Shields prior to 750 m.y. (see Fig. 8) because if the Australian data of Figure 12 are plotted on the reconstruction of Figure 15 the Upper Proterozoic poles plot remote from contemporaneous data from elsewhere (Fig. 14B). It is clear that the Australian Shield was only welded to the remainder of Gondwanaland in the conventional Smith-Hallam (1970) reconstruction at some time prior to 750 m.y. It remains to be resolved whether this shield was entirely uncoupled from the

African-Laurentian Shield in earlier Proterozoic times: Initial examination of Figure 12 suggests that there is little comparison between
the a.p.w. curves of the two regions, but this may reflect the paucity
of data in the case of Australia. The age control of Proterozoic palaeomagnetic data from India is in general so poor that it cannot yet be
used to support or refute the conventional Gondwanaland reconstruction
prior to 750 m.y.

7. Proterozoic Supercontinent

Analysis of the foregoing sections shows that a picture is emerging of
Proterozoic tectonics which cannot be reconciled completely with either
Plate Tectonic mechanisms or with a single rigid continental unit. A
very large supercontinent incorporating the African, the Laurentian,
and probably the Baltic-Ukrainian Shields existed without loss of general coherence during most of this time (Fig. 16), but relative plate
movements also occurred between the Baltic and Laurentian Shields at
least between 1200 and 1000 m.y., and between Australia and Africa
prior to 750 m.y.

Bridgwater and Windley (1973) have shown that the structural and magmatic evolution of the crust in Proterozoic times was characterised
by a distinctive suite of rock associations. Long linear mobile belts
developed within reactivated older crust while massive anothosite plu-
tons were intruded across a broad belt of crust in association with
rapakivi-type granites and with alkaline complexes; the extrusive equiv-
alents of this magmatism are present as acid volcanics, ignimbrites,
and plateau basalts. Fracturing of the crust with the development of
grabens in which fluviatile sediments and basalt lavas accumulated,
was also widespread in Middle to Upper Proterozoic times. On the palaeomagnetic reconstruction of the land masses all the Upper Proterozoic
mobile belts are colinear (Fig. 16) and parallel to the axis of the
continental body. Massive anorthosites and rapakivi-type anorogenic
granites are also confined to a belt close to the southeast leading
edge of the supercontinent. Intrusion of these bodies commenced within
the stabilised cratonic areas at about 2200 m.y., and their intrusion
became progressively more restricted towards this leading edge during
Proterozoic times. This probably represents the decrease in crustal
temperature gradients during Proterozoic times, and anorthosites were
still being intruded at present crustal levels in Labrador at about
1300 m.y., at shallow depths in south Greenland at 1200 m.y., and in
southwest Norway at about 1100 m.y.

Earlier components of the Proterozoic crust include long linear shear
belts exposed either as ductile or brittle shear zones (e.g., Bak et
al., 1975) or inferred indirectly from aeromagnetic anomalies (Watson,
1973), and formed between 2700 and 1600 m.y. (Davies and Windley, 1976).
These lineaments become parallel to one another and to the younger
mobile belts on the supercontinent reconstruction, and it is clear
that this great circle belt of tectonic-magmatic activity parallel to
the axis of the continental body was of fundamental importance for
most of Proterozoic times. Early (> 2500 m.y.) deformed calcic anorthosites (Windley and Bridgwater, 1971) occur within a number of
high-grade terrains which also lie near a great circle (Fig. 16) and
close to the leading edge of the supercontinent; they suggest that
this distribution of continental crust was already established in
Archaean times, but there is currently insufficient palaeomagnetic
evidence to test this.

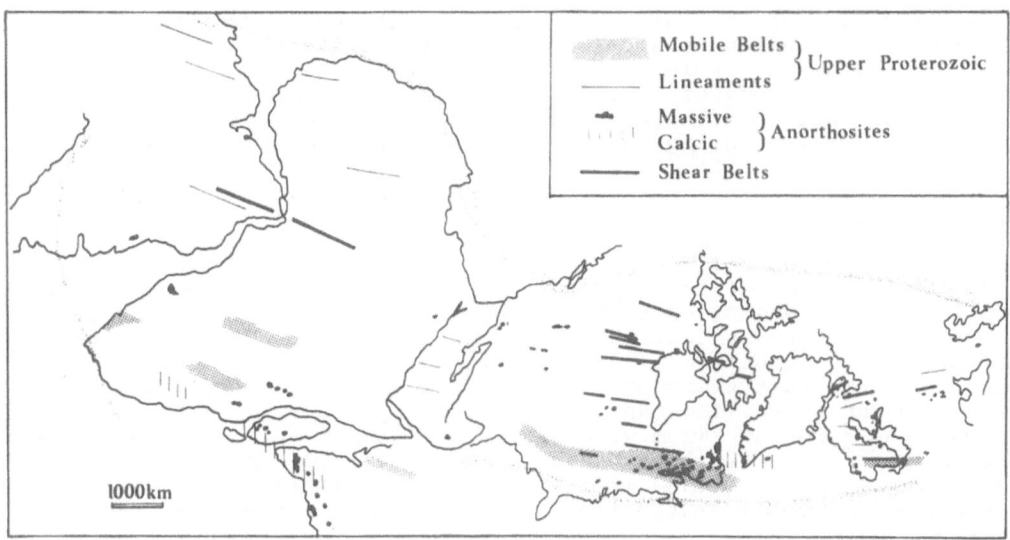

Fig. 16. The Proterozoic supercontinent as defined by palaeomagnetic data. Note the alignment of Upper Proterozoic mobile belts, Lower Proterozoic tectonic lineaments, and the restriction of magmatic activity at exposed crustal levels to the south east margin of the Shield. The age of the latest anorthosite-rapakivi granite type magmatism also decreases towards this margin of the supercontinent (Bridgwater and Windley, 1973); lineaments are taken in part from Davies and Windley (1976)

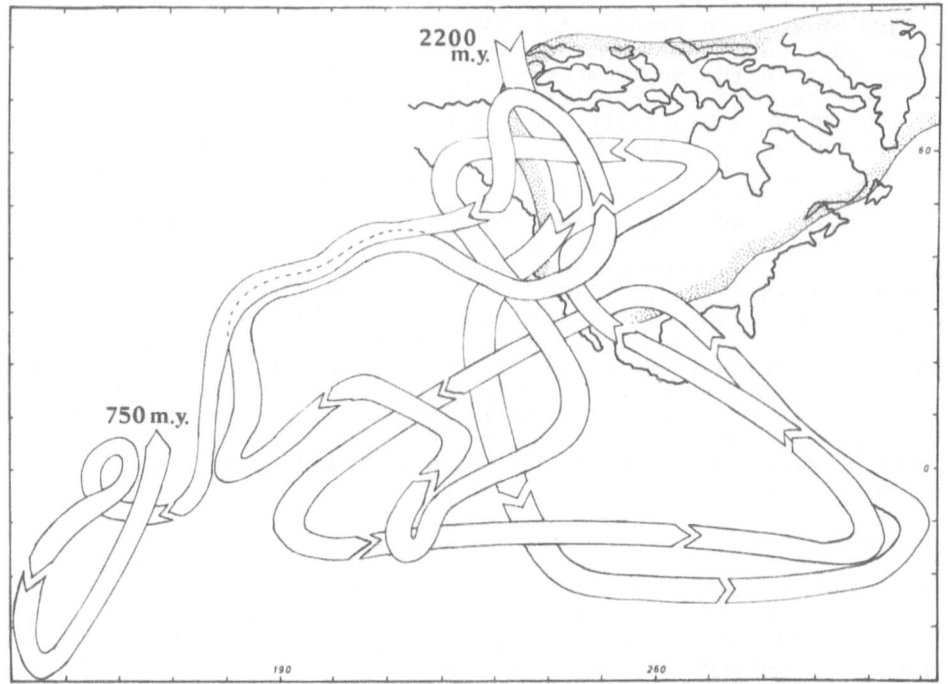

Fig. 17. Apparent polar wander path for the interval 2200 - 750 m.y. as derived from the Laurentian craton

The distribution of tectonic and magmatic features in Proterozoic times indicates that continuous accession of heat from the mantle took place along a broad great circle belt of continental crust. The increasing coherence of this body of continental crust with time is reflected both in the localisation of thermal/magmatic activity following widespread internal deformation along shear belts in Lower Proterozoic times, and in the gradual reduction of temperature gradients which localised and eventually precluded the intrusion of the massive anorthosite-rapakivi granite suite. This great circle belt must correspond to a very simple and more fluid (Sect. 5) large-scale mantle convection system possibly comprising only two cells. The downwelling of the oceanic lithosphere at the leading edge of the supercontinent could then have produced the range of mantle igneous derivatives to give the various Proterozoic analogues of modern subducting margins while preserving the general coherence of the continental unit. It is evident from the pattern of polar movements that the supercontinent had an inherent stability with respect to the prevailing mantle convection system (Fig. 17). The a.p.w. path between 2200 and 750 m.y. exhibits a sequence of loops; some of them are nearly identical although separated in time and all of them are closed. The continental body was apparently returned to a mean stable position on the globe over time intervals of the order of 100 - 200 m.y.

8. Age of the Earth-Moon System

The close approach of the Moon to the Earth would be recognised as a catastrophic event in Earth history (Lamar et al., 1970; Tarling, 1975), initiating widespread partial or complete melting of crust and mantle. The absence of such an event after 3000 m.y. provides additional confirmation that the Moon was already a satellite at this time. The oldest dated event in the Archaean crust of West Greenland is a thorough reworking of older granitic crust formed within the first 800 m.y. of the history of the solid Earth, and is only marginally older than widespread melting events recorded in several continents. Until about a decade ago it was possible to argue that the Archaean high-grade crust had been through a single evolutionary stage. On the basis of available age data, Cloud (1968) suggested that widespread Archaean crustal melting was caused by lunar capture at ca. 3700 m.y., and this concept received considerable independent support when the Maria volcanic episode on the Moon was dated at ca. 3700 m.y. (although with a spectrum of ages ranging from 3700 - 3100 m.y.). However, it is now known that the Archaean terrains did not experience a unified thermal history and while most high-grade terrains yield ages in the range 2700 - 3100, they occur, sometimes in close association, with small relicts of crust yielding older ages in the range 3800 - 3400 m.y., in Rhodesia, Greenland, Minnessota, and elsewhere. In addition the broadly contemporaneous low-grade greenstone terrains were forming at various times between 3500 and 2200 m.y.

Thus the geological grounds for recognising a lunar capture event are less good than was originally believed, and it is probable that the Moon was an earth satellite prior to all recognisable events in the Archaean crust. The ca. 3800 m.y. Isua supracrustal sequence of West Greenland contains banded iron formations which formed extensively in the oxygen-poor environment prior to about 2000 m.y. and reached a climax between about 2500 and 2100 m.y. They are considered to have been deposited in alkaline lakes where alternating siliceous and ferruginous layers represent a cyclic change in processes controlling sedimentation (Trendall, 1968). We do not have a present-day analogue

to establish whether this cycle was tidal, but the ferrous iron in solution clearly required an episodic supply of oxygen to be precipitated
as ferric oxides or hydroxides. According to Cloud (1968) this oxygen
supply was provided by primitive organic activity, and if this model
is correct the Greenland occurrence requires that blue-green algae
were established at 3800 m.y.; they could not have survived subsequent
close lunar approach. Models postulating the presence of the Moon in
a prograde orbit shortly after formation of the solid Earth (ca. 4600
m.y.) are thus favoured (e.g., Singer, 1970).

9. Stromatolite Evidence for Precambrian Tidal Parameters

Stromatolites are the fossilised remains of structures produced by
blue-green algae (photosynthetic oxygen producers) and bacteria; present-day algal mats form largely in intertidal zones but may also be
represented in supratidal and subtidal areas, and this is presumed to
have been the case in Precamrian times. Because this is a high-energy
environment microunconformities are widely developed in fossil stromatolites, and it is likely that only stromatolite sequences of de-
monstrable subtidal origin are suitable for analysing tidal rhythms
(Truswell and Eriksson, 1975). Studies of contemporaneous algal mats
suggest that the lamination here is diurnal because sediment trapping
during the day is followed by nocturnal algal binding of the particles.
Long-period cycles may reveal lunar (tidal) and seasonal variations
of sediment influx and biological activity but are more difficult to
evaluate both because the data accumulate more slowly and because continuous records of several periods of long-term fluctuations are seldom located. At the present time the surface environment is oxidising
and blue-green algae preciptate calcareous mats in all but a few environments. Prior to 2000 m.y., however, there was little or no free
oxygen and the algae frequently precipitated silica stromatolites. Indeed the oldest stromatolites of the Pongola Supergroup (3060 - 2870
m.y., Mason and Brunn, 1977), the Bulawayan Group (2700 - 2600 m.y.,
Pannella, 1972), the Ventersdorp (ca. 2300 m.y.) and Wolkberg Systems
were deposited in reducing environments, probably of quite restricted
vertical and lateral extent, and closely related to volcanicity. The
Transvaal Dolomites (> 2224 m.y., Truswell and Eriksson, 1975) appear
to be among the oldest algal stromatolites within a shelf carbonate
sequence.

Since they provide the only organic evidence for tidal rhythms in Precambrian times, stromatolites have been widely examined but currently
with only limited success (see Scrutton, this Volume). Mohr (1975)
derived a value of 25.6 days/synodic month from ca. 2000 m.y.-Biwabik
stromatolites by regarding observed periodicities of 12.8 as fortnightly. The basis of this argument is that the tidal effects on the algal
environment would have been in phase with the optimum sunlight twice
in each lunar cycle. Pannella (1972) derives a figure of 32 days/syn-
odic month and 448 days/year for stromatolites of the Gunflint Formation of broadly comparable age. The significance of Pannella's results
from the Bulawayan Group is more difficult to evaluate because, if the
total angular momentum of the Earth-Moon system is assumed to be conserved, it can be shown that the number of days/synodic month can never
approach 39 (Lamar et al., 1970, Fig. 2; Mohr, 1975, Fig. 3); Mohr suggests that the Bulawayan data represent a synodic month of about 20
days during Bulawayan times. These very sparse data suggest that a
gradual increase in the number of days per year took place in Precambrian times comparable to the trend found from Phanerozoic data

(Scrutton, this Volume). It is generally assumed that the sidereal year has remained constant; hence these data imply a gradual increase in the length of the day.

It was originally considered that the amplitude of domal stromatolites reflected the tidal variation and could thus be used as a measure of this parameter in the geologic past (Cloud, 1968). However, the observational data were found to be highly inconsistent (Awramik, 1971), and it is now considered that most columnar stromatolites grew below the intertidal zone so that the amplitude of the domal morphology cannot be used to indicate tidal amplitudes. From a sedimentological study of the Transvaal stromatolites, Truswell and Eriksson (1975) inferred that the tidal amplitude was about 1/2 - 1 1/2 metres at ca. 2300 m.y.

10. Sedimentologic Evidence for Precambrian Tidal Parameters

This evidence is of a much more subjective nature than that from stromatolites. Indeed sedimentologic of the past two decades has shown that many reasons advanced in the last century for strong tidal currents in the geological past can no longer be substantiated (see historical summary in Merifield and Lamar, 1970). Currents with velocities in the range 1 to 6 km/hr produce sand waves at the sediment-fluid interface which are widely reported from the present continental shelf (summary in Merifield and Lamar, 1968) where local topographic effects (apparently ∿ 100 km or less in scale) accentuate ambient tidal velocities which are normally less than about 1 km/h. These features would be preserved in the geological record as large-scale cross bedding and are distinguished from ripple marks (formed by currents as low as 0.9 km/h) by their much higher amplitudes. Hargraves (1961) and Stewart (1966) cite examples of high-amplitude cross bedding in well-sorted sandstone formations ca. 2500 m.y. and 600 m.y. old, respectively. However, to prove that these features represent high global tidal velocities and are not (like contemporaneous sand waves) a product of local conditions, requires that they be shown to have both a very wide lateral extent and a high average amplitude. This appears to be the case with the example documented by Hargraves (∿ 400 km lateral extent) and may apply to the widespread Lower Cambrian quartzites of Europe and North America (Olsen, 1966). However, abundant observational data are lacking, and it is more satisfactory to draw conclusions from a range of sedimentary characteristics that can only result from strong currents, and then endeavour to resolve whether these currents were tidal. Evidence of this nature relating to the Witwatersrand System (2500 m.y.) and in particular to the auriferous conglomerates (banket deposits) contained therein has been analysed by Hargraves (1961). The range of sedimentary features considered to be indicative of strong currents of tidal origin may be summarised under three headings:

1. The deposits are well sorted; grains of heavy minerals are associated with the conglomerates; the sand + conglomerate:shale ratio is high in general, and in particular quarzites are coarse grained.

2. There is a narrow azimuthal distribution of current directions indicated by large and small-scale cross bedding. Sand ripples are parallel and not perpendicular to current directions inferred from cross bedding.

3. Individual beds exhibit wide lateral persistence over large distances but have characteristics of shallow water deposition.

Hargraves' conclusions are indirectly supported by comparisons with other Lower Proterozoic auriferous conglomerates. For example the Tarkwaian Series (> 2150 m.y.) of West Africa exhibits a poorly sorted suite of conglomerates and other clastic sediments, with a wide azimuthal distribution of current indicators. Here a range of sedimentary features suggests that the environment of deposition was fluviatile rather than marine (Sestini, 1970).

Klein (1972) developed a sedimentation model for intertidal flats and employed it to estimate the tidal amplitude in Late Precambrian and Phanerozoic sequences. The derived amplitudes ranged from 17 to 0.5 metres, and Klein found no significant change in tidal amplitude since late Precambrian times. Using Klein's model, Brunn and Hobday (1976) estimated four tidal amplitudes of 12, 12, 20, and 25 metres in 3230 - 2870 m.y. sediments of the (late Archaean) Pongola Supergroup; two of these measurements are significantly higher than the Phanerozoic estimates of Klein and others. However, significantly higher tidal amplitudes in Lower Proterozoic times (∿ 2300 m.y.) seem to be ruled out by the work of Truswell and Eriksson (1975).

Thus the sedimentologic evidence is of a very fragmentary nature. The best that can be said is that it seems to indicate stronger tidal currents and possibly slightly higher tidal amplitudes in Late Archaean/Lower Proterozoic times, but since the force developing the tidal current is inversely proportional to the cube of the Earth-Moon distance, the Moon was probably only marginally closer to the Earth at that time.

11. The Tidal Couple and Changes in the Earth's Rotation in Precambrian Times

Contributions to the present Volume discuss the determination and magnitude of the present tidal retarding couple, and the palaeontological data suggest that a broadly comparable value is applicable over the past 400 m.y. since Silurian times (Hipkin, 1970; Scrutton, this Volume). The fundamental geological problem associated with these observations is that when extrapolated schematically back into Precambrian times, they imply a catastrophic approach of the Earth and Moon to the Roche limit (∿ 2.9 Earth radii), prior to about 1400 m.y. Since we now know that the Earth-Moon system has existed since at least 3800 m.y. with appreciable separation since at least 2500 m.y., either the tidal retarding couple or the Earth's moment of inertia, or both, must have been much lower in Proterozoic times.

The present coefficient of moment of inertia of the Earth is 0.33, which differs from the value of 0.40 for a homogeneous sphere. We may speculate that this difference results from differentiation within the Earth since its formation, and consider under four headings the possible changes that occurred in Precambrian times.

1. Fundamental differentiation into core and mantle. Runcorn (1964) considered that the Earth's core had continued to grow by accretion of iron through geological time; this would be a viable process if mantle convection extends to the base of the mantle. However, because of its high density and small radius the core makes a contribution of only about 10 % to the total moment of inertia of the Earth and the explanation of any appreciable accelerations in terms of this mechanism requires a very rapid core growth. Also formation of the core probably has no bearing on the present problem because the existence of the geomagnetic field prior to 2700 m.y., and arguments based on the chemical and thermal history of the Moon, imply that the core formed early in Earth history; indeed a strong case can be made for condensation of the core at a rapid phase of the Earth's development prior to condensation of the mantle constituents (Clark et al., 1970).

2. Thermal contraction. The decrease in radiogenic heat production in
the mantle would lead to progressive thermal contraction, a decrease
in the moment of inertia, and hence an increase in the angular rotation
rate. Calculations (Creer, 1975) show that this increase is an order
or more less than the probable decrease in the rotation rate since
Precambrian times.

3. Mantle convection and differentiation lead to addition of light
material to the crust and heavier depleted mantle to the mesosphere.
Again the effect would be to decrease the moment of inertia. Since
the bulk of the crust was formed by 2500 m.y. this effect is likely
to have been very small.

4. Elevation of mantle layers. The diminishing radiogenic heat production will have a more important consequence in elevating the transition
zones within the mantle, leading to a decrease in moment of inertia
and a decrease in the length of the day. In this context the most important interface is the (C) transition zone currently lying between
about 350 and 650 km depth (Stacey, 1969) with a density increase of
about 25 % probably representing an olivine → spinel transition. Creer
(1975) analysed the consequences of a temperature drop of 200°C on the
moment of inertia. Distributed over 2000 m.y. this would decrease the
length of the day by about 5 % of the increase in length actually suggested by the palaeontological data. Of course the temperature decrease
since Archaean times has probably been significantly greater than this,
but the resultant accelerating couple would remain only a small fraction of the tidal decelerating couple.

In addition to these changes in the moment of inertia, a number of
miscellaneous phenomena discussed by Tarling (1975) also produce accelerating moments, but the total effect of geological changes in Precambrian times is only a fraction (up to about a quarter) of the tidal
retarding couple, and we are forced to conclude that it is changes in
this parameter that have been most important in controlling the history of the Earth-Moon system.

The magnitude of tidal dissipation due to bottom-friction of the ocean
tides depends on the distribution of the continental crust over the
globe. The present global tectonic situation including many small continental plates is, however, seen to be the exception rather than the
rule, and Sundermann (this Volume) shows how the magnitude of the tidal retarding couple is appreciably lower when the present continents
were agglomerated together as the supercontinent Pangaea in Permian
times (225 m.y.). Prior to the welding of this supercontinent along
the Acadian-Hercynian-Caledonian orogenic belts, four large crustal
plates (Gondwanaland, Laurentia, Eurasia, and Siberia) appear to have
existed in Lower Palaeozoic times, with an unspecified number of minor
plates which became incorporated in these belts. The foregoing analysis of Proterozoic palaeomagnetism suggests that only a small number
of very large crustal plates, which may well have been closely coupled
together, existed during those times. The total tidal dissipation will
of course depend on the disposition of these crustal plates over the
globe, but it is likely to have been considerably lower, at least in
Proterozoic times, because the preponderance of palaeomagnetic poles
from Africa and Laurentia show that these regions at least, lay at in-
termediate or high latitudes during most of this time (Fig. 17). The
dissipative tidal system was probably much more complex in Archaean
times when the global surface comprised a large number of loosely
coupled small plates (although it is probable that the hydrosphere
had then not reached its present volume).

The mass and lithological spectrum of marine sediments deposited during
geological time may be used indirectly to estimate the importance of

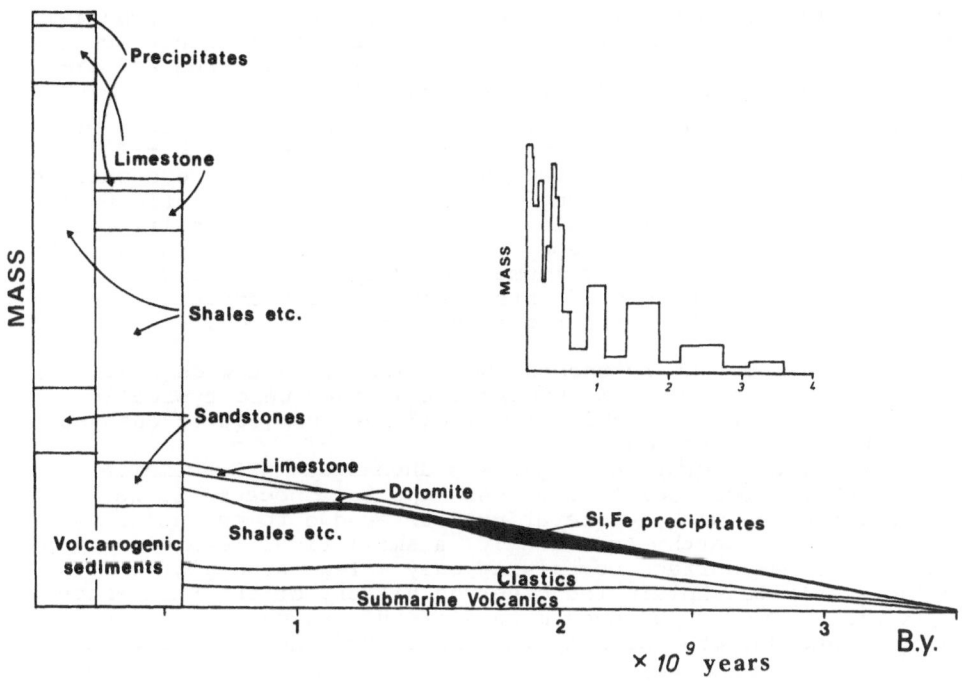

Fig. 18. Schematic diagram showing the relative masses of the sedimentary rocks and their main lithological subdivisions as a function of age (based on data of Ronov, 1968). The *inset diagram* shows the mass-age distribution of existing sedimentary rocks as estimated by Garrels and Mackenzie (1971)

shelf seas in the past. These parameters have been examined by Garrels and Mackenzie (1971), who found that the total mass of Precambrian sediments is less than that of all Phanerozoic sediments. This figure only in part reflects the greater time available for erosion and recycling because several features characterise the Precambrian record (Fig. 18):

1. In addition to the great thicknesses of basic lavas deposited in the Archaean greenstone belts, a significant proportion of Precambrian sediments are chemical precipitates of silica and iron deposited in shallow water in environments essentially free of clastic deposits. The bulk of the Proterozoic sedimentary record is fluviatile or shallow marine.

2. Deposition was apparently irregular with a maximum at about 1000 m.y. and a minimum at 800 - 600 m.y.; however, there are no records of any regressions comparable to the Permian or transgressions comparable with the Cretaceous. This is to be expected because long-term sea level fluctuations are largely controlled by changes in the length of the mid-ocean ridge system (Valentine and Moores, 1970). Not only would this have been largely constant in a tectonic regime not characterised by repeated break-up and rewelding of plates, but ridges of higher topography are associated with slow spreading, and higher ocean lithosphere spreading rates in the Proterozoic would mitigate against large ocean volume fluctuations caused by changing ridge geometry even if such changes took place.

3. There was a marked change in sedimentation reflecting the transition from unstable Archaean environments to stable Proterozoic environments, and defined by a change from immature greywackes and arkoses to mature quartz sediments (Ronov, 1968, 1972).

4. Proterozoic erosion and deposition rates were much lower than present rates. There are several reasons for this: Outside of restricted geosynclinal belts and granulite regions, rates of uplift were extremely slow such that many cratonic regions have experienced little uplift since 2000 m.y. (Watson, 1976). Also continent-continent collisions were absent or rare, and the steeper temperature gradients in the lithosphere did not permit the development of deep root zones to fold belts; hence the topographic gradients favouring high deposition and erosion rates could not have existed in Proterozoic times.

Thus although the sedimentary record shows clearly that wide shelf seas existed in Proterozoic times, it suggests that they were typically shallower, of lower gradient and volume, and less subject to changes in level than Phanerozoic seas.
Even if the tidal decelerating couple was unchanged in Precambrian times it would have been relatively less important because it was partly counterbalanced by week accelerating couples due to evolutionary processes within the Earth. Accordingly, a small tidal retarding couple acting over a global surface comprising only a few large plates is qualitatively able to explain the observed history of the Earth-Moon system. It will, however, be much more difficult to derive a quantitative model to accommodate the many variables producing the resultant couple.

References

Anhaeusser, C.R.: The evolution of the early Precambrian crust of southern Africa. Philos. Trans. R. Soc. Lond. A, 273, 359-388 (1973)

Anhaeusser, C.R., Mason, R., Viljoen, M.J., Viljoen, R.P.: A reappraisal of some aspects of Precambrian Shield geology. Bull. Geol. Soc. Am. 80, 2175-2220 (1969)

Armstrong, R.L.: A model for the evolution of strontium and lead isotopes in a dynamic earth. Rev. Geophys. 6, 175-199 (1968)

Awramik, S.M.: Precambrian columnar stromatolite diversity: reflection of Metazoan appearance. Science 174, 825-826 (1971)

Bak, J., Sørensen, K., Grocott, J., Korstgaard, J.A., Nash, D., Watterson, J.: Tectonic implications of Precambrian shear belts in western Greenland. Nature (London) 254, 566-569 (1975)

Barr, M.W.C.: Crustal shortening in the Zambezi Belt. Philos. Trans. R. Soc. Lond. A 280, 555-568 (1976)

Beckmann, G.E.J., Olesen, N., Sørensen, K.: A palaeomagnetic experiment on crustal uplift in West Greenland, Earth Planet. Sci. Lett. 34, 592-599 (1977)

Bridgwater, D., Escher, A., Watterson, J.S.: Tectonic displacements and thermal activity in two contrasting Proterozoic mobile belts from Greenland. Philos. Trans. R. Soc. Lond. A 273, 513-533 (1973)

Bridgwater, D., Windley, B.F.: Anorthosites, post-orogenic granites, acid volcanic rocks and crustal development in the North Atlantic Shield during the mid-Proterozoic. Spec. Publ. Geol. Soc. S. Afr. 3, 307-317 (1973)

Brown, G.C.: Mantle origin of Cordilleran granites. Nature (London) 265, 21-24 (1977)

Brunn, V. von, Hobday, D.K.: Early Precambrian tidal sedimentation in the Pongola Supergroup of South Africa. J. Sed. Petrol. 46, 670-679 (1976)

Buetner, E.K.: The origin and evolution of free oxygen in the atmosphere. IAGA/IAMAP Joint Assembly (abst.) 108 (1977)

Bullard, E., Everett, J.E., Smith, A.G.: The fit of the continents around the Atlantic. Philos. Trans. R. Soc. Lond. A 258, 41-51 (1965)

Bullen, K.E., Haddon, R.A.W.: Derivation of an Earth model from free oscillation data. Proc. Natl. Acad. Sci. USA 58, 846-852 (1967)

Burke, K., Dewey, J.F.: An outline of Precambrian plate development. In: Implications of Continental Drift to the Earth Sciences, Vol. 2, pp. 1035-1045. Tarling, D.H., Runcorn, S.K. (eds.). London: Academic Press, 1973

Cavenaugh, M.D., Seyfert, C.K.: Apparent polar wander paths and the joining of the Superior and Slave Provinces during early Proterozoic time. Geology 5, 207-211 (1977)

Chase, C.G., Perry, E.C.: The ocean: Growth and oxygen isotope evolution. Science 177, 992-994 (1972)

Clark, S.P., Turekian, K.K., Grossman, L.: Model for the early history of the Earth. In: The Nature of the Solid Earth. Robertson, E.C. (ed.). New York: McGraw Hill, 1970, pp. 3-18

Clifford, T.N.: The structural framework of Africa. In: African Magmatism and Tectonics. Clifford, T.N., Gass, I.G. (eds.). London: Oliver and Boyd, 1970, pp. 1-26

Cloud, P.: Atmospheric and hydrospheric evolution on the primitive Earth. Science 160, 729-736 (1968)

Condie, K.C.: Archaean magmatism and crustal thickening. Bull. Geol. Soc. Amer. 84, 2981-2992 (1973)

Condie, K.C., Potts, M.T.: Calc-alkaline volcanism and the thickness of the early Precambrian crust in North America. Can. J. Earth Sci. 6, 1179-1184 (1969)

Coward, M.P., Lintern, B.C., Wright, L.I.: The pre-cleavage deformation of the sediments and gneisses of the northern part of the Limpopo belt. In: The Early History of the Earth. Windley, B.F. (ed.). London: Wiley, 1976, pp. 323-330

Creer, K.M.: On a tentative correlation between changes in the geomagnetic polarity bias and reversal frequency and the Earth's rotation through Phanerozoic time. In: Growth Rhythms and the History of the Earth's Rotation. Rosenburg, G.D., Runcorn, S.K. (eds.). London: Wiley, 1975, pp. 293-318

Davies, F.B., Windley, B.F.: The significance of major Proterozoic high grade linear belts in continental evolution. Nature (London) 263, 383-385 (1976)

Dearnley, R.: Orogenic fold belts and a hypothesis of Earth evolution. In: Physics and Chemistry of the Earth, 7. Ahrens, L.H. et al. (eds.). London: Pergamon Press, 1965, pp. 1-114

Dewey, J.F.: Evolution of the Appalachian/Caledonian orogen. Nature (London) 222, 124-129 (1969)

Dickinson, B.B., Watson, J.: Variations in crustal level and geothermal gradient during the evolution of the Lewisian complex of northwest Scotland. Precambrian Res. 3, 363-374 (1976)

Dickinson, W.R., Luth, W.C.: A model for plate tectonic evolution of mantle layers. Science 174, 400-404 (1971)

Dietz, R.S.: Continent and ocean basin evolution by spreading of the sea floor. Nature (London) 190, 854-857 (1961)

Donaldson, J.A., McGlynn, J.C., Irving, E., Park, J.K.: Drift of the Canadian Shield. In: Implications of Continental Drift to the Earth Sciences. Tarling, D.H., Runcorn, S.K. (eds.). London: Academic Press, 1973, pp. 3-18

DuBois, P.M.: Palaeomagnetism and correlation of Keweenawan rocks. Geol. Surv. Can. Bull. 71, 75p. (1962)

Eade, K.E., Fahrig, W.F.: Geochemical evolutionary trends of continental plate — a preliminary study of the Canadian Shield. Bull. Can. Geol. Surv. 179, 1-51 (1971)

Engel, A.E.J., Itson, S.P., Engel, C.G., Stickney, D.M., Cray, E.J.: Crustal evolution and global tectonics. A petrogenic view. Bull. Geol. Soc. Am. 85, 843-858 (1974)

Ernst, W.G.: Occurrence and mineralogical evolution of blueschist belts with time. Am. J. Sci. 272, 657-668 (1972)

Fahrig, W.F., Bridgwater, D.: Late Archaean — Early Proterozoic palaeomagnetic pole positions from West Greenland. In: The Early History of the Earth. Windley, B.F. (ed.). London: Wiley, 1975, pp. 427-439

Fisher, D.E.: Rare gas clues to the origin of the terrestrial atmosphere. In: The Early History of the Earth. Windley, B.F. (ed.). London: Wiley, 1976, pp. 547-556

Fyfe, W.S.: Archaean tectonics. Nature (London) 249, 338 (1974)

Garrels, R.M., Mackenzie, F.T.: Evolution of Sedimentary Rocks. New York: Norton, 1971, p. 397

Gastil, R.G.: Continents and mobile belts in the light of mineral dating. Int. Geol. Congr. 21(a), 162-169 (1960)

Glikson, A.Y.: Early Precambrian evidence of a primitive ocean crust and island nuclei of sodic granite. Bull. Geol. Soc. Am. 83, 3323-3344 (1972)

Glikson, A.Y., Lambert, I.B.: Relations in space and time between major Precambrian Shield units. An interpretation of Western Australian data. Earth Planet. Sci. Lett. 20, 395-403 (1976)

Goodwin, A.M.: Giant impacting and the development of continental crust. In: The Early History of the Earth. Windley, B.F. (ed.). London: Wiley, 1976, pp. 77-98

Greenwood, W.R., Hadley, D.G., Anderson, R.E., Fleck, R.J., Schmidt, D.L.: Late Proterozoic cratonisation in south western Saudi Arabia. Philos. Trans. R. Soc. Lond. A 280, 499-516 (1976)

Hargraves, R.B.: Cross-bedding and ripple-marking in the Main Bird Series of the Witwatersrand System of the East Rand Area. Trans. Geol. Soc. S. Afr. 65, 264-275 (1961)

Hargraves, R.B.: Precambrian geological history. Science 193, 363-371 (1977)

Heier, K.S.: Geochemistry of granulite facies rocks and problems of their origin. Philos. Trans. R. Soc. Lond. A 273, 429-442 (1973)

Hess, H.H.: History of ocean basins. In: Petrologic Studies (Buddington Volume), Engel, A.E.J., James, H.L., Leonard, B.F. (eds.). Geol. Soc. Amer. 1962, pp. 599-620

Hipkin, R.G.: A review of theories of the Earth's rotation. In: Palaeogeophysics. Runcorn, S.K. (ed.). London: Academic Press, 1970, pp. 53-59

Hoffman, P.: Evolution of an early Proterozoic continental margin: the Coronation geosyncline and associated aulacogens in the north western Canadian Shield. Philos. Trans. R. Soc. Lond. A 273, 547-581 (1973)

Hurley, P.M., Rand, J.R.: Pre-drift continental nuclei. Science 164, 1229-1242 (1969)

Irving, E., McGlynn, J.C.: Proterozoic magnetostratigraphy and the tectonic evolution of Laurentia. Philos. Trans. R. Soc. Lond. A 280, 433-468 (1976)

Irving, E., Naldrett, A.J.: Palaeomagnetism in Abitibi Greenstone Belt, and Abitibi and Matachewan diabase dikes: Evidence of the Archaean geomagnetic field. J. Geol. 85, 157-176 (1977)

Irving, E., Park, J.K., Roy, J.L.: Palaeomagnetism and the origin of the Grenville Front. Nature (London) 236, 344-346 (1972)

Isacks, B., Oliver, J., Sykes, L.R.: Seismology and the new global tectonics. J. Geophys. Res. 73, 5855-5899 (1968).

Jahn, B.M., Nyquist, L.E.: Crustal evolution in the early Earth - Moon system: constraints from Rb-Sr studies. In: The Early History of the Earth. Windley, B.F. (ed.). London: Wiley, 1976, pp. 55-76

Jones, D.L., Robertson, I.D.M., McFadden, P.L.: A palaeomagnetic study of Precambrian dyke swarms associated with the Great Dyke of Rhodesia. Trans. Geol. Soc. S. Afr. 78, 57-65 (1977)

Kalsbeek, F.: Metamorphism of Archaean rocks of West Greenland. In: The Early History of the Earth. Windley, B.F. (ed.). London: Wiley, 1976, pp. 225-236

Klein, G.de V.: Determination of palaeotidal range in clastic sedimentary rocks. Int. Geol. Congr. 24 (b), 397-405 (1972)

Kruglyakova, G.I.: Results of palaeomagnetic research on the Ukrainian crystalline massif and adjacent regions. Akad. Nauk. SSSR, Izv. Earth Phys. Ser. (1961), pp. 1-238

Kröner, A.: Precambrian·mobile belts in southern and eastern Africa — ancient sutures or sites of ensialic mobility? A case for crustal evolution towards plate tectonics. Tectonophysics 40, 101-135 (1977)

Lamar, D.L., McGann-Lamar, J.V., Merifield, P.M.: Age and origin of the Earth - Moon system. In: Palaeogeophysics. Runcorn, S.K. (ed.). London: Academic Press, 1970, pp. 41-52

Lambert, R.St.J.: Archaean thermal regimes, crustal and upper mantle temperatures and a progressive evolutionary model for the Earth. In: The Early History of the Earth. Windley, B.F. (ed.). London: Wiley, 1976, 363-387

Leblanc, M.: Proterozoic ocean crust at Bon Azzes. Nature (London) 261, 34-35 (1976)

MacIntyre, F.: Why the Sea is salt. Sci. Am. 223, 104-115 (1970)

Mason, T.R., Brunn, V. von: 3 Gyr-old stromatolites from South Africa. Nature (London) 266, 47-56 (1977)

McBirney, A.R.: Factors governing the nature of submarine volcanism. Bull. Volcanol. 26, 455-469 (1963)

McElhinny, M.W.: Palaeomagnetism and Plate Tectonics. Cambridge: University Press, 1973, p. 358

McElhinny, M.W., Embleton, B.J.J.: Precambrian and early Palaeozoic palaeomagnetism in Australia. Philos. Trans. R. Soc. Lond. A 280, 417-431 (1976)

McElhinny, M.W., Giddings, J.W., Embleton, B.J.J.: Palaeomagnetic results and Late Precambrian glaciations. Nature (London) 248, 557-561 (1974)

McKenzie, D., Weiss, N.: Speculations on the thermal and tectonic history of the Earth. Geophys. J. R. Astr. Soc. 42, 131-174 (1975)

McWilliams, M.O., Dunlop, D.J.: Precambrian palaeomagnetism: Magnetisations reset by the Grenville orogeny. Science 190, 269-272 (1975)

Merifield, P.M., Lamar, D.L.: Sand waves and early Earth - Moon history. J. Geophys. Res. 73, 4767-4774 (1968)

Merifield, P.M., Lamar, D.L.: Palaeotides and the geologic record. In: Palaeogeophysics. Runcorn, S.K. (ed.). London: Academic Press, 1970, pp. 31-40

Mohr, R.E.: Measured periodicities of the Biwabik (Precambrian) stromatolites and their geophysical significance. In: Growth Rhythms and the History of the Earth's Rotation. Rosenburg, G.D., Runcorn, S.K. (eds.). London: Wiley, 1975, pp. 43-56

Moorbath, S.: Age and isotope constraints for the evolution of Archaean crust. In: The Early History of the Earth. Windley, B.F. (ed.). London: Wiley, 1976, pp. 351-360

Moorbath, S., O'Nions, R.K., Pankhurst, R.J.: Early Archaean age for the Isua iron formation, West Greenland. Nature (London) 245, 138-139 (1973)

Moorbath, S., O'Nions, R.K., Pankhurst, R.J.: The evolution of early Precambrian crustal rocks at Isua, West Greenland — geochemical and isotopic evidence. Earth Planet. Sci. Lett. 27, 229-239 (1975)

Morgan, G.E.: Palaeomagnetism of a slowly cooled plutonic terrain in western Greenland. Nature (London) 259, 382-385 (1976)

Morris, W.A., Roy, J.L.: Discovery of the Hadrynian Polar Track and further study of the Grenville Problem. Nature (London) 266, 689-692 (1977)

Naqvi, S.M., Rao, V.D., Narain, H.: The protocontinental growth of the Indian shield and the antiquity of the rift valleys: Precambrian Res. 1, 345-398 (1974)

O'Hara, M.J.: Thermal history of excavation of Archaean gneisses from the base of the continental crust. J. Geol. Soc. 134, 185-200 (1977)

Olsen, W.S.: Origin of the Cambrian-Precambrian unconformity. Amer. Sci. 54, 458-464 (1966)

Pannella, G.: Palaeontological evidence on the Earth's rotational history since the Early Precambrian. Astrophys. Space Sci. 16, 212-237 (1972)

Park, J.K.: Palaeomagnetism of the Flin Flon - Snow Lake greenstone belt, Manitoba and Saskatchewan. Can. J. Earth Sci. 12, 1272-1290 (1975)

Patchett, P.J., Bylund, G.: Age of Grenville belt magnetisation: Rb-Sr and palaeomagnetic evidence from Swedish dolerites. Earth Planet. Sci. Lett. 35, 92-104 (1977)

Perry, E.C., Tan, F.C.: Significance of oxygen and carbon isotope variations in early Precambrian cherts and carbonate rocks of southern Africa. Bull. Geol. Soc. Am. 83, 647-664 (1972)

Poorter, R.P.E.: Palaeomagnetism of the Svecofennian Loftahammar gabbro and some Jotnian dolerites in the Swedish part of the Baltic Shield. Phys. Earth Planet. Int. 12, 51-64 (1976)

Piper, J.D.A.: Proterozoic crustal distribution mobile belts and apparent polar movements. Nature (London) 251, 381-384 (1974)

Piper, J.D.A.: Palaeomagnetic evidence for a Proterozoic Supercontinent. Philos. Trans. R. Soc. Lond. A 280, 469-490 (1976a)

Piper, J.D.A.: Definition of pre-2000 m.y. Apparent polar movements. Earth Planet. Sci. Lett. 28, 470-478 (1976b)

Piper, J.D.A.: Magnetic stratigraphy and magnetic-petrologic properties of Precambrian Gardar lavas, South Greenland. Earth Planet. Sci. Lett. 34, 247-263 (1977)

Piper, J.D.A., Briden, J.C., Lomax, K.: Precambrian Africa and South America as a single continent. Nature (London) 245, 244-248 (1973)

Piper, J.D.A., Stearn, J.E.F.: Palaeomagnetism of Ketilidian metamorphic rocks of
  SW Greenland and 1850 - 1650 m.y. apparent polar movements. Phys. Earth Planet.
  Int. 13, 143-156 (1976)
Pullaiah, G., Irving, E., Buchan, K.L., Dunlop, D.J.: Magnetisation changes caused
  by burial and uplift. Earth Planet Sci. Lett. 28, 133-143 (1975)
Rankin, D.W.: Appalachian Salients and Recesses: Late Precambrian Continental break-
  up and the opening of the Iapetus Ocean. J. Geophys. Res. 81, 5605-5619 (1976)
Robertson, W.A., Fahrig, W.F.: The Great Logan Palaeomagnetic Loop — the polar
  wandering path from Canadian Shield rocks during the Neohelikian Era. Can. J.
  Earth Sci. 8, 1355-1371 (1971)
Rogers, J.W., Novitsky-Evans, J.M.: Evolution from oceanic to continental crust in
  northwestern U.S.A. Geophys. Res. Lett. 4, 347-350 (1977)
Ronov, A.B.: Probable changes in the composition of sea water during the course of
  geological time. Sedimentology 10, 25-43 (1968)
Ronov, A.B.: Evolution of rock composition and geochemical processes in the sedimentary shell of the Earth. Sedimentology 19, 157-172 (1972)
Roy, J.L., Lapointe, P.L.: The palaeomagnetism of Huronian redbeds and Nipissing
  diabase; post-Huronian igneous events and apparent polar path for the interval
  -2300 to -1500 Ma for Laurentia. Can. J. Earth Sci. 13, 749-773 (1976)
Runcorn, S.K.: Changes in the Earth's moment of inertia. Nature (London) 204, 823-
  825 (1964)
Schidlowski, M., Eichmann, R., Junge, C.E.: Precambrian sedimentary carbonates:
  carbon and oxygen isotope geochemistry and implications for the terrestrial oxygen
  budget. Precambrian Res. 2, 1-69 (1975)
Sestini, G.: Sedimentological study of the Tarkwaian gold deposits, Ghana: 14th
  Ann. Rep. Res. Inst. Afr. Geol., Univ. Leeds, 23-26 (1970)
Singer, S.F.: Origin of the Moon by capture and its consequences. Am. Geophys. Union
  Trans. 51, 637-641 (1970)
Shackleton, R.M.: Correlation of structures across Precambrian orogenic belts in
  Africa. In: Implications of Continental Drift to the Earth Sciences. Tarling, D.H.,
  Runcorn, S.K. (eds.). London: Academic Press, 1973, pp. 1091-1095
Shackleton, R.M.: Shallow and deep level exposures of Archaean crust in India and
  Africa. In: The Early History of the Earth. Windley, B.F. (ed.). London: Wiley,
  1976, pp. 317-321
Smith, A.G., Hallam, A.: The fit of the southern continents. Nature (London) 225,
  139-144 (1970)
Stacey, F.: Physics of the Earth. New York: Wiley, 1969, p. 324
Stewart, J.H.: Regional sedimentary characteristics of Upper Precambrian and Lower
  Cambrian strata in the southern Great Basin. Geol. Soc. Am., Spec. Paper 101 (1966)
  p. 56
Strong, D.F., Stevens, R.K.: Possible thermal explanation of contrasting Archaean
  and Proterozoic geological regimes. Nature (London) 249, 545-546 (1974)
Sutton, J.: Some changes in continental structure since early Precambrian time.
  In: Implications of Continental Drift to the Earth Sciences. Tarling, D.H.,
  Runcorn, S.K. (eds.). London: Academic Press, 1973, pp. 1071-1081
Sutton, J.: Some consequences of horizontal displacements in the Precambrian. Tec-
  tonophysics 40, 161-181 (1977)
Tanner, P.W.G.: Orogenic cycles in East Africa. Bull. Geol. Soc. Amer. 84, 2839-
  2850 (1973)
Tarling, D.H.: Geological processes and the Earth's rotation in the past. In:
  Growth Rhythms and the History of the Earth's Rotation. Rosenberg, G.D., Runcorn,
  S.K. (eds.). London: Wiley, 1975, pp. 397-412
Tarney, J., Dalziel, I.W.D., Dewit, M.J.: Marginal basin "Rocas Verdes" complex
  from S. Chile: a model for Archaean greenstone belt formation. In: The Early
  History of the Earth. Windley, B.F. (ed.). London: Wiley, 1976, pp. 131-146
Trendall, A.F.: Three great basins of Precambrian banded iron formation. A system-
  atic comparison. Bull. Geol. Soc. Amer. 79, 1527-1544 (1968)
Truswell, J.F., Eriksson, K.A.: Facies and laminations in the Lower Proterozoic
  Transvaal Dolomite, South Africa. In: Growth Rhythms and the History of the Earth's
  Rotation. Rosenberg, G.D., Runcorn, S.K. (eds.). London: Wiley, 1975, pp. 57-74
Valentine, J.W., Moores, E.M.: Plate tectonic regulation of faunal diversity and
  sea level: A model. Nature (London) 228, 657-659 (1970)

Veizer, J.: $^{87}Sr/^{86}Sr$ evolution of sea water during geologic history and its sig-
nificance as an index of crustal evolution. In: The Early History of the Earth.
Windley, B.F. (ed.). London: Wiley, 1976, pp. 569-578
Walker, J.C.G.: Implications for atmospheric evolution of the inhomogeneous accre-
tion model of the origin of the Earth. In: The Early History of the Earth. Windley,
B.F. (ed.). London: Wiley, 1976, pp. 535-546
Watson, J.V.: Effects of reworking on high grade gneiss complexes: Philos. Trans.
R. Soc. London Ser. A 273, 443-455 (1973)
Watson, J.V.: Vertical movements in Proterozoic structural provinces. Philos. Trans.
R. Soc. London Ser. A. 280, 629-640 (1976)
Windley, B.F.: Crustal development in the Precambrian. Philos. Trans. R. Soc. London
A. 273, 321-341 (1973)
Windley, B.F., Bridgwater, D.: The evolution of Archaean low and high-grade terrains.
Geol. Soc. Austr. Spec. Publ. 3, 33-46 (1971)
Windley, B.F., Smith, J.V.: Archaean high grade complexes and modern continental
margins. Nature (London) 260, 671-675 (1976)
Wynne-Edwards, H.R.: The Grenville Province. In: Variations in Tectonic Styles in
Canada. Geol. Assoc. Can. Spec. Paper 11. Price, R.A., Douglas, R.J.W. (eds.).
263-334 (1972)

Glossary of Key Technical Terms

Definitions given here apply to the content of this paper and do not
in all cases embrace the full usage of the term.

Amphibolite — metamorphic rock consisting predominantly of the minerals
amphibole and feldspar; formed under conditions of moderate temperature
and pressure.
Anorogenic — descriptive term applied to areas and intervals of time
when major tectonic and magmatic activity did not take place, and pertaining predominantly to igneous activity in stabilised crustal regions.
Anorthosite — a plutonic rock comprising 90 % or more of feldspar and
formed under high temperature and/or fluid pressures in Archaean and
Proterozoic times.
Asthenosphere — a shell of relative weakness within the earth underlying the rigid lithosphere where yielding permits the lateral and
vertical motions of the latter.
Banded Iron Formation — a chemical precipitate formed in shallow lakes
or shelf seas prior to about 2000 m.y. usually in the form of alternating bands of silica and iron silicates and oxides.
Blueschist — a metamorphic rock characterised by presence of the mineral glaucophane and produced under conditions of high pressure and
low temperature; with a few exceptions it is not known to have formed
prior to Phanerozoic times and characterises zones where ocean crust
has been subducted.
Calcalkaline — a series of igneous rocks having a weight percentage
of silica in the range 55 - 61 where the weight percentages of CaO and
K_2O + Na_2O are equal.
Chondritic — pertaining to a meteorite containing silicates in the
form of small spherical chondrules.
Chrontour — a line connecting localities where the rocks yield the
same radiometric age.
Clastic — a textural term applied to rocks composed of detrital frag-
ments of pre-existing rocks or minerals.
Craton — an ancient nucleus of crust which stabilised in Archaean times;
the term "Shield" is usually applied to broader tracts of Precambrian
terrain comprising several cratons separated by later mobile belts.
Diapir — a body of lower density material of cylindrical or spherical

shape which forcibly penetrates through higher material of greater density.

Fractionation — the separation of a melt into two or more phases.

Geosyncline — a downwarped belt of crust usually margining a continent and acting as a site of deposition; geosynclines containing abundant volcanic rocks are termed "eugeosynclines" and those containing little or no volcanic rocks are termed "miogeosynclines".

Gneiss — a general descriptive term applied to banded and coarse-grained metamorphic rock.

Granodiorite — a granitic plutonic rock defined by its content of quartz and sodic feldspars.

Granulite — a medium to high-grade metamorphic rock composed of even-sized interlocking granular minerals.

Greenstone — classical field term applied to basic igneous rocks which have been subjected to low-grade metamorphism.

Greywacke — rapidly deposited and poorly sorted clastic rock; usually an important component of the infill of geosynclines.

Grade — the state of metamorphism achieved in a rock which has been subjected to low (termperatures ca. 150 - 300°C, pressures up to ca. 4000 kg/cm^2), medium (temperatures ca. 300 - 450°C, pressures up to ca. 7000 kg/cm^2) and high (temperatures > 450°C, pressures up to 10,000 kg/cm^2) conditions of metamorphism.

Hydrothermal system — a hot emanation rich in water which is responsible for chemical deposition when circulating into different temperature and pressure conditions.

Lithosphere — the solid outer part of the earth comprising the crust and the outer part of the mantle rigidly attached to it.

Magmatic — pertaining to or derived from an igneous melt.

Mesosphere — that part of the mantle lying below the asthenosphere.

Mobile Belt — a coherent zone of crust subjected to folding, metamorphism, and (sometimes) igneous activity; with few exceptions, all mobile belts are Precambrian in age.

Molasse — a detrital deposit formed from the erosion of a newly elevated orogenic belt.

Nappe — a large overturned fold translated from its original position by overthrusting or by recumbent folding.

Obduction — the emplacement of a slice of ocean crust onto a continental block during the collision between the two slabs of continental crust.

Ophiolite — applied to the sequence of rocks forming oceanic crust after that sequence has been emplaced by continent-continent or continent-ocean collision.

Orogenic Belt — a complex zone of folded rocks, often including slices of oceanic crust, that has formed in the collision zone between two plates.

Plate — a rigid slab of lithosphere including oceanic or continental crust or both that behaves as a single entity in the scheme of global tectonics; plates grow by the addition of ocean crust at "constructive" margins along the Mid-Ocean Ridge system and are consumed by near normal collisions with adjacent plates at "destructive" margins.

Pluton — an igneous intrusion.

Radiogenic — formed as a consequence of radioactive decay; U, K, and Th are the most important sources of radiogenic heat on a geological time scale although short-lived isotopes may have been important in the early history of the earth.

Rapakivi — applied in this context to a large mushroom-shaped intrusion emplaced into crust which had already cooled and stabilised; Rapakivi-type intrusions are all Proterozoic in age and the type locality is in Finland.

Shield — see Craton.

Sial — classical term applied to the continental crust.

Sima — equivalent term for rocks forming the oceanic crust.
Solidus — locus of points at temperatures above which solid and liquid phases are in equilibrium and below which the system is completely solid.
Spilite — a basaltic volcanic rock usually erupted in a submarine environment and subject to low temperature alteration.
Subduction — the consumption of oceanic crust by the mantle beneath a continent at zones of plate collisions.
Supracrustal — pertaining to rocks formed, folded, and metamorphosed at high levels in the continental crust.
Tonalite — a plutonic granitic rock defined by its content of quartz and sodic plagioclase feldspar.
Ultrabasic — rocks containing abundant Fe and Mg and less than 45 % silica.

Tidal Friction
and the Earth's Rotation

Ed. by P. Brosche · J. Sündermann

Proceedings of a Workshop Held at the
Centre for Interdisciplinary Research (ZiF)
of the University of Bielefeld,
26 – 30 September, 1977

Springer-Verlag
Berlin Heidelberg New York 1978

Concluding Remarks

The excellent reviews and new information presented at this meeting
do not necessitate an extra summary. However, answers to the questions
posed at the beginning seem to be desirable and also, to some extent,
possible.

1. We clearly are not obliged by the results of human astronomical
observations to accept any change of the tidal friction interaction
between Earth and Moon during the last two millenia. Even the palaeontological values for 5,00 million years (perhaps 2000 m.y.) can be
represented by the same quantity of the torque, but with larger error
margins, hence admitting the possibility of noteworthy variations.

2. It seems that oceanic tides dominate the tidal friction process.
However, the interaction of the hydrosphere with the solid Earth may
change the oceanic tides considerably. Since oceanic tide computations
which include this interaction have now been started, the comparison
with earlier computations assuming a rigid bottom of the ocean will
become possible very soon.

3. First computations for an ancient ocean configuration show that the
torque between Earth and Moon may have changed within geological times
in a way which was not envisaged by the schematic backwards-integrations, namely, that it might have been smaller than now in some periods.

4. The distance and the time of the nearest approach between Earth and
Moon could not — as expected — be estimated at our meeting. Nevertheless, the geological record tells us that there was no dramatic approach later then $3.5 \cdot 10^9$ years ago. This seems to be understandable
with regard to 3.

Our data and our reasoning on that basis can and should be improved
in many respects. In particular, two possibilities of progress seem
to be dictated by the circumstances: (1) the re-establishment and the
new evaluation of the palaeontological evidence, (2) the tide computations for ancient oceans — with all their prerequisites. The results
of the latter are needed before the former can be interpreted, and an
iterative scheme of improvements should be the consequence.

P.B.

Journal
of Geophysics
Zeitschrift
für Geophysik

Edited for the Deutsche Geophysikalische Gesellschaft by W. Dieminger, J. Untiedt

Editorial Board: K.M. Creer, Edinburgh;
W. Dieminger, Lindau üb. Northeim/Hannover;
K. Fuchs, Karlsruhe; C. Kisslinger, Boulder, CO;
Th. Krey, Hannover; J. Untiedt, Münster/Westfalen;
S. Uyeda, Tokyo

Journal of Geophysics publishes articles predominantly in English from the entire field of geophysics and space research, including original essays, short reports, letters to the editor, book discussions, and review articles of current interest, on the invitation of the German Geophysical Association.
The following fields of geophysics have been treated in recent volumes: applied geophysics, geomagnetism, gravity, hydrology, physics of the solid earth, seismology, physics of the upper atmosphere including the magnetosphere, space physics and volcanology.

Springer-Verlag
Berlin
Heidelberg
New York

Fields of Interest: Geophysics, Seismology, Geomagnetism, Aeronomy, Extraterrestrical Physics, Space Research, Meteorology, Oceanography, Applied Geophysics, Theoretical Geophysics, Tectonics, Geochemistry, Petrology.

Subscription information and sample copies upon request.

Wave Propagation and Underwater Acoustics

Editors: J.B. Keller, J.S. Papadakis
1977. 32 figures. VIII, 287 pages
(Lecture Notes in Physics, Volume 70)
ISBN 3-540-08527-0

Contents: I. *Survey of wave propagation and underwater acoustics*. With contributions by J.B. Keller. II. *Exact and asymptotic representations of the sound field in a stratifield ocean*. With contributions by D.S. Ahluwalia, J.B. Keller. III. *Horizontal rays and vertical modes*. With contributions by R. Burridge, H. Weinberg. IV. *Wave propagation in a randomly inhomogeneous ocean*. With contributions by W. Kohler, G.C. Papanicolaou. V. *The parabolic approximation method*. With contributions by F.D. Tappert

R.K. Zeytounian

Notes sur les Ecoulements Rotationnels de Fluides Parfaits

1974. XIII, 407 pages
(Lecture Notes in Physics, Volume 27)
ISBN 3-540-06721-3

Waves on Water of Variable Depth

Proceedings of a Symposium Held Under the Auspices of the International Union of Theoretical and Applied Mechanics (IUTAM) and the Australian Academy of Science at the Academy in Canberra, 20 - 23 July 1976
Editors: D.G. Provis, R. Radok
1977. 134 figures, 10 tables. 231 pages
(Lecture Notes in Physics, Volume 64)
ISBN 3-540-08253-0

Contents: I. *Wave Propagation in Water of Variable Depth*. With contributions by E.O. Tuck, R.W. Preisendorfer, P.L. Christiansen, D.G. Provis. II. *Tsunami Generation and Propagation*. With contributions by Y. Nagata, A.S. Alexeev, V.K. Cusyakov, K. Kajirua, A.T. Chwang, T.Y. Wu. III. *Waves on Beaches*. With contributions by J.D. Fenton, D.A. Mills, A.J. Bowen, S. Hibbera, D.H. Peregrine, H. Hornung, P. Killen, M.S. Longuet-Higgins. IV. *Waves and Currents*. With contributions by I.G. Jonsson, D.H. Peregrine, G.P. Thomas, R. Smith. V. *Waves in a Rotating Stratified Medium*. With contributions by M. Roseau, C.N.K. Mooers, J.A. Helbig, L.A. Mysak. VI. *Long Period Barotropic Waves*. With contributions by V.T. Buchwald, N.G. Barton, W.K. Melville, R.H.J. Grimshaw, W.D. McKnee, B.V. Hamon, G. Krause, R. Radok

 Springer-Verlag Berlin Heidelberg New York